AF535077

AN INTRODUCTION TO THERMODYNAMICS FOR ENGINEERING TECHNOLOGISTS

AN INTRODUCTION TO THERMODYNAMICS FOR ENGINEERING TECHNOLOGISTS

James P. Todd
President
Vermont Technical College
Randolph Center, Vermont

Herbert B. Ellis
Project Engineer
Aerojet Electrosystems Co.
Azusa, California

JOHN WILEY & SONS
New York Chichester Brisbane Toronto

This book is dedicated to our students—past, present and future—and to my wife, Ginger, for her unwavering support, and to Herb's wife, Ellen, for her "Ishkabod."

J. P. T.

Cover art by Melissa A. Birrittella.
Text and cover design by Laura C. Ierardi.

Library of Congress Cataloging in Publication Data:

Todd, James P
An introduction to thermodynamics for engineering technologists.

Includes bibliographical references and index.
1. Thermodynamics. I. Ellis, Herbert, B., joint author. II. Title.
TJ265.T62 1981 621.402'1 80-24055
ISBN 0-471-05300-7

Printed in the United States of America

10 9 8 7 6 5 4 3 2 1

PREFACE

This introductory textbook has been written to present clearly the basic principles and equations of thermodynamics relevant to current industrial and scientific applications. This book has four important objectives. The first objective of the book is to provide the engineering technology student with a basic introduction to the vocabulary, concepts, and fundamental relationships of thermodynamics and steady state heat transfer. The second objective is to expose the student to a methodology for the solution of thermal energy problems that will be advantageous in a career of technology, engineering, or science. The third is to give the student an adequate background for more advanced studies if so desired. The fourth objective is to provide practicing technologists and engineers with a convenient and useful reference book. To accomplish these objectives, the authors have combined an extensive teaching and academic experience at the college level with over 40 years of practical engineering experience in industry.

The book is organized in a conventional manner. The presentation assumes that the reader has a basic knowledge of physics and mathematics. Calculus is not required as a prerequisite; however, it is advantageous to a more complete understanding of the second law of thermodynamics. The International System of Units (SI) is applied throughout in accordance with the current shift to SI (metric) units in this country.

An introduction to thermodynamics is presented, followed by two brief chapters to introduce the reader to steady state heat transfer. Sample problems in each chapter illustrate the use of basic equations and suggested methodology for the solution of problems. At the end of each chapter, problems underscoring the key points in the chapter are given. Answers to selected problems are included at the end of the book to aid the student in his or her study.

An appendix with five sections contains all the reference material necessary to solve both the problems in this text and a variety of problems encountered in actual engineering practice. Appendix A provides thermodynamic data. Appendix B presents tables of physical properties and heat transfer data of metals, building materials, insulation, other solids, common liquids including metals and cryogens, and common gases. Appendix C presents conversion factors (equations of equality). Appendix D suggests a methodology for linear interpolation and an approach to problem solving. Appendix E provides mathematical tables most commonly required for text problems.

J. P. T.
H. B. E.

CONTENTS

CHAPTER 1
INTRODUCTION

One of the most significant contributions to the development of our technological society has been man's ability to utilize thermal energy. Through the application of thermodynamic principles, modern heat engines have been developed. They power our vehicles; refrigerate perishables; provide air conditioning; power industrial machines; lift weights; move earth; and perform other much needed tasks.

Support of the high standard of living in the United States has dictated extensive growth in fuel (energy) consumption per capita. In the past, fossil fuels (coal, oil, and gas) were the primary source of thermal energy and their availability was considered "inexhaustible." Now we face the reality that fossil fuel reserves are diminishing and may be insufficient in the forseeable future. Consequently, to those who study thermodynamics, increased efficiency in the use of fossil fuels and the development of alternate sources of thermal energy are the real challenges to technology for today and tomorrow.

1-1 A DEFINITION OF THERMODYNAMICS

Thermodynamics deals with energy in its broadest form. It is an applied science that is involved with the relations between energy and the properties of substances. A basic definition is

> THERMODYNAMICS: That branch of science which deals with energy, its conversion from one form to another, and the movement of energy from one location to another.

Thermodynamics is involved with energy exchanges (heat and work), stored energy (chemical, potential, etc.), and the associated changes in the properties (state) of the working fluid or substance (e.g., pressure, temperature, and specific volume). Although thermodynamics deals with systems in motion, it does not concern itself with the speed at which such processes or energy exchanges occur.

1-2 THE APPLIED SCIENCE OF THERMODYNAMICS

Historically, the science of thermodynamics was developed through the investigation and experimentation conducted by numerous individuals. Their work falls into five main aspects of the science of thermodynamics. These include:

1. Basic concepts of temperature and heat.
2. The transformation of energy.
3. The behavior of gases.
4. Power production from heat.
5. Measurement of values of thermodynamic properties of substances.

1-2.1 BASIC CONCEPTS OF TEMPERATURE AND HEAT

The science of thermodynamics began with the measurement of temperature in 1592 by Galileo. Later investigators examined the nature of heat. For a period of time the generally accepted explanation of heat was the theory of a weightless fluid, or "caloric," which permeated matter. It was thought that friction and compression produced heat simply by squeezing some of the "caloric fluid" out of the bodies concerned.

Experiments by Rumford in 1798 and Davy in 1799 led to the concept that heat is thermal energy in the form of molecular motion in the structure of matter. This form of thermal energy is currently referred to as "internal" energy.

Temperature is a measurement of the magnitude of the internal energy. As the internal energy of a substance (matter) increases, its temperature also increases. The theory of motion in the internal structure of matter is still the accepted concept of thermal energy.

1-2.2 THE TRANSFORMATION OF ENERGY

In 1693 G. W. Liebnitz described the conservation of mechanical energy, both potential and kinetic. Carnot in 1830 published a treatise on the mechanical theory of heat. In 1840 J. P. Joule's experiments proved the equivalence of heat and work. In 1847 Helmholtz wrote a landmark paper "On the Conservation of Force," wherein he discussed all the known cases of transformation of energy. In 1850 Clausius and Rankine established that when heat is applied to a body, part of the heat remains in the body as intrinsic energy and part of the heat is converted to the external work of expansion and ceases to exist as heat (see Section 1-4).

1-2.3 THE BEHAVIOR OF GASES

In the study of the behavior of gases, Boyle in 1662 investigated the resistance of gas to compression. Subsequent investigations such as by

Laplace, Dalton, Gay-Lussac, and others contributed to the development of the perfect (ideal) gas laws in the early 1800s (see Chapter 6).

1-2.4 POWER PRODUCTION FROM HEAT

Sadi Carnot addressed himself to the relationship between heat and power production. In 1824 he published "Reflections on the Motive Power of Heat" in which he developed the optimum theoretical power-producing cycle, which is known as the Carnot cycle. It is used today as a comparative measure for the efficiency of a heat engine's performance (see Chapter 9).

1-2.5 MEASUREMENTS OF THERMODYNAMIC PROPERTIES

Even before the start of the nineteenth century, thermodynamic properties of various substances were being measured experimentally. In 1764 James Watt measured the latent heat of vaporization of steam. More refined measurements were made in 1780 by Laplace and Lavoisier.

Tables of values for the thermodynamic properties of various substances were developed and published. Measurements of thermodynamic properties are continually being made as new substances of interest appear, adding to the library of thermodynamic reference data. These thermodynamic data are basic information needed in the design or evaluation of any thermodynamic system (see Appendix A).

1-3 THE DEVELOPMENT OF THERMODYNAMIC MACHINES

In general, thermodynamic machines have been invented and developed in response to the needs of society. Often their development was restricted by the available materials, manufacturing processes, and current design technology.

1-3.1 ROCKETS

The first historically recorded thermodynamic machine was the rocket, which was applied to arrows by the Chinese. Chinese arrows are mentioned in writings of A.D. 1232 and subsequently in the Chronicle of Cologne[(1)] in 1258. By 1400 experiments of military engineers of various countries had produced many types of rockets. In 1800 Sir William Congreve revived the rocket projectile only to have it supplanted by the more accurate rifled cannon. In 1915 Robert H. Goddard began experiments leading to the rocket engine of today's satellites and space vehicles. During World War II, rockets were revived and new technology applied. Extensive use of rockets was made for propelling warheads and assisting heavily laden aircraft on takeoff. Subsequently, sophisticated rocket-propelled missiles have been added to military arsenals.

1-3.2 STEAM PUMPS

In 1698 Thomas Savery perfected the first commercially successful, water-raising engine using steam for pumping mines. His invention was perhaps sparked by published treatises by Giovanni Battista della Porta in 1601 and the steam fountain concept of Solomon de Caus in 1615. Steam was injected into vessels filled with water until the water was discharged. The vessel was then cooled to cause condensation with a resulting vacuum that would allow water to be inserted by atmospheric pressure into the vessel for the next cycle.[2]

In 1705 Thomas Newcomen applied the piston and cylinder, used in water pumps, to separate the steam from the water being pumped so as to reduce the steam consumption. This engine introduced atmospheric pressure steam to fill the cylinder. Then the steam was condensed in the cylinder to produce a vacuum. The atmospheric air pressure on the other side of the piston produced the power stroke. This engine remained basically unchanged for 60 years.

In 1765 James Watt improved the Newcomen engine efficiency by introducing a separate cold condensing chamber so that the steam cylinder could remain hot, thereby significantly reducing the steam consumption. In 1782 James Watt applied steam alternately to both sides of the piston making a double-acting pump engine. He also introduced the concept of using higher than atmospheric pressure steam and making use of an expansion of the steam to obtain more work out of the steam.

1-3.3 STEAMBOAT

During 1807 Robert Fulton modified the double-acting Watt pumping engine, combined it with a crankshaft, and applied it successfully to driving a steamboat on the Hudson River. A rapid development of the steamboat followed.

1-3.4 STEAM LOCOMOTIVE

In 1804 Richard Trevithick built the first railway locomotive to run on a horse tramway in Wales. It was not particularly successful and was retired after a short period of use.

Twenty-five years later, in 1829, George Stephenson built the "Rocket," which by its outstanding performance established that steam locomotives (rather than horses) would pull trains. The principal features of the Rocket were followed in all essentials in subsequent steam locomotive development. Rapid advances in boiler and steam engine design occurred. The world was waiting and a rapid worldwide development of the steam locomotive followed.

1-3.5 STEAM TURBINE

The steam turbine introduced in the latter part of the nineteenth century provided a more efficient and compact engine than the reciprocating engine.

In 1884 Parsons introduced a successful steam turbine that divided the total expansion of the steam into a large number of successive and separate stages, making it easier to extract the kinetic energy.

Five years later in 1889, Gustaf de Laval introduced a nondispersing steam expansion nozzle and a turbine design permitting the high speeds necessary for efficiently extracting the kinetic energy from the high velocity jet of the expanded steam.

1-3.6 INTERNAL COMBUSTION ENGINES

Nicholas A. Otto (1832–1891) patented the first successful internal combustion engine in 1876 (Figure 1-1). His disclosure served as the basis for much subsequent oil and gas spark-ignition engine development. The common automobile engine is a prime example of an Otto cycle engine.

The Otto cycle engine has been applied to a wide variety of applications both large and small. Its initial development was along the lines of large,

Figure 1-1 Automobile engine 1979. (Photo courtesy of the Oldsmobile Division of General Motors.)

heavy slow-speed engines. In 1887 Daimler demonstrated a motorcycle with a light, high-speed engine; and shortly thereafter demonstrated the first automobile. To obtain a greater power output from the same-size engine, the two-stroke cycle was introduced.

The compression ignition engine with its higher compression ratios is capable of higher efficiencies. In 1893 Rudolf Diesel obtained a patent on the type of compression ignition engine that bears his name. This engine has been developed for many applications and is widely used on ships, railroad locomotives, and engines for trucks, buses, earth-moving equipment, and the like.

1-3.7 GAS TURBINES

George B. Brayton, a Boston engineer, proposed the thermodynamic cycle that bears his name. Independently, James Joule proposed a similar cycle. The earliest patent on a gas turbine was that of the Englishman John Barber in 1791. The first practical turbojet engine design is attributed to Sir Frank Whittle of England,(3) who filed his patent in 1930. His first operating turbine engine was demonstrated in 1937. The Germans, however, were the first to fly a turbojet-powered airplane—the Heinkel HE-178—on August 27, 1939. Rapid progress in the development of gas turbine engines for aircraft, auxiliary power generation, and other applications occurred during the post-World War II period and is still in process. Examples of the latest gas turbine engines used in a variety of applications are shown in Figure 1-2.

1-3.8 REFRIGERATION AND AIR CONDITIONING

Refrigeration is the effect produced by a thermodynamic machine that results in the cooling of a substance such as air. In our everyday existence we encounter it in our kitchens (refrigerator) and in buildings and vehicles (air conditioners). There are four basic types of refrigeration systems: vapor compression, air compression, absorption, and desicant. All of these systems require an energy input to provide the refrigeration desired.

Air conditioning consists primarily of the simultaneous control of temperature, humidity, flow (motion), and purity of air to meet the requirements of human comfort.(4) Most modern buildings are designed with a central system(s) for maximum comfort and minimum energy consumption. It was not until 1917, when Willis H. Carrier began a systematic study of air–water vapor mixtures, that suitable air conditioning equipment could be designed. During the 1930s the production of air conditioning units started, and in the 1940s the air conditioning industry became established. Today air conditioning is no longer a luxury but an essential part of modern living.

Figure 1-2*a–d* Gas turbine applications. (Photo courtesy of Solar Turbines International, an Operating Group of International Harvester.)

Figure 1-2 Gas turbine applications (continued)

1-4 DEFINITIONS AND BASIC LAWS

Thermodynamics, as with any other engineering subject, has various basic definitions that the student should learn and understand. Thermodynamics is involved with the relationship between energy, work, and heat, and changes in the condition or state of the thermodynamic system's working fluid or substance.

1-4.1 ENERGY

Elementary physics acquaints us with at least two basic forms of energy, namely potential and kinetic. In the study of thermodynamics we must add another form, internal energy, which helps to explain the "hotness" and "coldness" of a body. These forms of energy are discussed in subsequent sections, but first we need some definitions. One very basic definition of energy is

ENERGY: The capacity for doing work.

All forms of energy are ultimately measured in terms of work, which has the dimensions of force times length. In this book the symbol E is used to identify energy, and the units of energy are expressed in joules [J]. A joule is equivalent to a newton·meter [N·m], that is, 1 J=1 N·m.

Energy takes many forms: thermal, chemical, electrical, strain, heat, light, and nuclear. The types of energy that are most easily recognized are, of course, those in which the energy can be most effectively used in doing mechanical work. Obvious examples include springs, compressed gases, flywheels, explosive substances, internal combustion engines, and the like. Lord Kelvin made a fundamental distinction in energy between "available energy," which can be converted to mechanical effects, and "diffuse or unavailable energy," which is useless for that purpose.

1-4.2 WORK

The object of many thermodynamic machines (heat engines) is to produce mechanical power from heat power. Power is simply the rate of producing or consuming work. A basic definition of work is

WORK: The product of (1) the component of force in the direction of motion, and (2) the distance through which the point of application of the force moves during its action.

This concept of work was introduced in 1826 by the French mathematician J. V. Poncelet and has been accepted as a basic definition. In this book the symbol W_k is used to identify work. The units of work, like energy, are expressed in joules (J).

1-4.3 HEAT

Francis Bacon (1561–1626), in *Novum Organum*, said "The very essence of heat is motion and nothing else." In thermodynamics, heat is defined in a somewhat different manner from the way it is commonly used. It is therefore necessary to present a very clear and concise definition because heat is involved in the solution of many thermodynamic problems. Heat is one form of thermal energy, and in the study of thermodynamics is defined as:

HEAT: A form of energy in transit across a system boundary.

When heat has crossed the thermodynamic system boundary, it is transformed into another form of energy. This transformed (heat) energy becomes internal energy or work as discussed later in this book. Consequently, as soon as heat passes through the system boundary it ceases to exist as heat per se. In this book the symbol Q is used to designate heat. Since it is a form of energy, its units are expressed in joules [J].

1-4.4 THERMODYNAMIC SYSTEMS

In addition to energy, work, and heat, the definition of a thermodynamic system is important. A definition of the thermodynamic system is

THERMODYNAMIC SYSTEM: A region in space that occupies a given volume, has a specific boundary, and contains a thermodynamic substance.

Every thermodynamic analysis must begin with the specification of the thermodynamic system. The thermodynamic system, as defined, is that region in space containing a quantity of thermodynamic substance (matter) which is to be studied. The system is distinguished from its surroundings by system boundaries. These boundaries may be either fixed or flexible. The concept of a thermodynamic system is shown in Figure 1-3.

The boundaries of a thermodynamic system may be drawn to include a group of interrelated system components, forming a relative complex system, or around a subgroup of such components, or even around a single component.

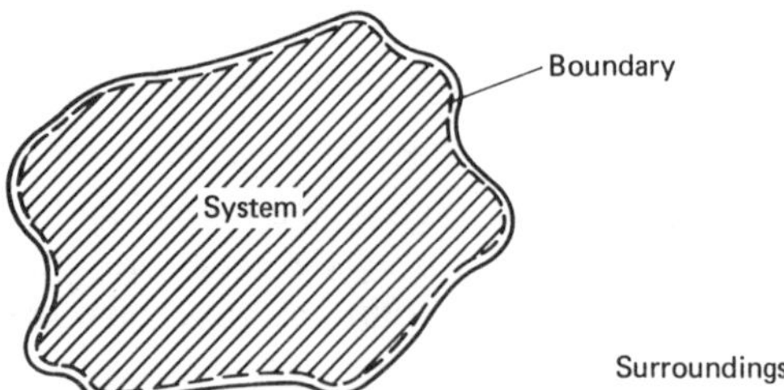

Figure 1-3 The concept of a thermodynamic system.

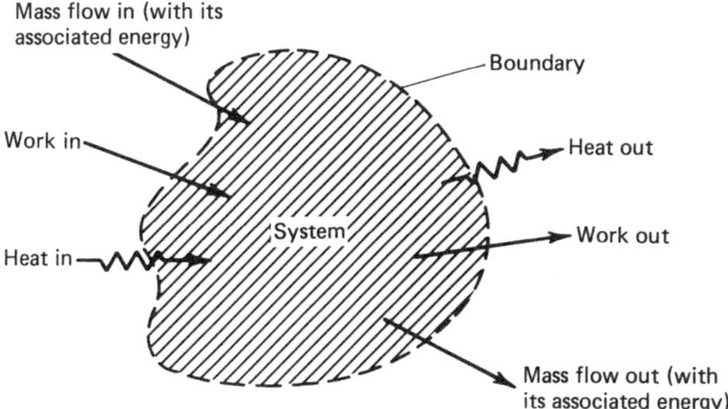

Figure 1-4 Diagram of an OPEN system.

1-4.5 TYPES OF THERMODYNAMIC SYSTEMS

There are three basic types of thermodynamic systems:

- Open system.
- Closed system.
- Isolated system.

Open System

In the open system, *heat*, *work*, and *mass* (with its associated energy) all cross the system boundary, as shown in Figure 1-4.

A simple example of an open system is a steam turbine. Such an open system is shown in Figure 1-5. Another example of an open system is the automobile engine.

Since the system boundaries can be drawn in various manners, the example of a steam turbine shown in Figure 1-5 may be a part of a larger

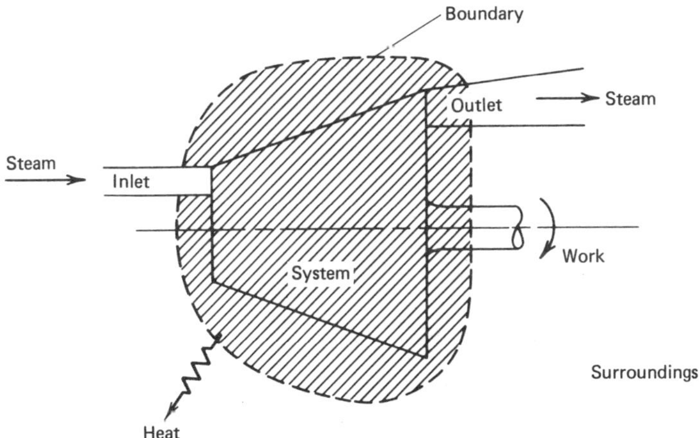

Figure 1-5 Open system (steam turbine).

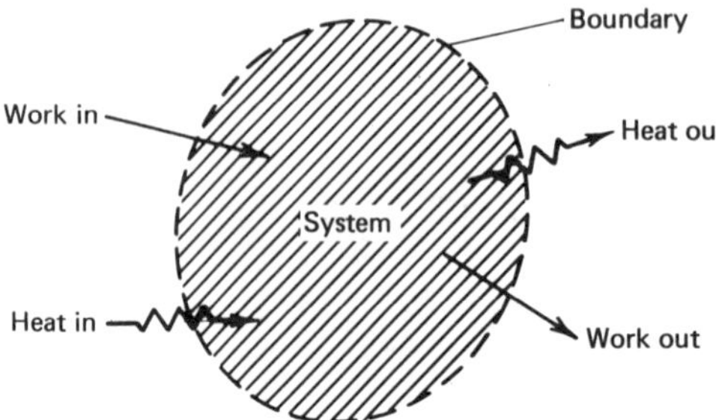

Figure 1-6 Diagram of a CLOSED system.

system comprising boilers, condensers, and pumps, which in its entirety may be a closed system. Still, it is valid to analyze a component of such a larger system, like the turbine, as a separate open system.

Closed System

In the closed system, only *heat* and *work* can cross the system boundary as shown in Figure 1-6. There is no transfer of mass (or mass flow) across the boundary.

An example of a simple closed system wherein only heat crosses the boundary is the mercury or alcohol thermometer. A closed thermodynamic system may vary from a simple closed tank containing a substance to a more complex system such as the sealed Freon refrigerator system shown in Figure 1-7, which has both heat and work crossing its system boundary.

Isolated System

In the isolated system, nothing crosses the system boundary (no heat, no work, no mass). Consequently, an isolated system is not affected by its

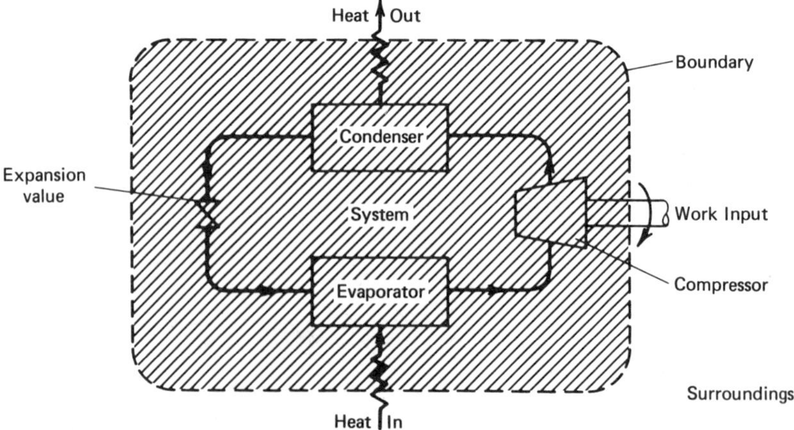

Figure 1-7 Closed system (refrigerator).

surroundings (and in turn has no affect upon its surroundings). The isolated system will not be discussed in this book.

1-4.6 BASIC LAWS OF THERMODYNAMICS

During the development of the applied science of thermodynamics, four basic laws evolved. After the first three laws were established, another law evolved. Because of its nature it was ranked first in the order of the prior laws. Because the first three laws were well established as the first, second, and third laws, the new law was termed the zeroth law to list it first. These four laws are described briefly as follows:

- The zeroth law is a statement concerning thermal equilibrium and is the basis for temperature measurement. It states that when two bodies have equality of temperature with a third body, they in turn have equality of temperature with each other; the three systems are in thermal equilibrium.
- The first law is often called the law of conservation of energy. It relates the net amount of heat transferred into, and the net work output from, a thermodynamic system to the change in total energy. Energy itself cannot be created or destroyed but only transformed into other forms.
- The second law is concerned with the availability of energy from a thermodynamic cycle and demonstrates the impossibility of a perpetual motion machine. A basis is provided for a relationship between the temperature levels of heat input and rejection and the relative amount of the heat input energy that can be transformed into available energy.
- The third law represents a restriction of all physical systems to the temperature regime that excludes absolute zero.

The first and second laws are of primary interest in this text. Further definition and discussions of applications of these laws for thermodynamic processes and systems are given in Chapters 3 and 5.

1-5 PROBLEMS

1-1. Define energy.
1-2. Define work.
1-3. Define heat.
1-4. Define a thermodynamic system.
1-5. What are the different types of thermodynamic systems? In what way do they differ?
1-6. Give two examples each of (a) open, and (b) closed systems. Explain using sketches (block diagrams) why your examples are the types of thermodynamic system indicated.
1-7. State the four laws of thermodynamics.

CHAPTER 2

PROPERTIES OF THERMODYNAMIC SUBSTANCES

Properties are the quantities that describe and define the thermodynamic substance. In order to analyze a thermodynamic system, we must know some of these quantities (properties). In order to know such quantities (properties), dimensions and units are used to describe them. Properties are usually divided into two general classifications:

- Intensive.
- Extensive.

The symbols for these properties in this book will follow the conventional designation of lowercase letters for intensive properties (except temperature) and uppercase letters for extensive properties.

2-1 DIMENSIONS AND UNITS

Before attempting to study any engineering subject, dimensions and the units for these dimensions must be understood. Dimensions describe complex physical parameters in terms of simple or basic definitions. Units are the basis for measurements that evaluate or "quantize" the dimensions of the physical parameter.

2-1.2 DIMENSIONS

A dimension can be defined as a term that describes certain qualities or characteristics of an entity (such as mass, M; length, L; area, A; etc.). In this context, a large number of dimensions is possible. To reduce such a large number of dimensions, many can be expressed in terms of basic dimensions. For example, length, area, and volume are dimensions describing certain characteristics of an object. But area $[A]$ can be defined as a length squared $[L^2]$ and volume $[V]$ as a length cubed $[L^3]$. These dimensions can be stated in terms of the basic dimension length $[L]$ as follows:

$$[A]=[L^2]$$
$$[V]=[L^3]$$

Note the use of brackets, which are often used to identify a dimension as in the above equations. These should be read as: The dimension of area is equivalent to the dimension of length squared; or the dimension of volume is equivalent to the dimension of length cubed.

By using this technique, a large number of dimensions can be reduced to a smaller number of basic dimensions. All other dimensions expressed in terms of these basic dimensions are known as secondary or derived dimensions. Thus, in the previous examples, area and volume are derived (dimensions) in terms of the basic dimension of length.

2-1.2 UNITS

Whereas a dimension provides a description, a *unit* is a definite standard or measure of a dimension. A unit may be defined as a particular amount (or quantity) to be measured. For example, kilometer (km), meter (m), and millimeter (mm) are all different units used to specify definite lengths. The common dimension of all these units is the dimension of length $[L]$.

2-1.3 SI (METRIC) UNITS

This book is written using SI (metric) units because of the current adoption of SI (metric) units for academic and industrial use in the United States. Usage has shown the SI units to be very convenient, simplifying many calculations.

The SI (metric) system is designated SI for System International d'unités[5], and (metric) is usually added to the SI designation. A set of seven basic units, listed in Table 2-1, was officially adopted in a resolution of the Eleventh General Conference on Weights and Measures in Paris, in 1960.

Table 2-1 Basic SI (Metric) Units.

Dimension		Units	
Physical Parameter	Symbol	Name	Abbreviation
Length	L	meter	m
Mass	M	kilogram	kg
Time	τ	second	s
Electric current	I	ampere	A
Temperature	T	kelvin	K
Amount of substance	$\mathfrak{M}$	mole	mol
Luminous intensity	I	candela	cd

Table 2-2 Secondary SI (Metric) Units Used in Thermodynamics

Dimension	Units	
Physical Parameter	Abbreviation	Name Equivalences
Force	N	newton = [kg·m/s²]
Pressure	Pa	pascal = [N/m²] = [kg/m·s²]
Energy, heat, work, enthalpy	J	joule = [N·m] = [kg·m²/s²]
Power, heat transfer rate	W	watt = [J/s] = [N·m/s] = [kg·m²/s³]

The secondary SI (metric) units commonly used are listed in Table 2-2.

These basic units measure quantities that could vary considerably in magnitude. To avoid awkwardly large or small figures, common prefixes representing multiples and submultiples are given in Table 2-3.

Multiples and submultiples are generally in increments of three orders of magnitude. Other orders of magnitude that do not follow this pattern are not part of the SI system.

Thermal energy problems are involved with parameters that have various dimensions, such as heat, mass, force, area, length, time, and temperature, as well as various combinations of these dimensions. Each of these parameters is characterized by its dimensions. Each dimension is evaluated by a system of units. The symbols, dimensions, and SI (metric) units for the common parameters encountered in thermodynamics problems are listed in Appendix A, Table A-1.

In the solution of any thermodynamics or heat transfer equation, it is mandatory that a consistent set of dimensional units be used throughout. For example, the units of time for all the parameters in the equation must be consistently seconds, minutes, or hours; units of length must be consistently centimeters or meters, and so forth.

Table 2-3 Prefixes to SI (Metric) Units

Name	Symbol	Multiply by
	Multiples	
tera	T	10^{12}
giga	G	10^{9}
mega	M	10^{6}
kilo	k	10^{3}
	Submultiples	
centi	c	10^{-2}
milli	m	10^{-3}
micro	μ	10^{-6}
nano	n	10^{-9}
pico	p	10^{-12}
femto	f	10^{-15}
atto	a	10^{-18}

2-1.4 EQUATIONS

For the ease of identifying related equations in this book, the authors have adopted the use of adding a letter to such equations. The original numbered equation will generally be for the first statement of the equation. Subsequent permutations of the same equation will have the basic equation number plus a letter suffix (e.g., Equations 2-25 and 2-25a on page 48). If a rearrangement (permutation) of a basic equation is used, the authors often use the letter "x" as a suffix.

2-1.5 CONVERSION OF UNITS

This book was written in a transition period in which a large quantity of published and manufacturers' data were in units other than SI (metric). Consequently, the student can expect to be faced with the need to make numerous conversions of units. Conversions from one unit system to another are simple if the following procedure is used: Two equations are involved, namely the general conversion equation and the equation of equality.

The general conversion equation is

$$\psi = \psi' F_c \qquad [\text{converted units}] \tag{2-1}$$

where

ψ = value of any parameter in converted units
ψ' = value of the parameter in initial units
F_c = parameter conversion factor between units

It is important to note that in the general conversion equation the units on both sides of the equation equals sign (=) are the same.

The equation of equality numerically relates one set of units to another set of units:

$$100\,[\text{cm}] = 1\,[\text{m}]$$

It is important to note that the units are different on the two sides of the equation equals sign (=).

The equation of equality is used only to obtain the parameter conversion factor (F_c) between units that is used in the general conversion equation. For example, to obtain a conversion factor F_c with the units of [cm/m], divide both sides of the example equation of equality above by 1 [m].

$$F_c = \frac{100\,[\text{cm}]}{1\,[\text{m}]} = 100\left[\frac{\text{cm}}{\text{m}}\right]$$

A conversion of units is made by taking the following steps:

1. The first step is to write Equation 2-1 indicating the desired units for the parameter ψ.
2. The next step is to list the given value of the parameter ψ' and its units.
3. The next step is to list the equation of equality between the given and desired units of the parameter. Equations of equality for various conversions are given in Appendix C.
4. The next step is to derive from the equation of equality the conversion factor (F_c) with the required ratio of units.
5. The values of the parameter ψ' and the conversion factor F_c are then substituted into Equation 2-1; the numerical values in one substitution, and the units in a separate substitution.

The use of the general conversion equation is shown by the following illustrative problems.

Example Problem 2-1

Convert a mass of 77 lb_m to kg

Equation

$$M = M^1 F_c \qquad [kg] \qquad (Eq. 2\text{-}1)$$

Parameters

$$M^1 = 77\ lb_m \qquad (given)$$

$$1\ [lb_m] = 0.4536\ [kg] \qquad (Appendix\ C\text{-}1.2)$$

$$F_c = 0.4536\ kg/lb_m$$

Substitution

$$M = (77)(0.4536) \qquad [lb_m][kg/lb_m] = [kg]$$

Answer

$$\mathbf{M = 35\ kg}$$

Example Problem 2-2

Convert a pressure of 1.48 atm to MPa.

Equation

$$p = p^1 F_c \qquad [MPa] \qquad (Eq. 2\text{-}1)$$

Parameters

$$p^1 = 1.48\ atm \qquad (given)$$

$$1\ atm = 0.10135\ MPa \qquad (Appendix\ C\text{-}1.2)$$

$$F_c = 0.10135\ MPa/atm$$

Substitution

$$p=(1.48)(0.10135) \qquad [\text{atm}][\text{MPa/atm}]=[\text{MPa}]$$

Answer

$$\mathbf{p=0.15\ MPa}$$

Example Problem 2-3

Convert a density of 62.42 lb_m/ft^3 to kg/m^3

Equation

$$\rho=\rho^1 F_c \qquad [\text{kg/m}^3] \qquad \text{(Eq. 2-1)}$$

Parameters

$$\rho^1=62.42\ \text{lb}_\text{m}/\text{ft}^3 \qquad \text{(given)}$$

$$1\ [\text{lb}_\text{m}/\text{ft}^3]=16.0185\ \text{kg/m}^3 \qquad \text{(Appendix C-1.2)}$$

$$F_c=16.0185\left[\frac{\text{kg/m}^3}{\text{lb}_\text{m}/\text{ft}^3}\right]$$

Substitution

$$\rho=(62.42)(16.0185) \qquad [\text{lb}_\text{m}/\text{ft}^3]\left[\frac{\text{kg/m}^3}{\text{lb}_\text{m}/\text{ft}^3}\right]=[\text{kg/m}^3]$$

Answer

$$\boldsymbol{\rho}\mathbf{=1000\ kg/m^3}$$

Example Problem 2-4

Convert an absolute viscosity of 1.15 $lb_m/ft \cdot s$ to $Pa \cdot s$

Equation

$$\mu=\mu^1 F_c \qquad [\text{Pa}\cdot\text{s}] \qquad \text{(Eq. 2-1)}$$

Parameters

$$\mu^1=1.15\ \text{lb}_\text{m}/\text{ft}\cdot\text{s} \qquad \text{(given)}$$

$$1\ [\text{lb}_\text{m}\cdot\text{ft}\cdot\text{s}]=1.488[\text{Pa}\cdot\text{s}] \qquad \text{(Appendix C-1.2)}$$

$$F_c=1.488\left[\frac{\text{Pa}\cdot\text{s}}{\text{lb}_\text{m}/\text{ft}\cdot\text{s}}\right]$$

Substitution

$$\mu=(1.15)(1.488) \qquad [\text{lb}_\text{m}/\text{ft}\cdot\text{s}]\left[\frac{\text{Pa}\cdot\text{s}}{\text{lb}_\text{m}/\text{ft}\cdot\text{s}}\right]=[\text{Pa}\cdot\text{s}]$$

Answer

$$\boldsymbol{\mu=1.71\ \text{Pa}\cdot\text{s}}$$

2-2 THE THERMODYNAMIC SUBSTANCE

In a thermodynamic system the fluid that receives, transports, and transfers energy (i.e., the working fluid) is called a substance. The definition of a pure thermodynamic substance is different from the definition of a chemically pure substance in that the thermodynamic pure substance, in addition to being chemically pure, is composed of only one type of molecule. For example, chemically pure water (or H_2O), which is composed entirely of molecules consisting of two atoms of hydrogen and one atom of oxygen, is an example of a pure substance.

A thermodynamic substance is described and defined by properties that have quantitative values. Properties are usually divided into two general classifications, namely

- Intensive.
- Extensive.

These two classifications and properties are discussed in Sections 2-3 and 2-4.

Substances have three phases of physical being, namely

- Solid.
- Liquid.
- Gaseous.

2-2.1 SOLID PHASE

The solid phase is characterized by rigid bonds between atoms or molecules. A solid substance possesses both definite volume and definite shape, resisting any force that tends to alter the volume or shape. For example, ice is the solid phase of H_2O. The solid phase of a substance is often composed of a crystal structure. A typical lattice-type crystal structure of a solid is shown in Figure 2-1.

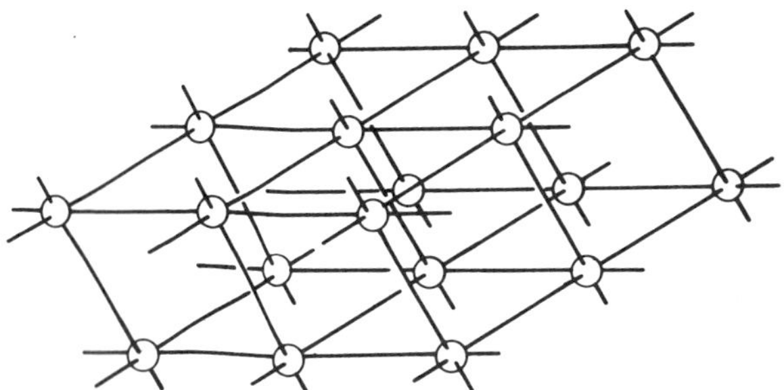

Figure 2-1 Solid phase—typical lattice structure.

2-2.2 LIQUID PHASE

The liquid phase is characterized by being fluid but relatively incompressible, having no permanent resistance to shear stress. Liquids cannot retain an unconstrained shape, but conform to the shape of a container. The molecules of a liquid adhere to each other under minimal forces and can easily slide apart and shift arrangements. For example, water is the liquid phase of H_2O. An example of the liquid phase is shown in Figure 2-2.

2-2.3 THE GASEOUS PHASE

The gaseous phase is characterized by a free movement of the molecules. Consequently, the entire mass tends to expand indefinitely, completely occupying the volume of any container. A gas does not have an independent shape or volume. For example, steam is the gaseous phase of H_2O. An example of the gaseous phase is shown in Figure 2-3.

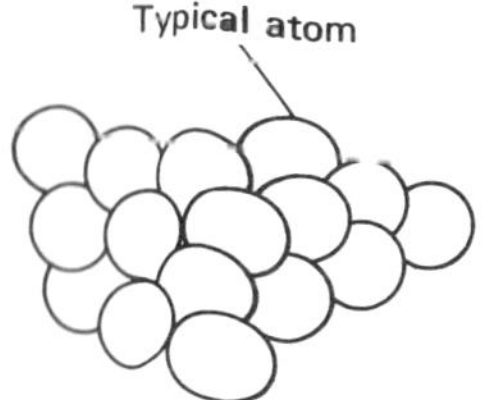

Figure 2-2 Liquid phase—compact structure.

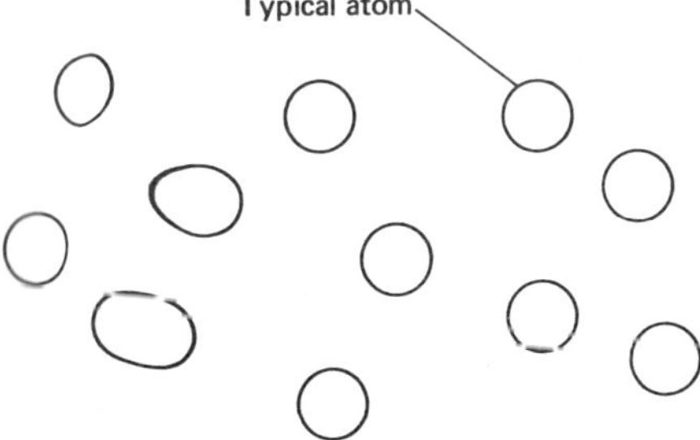

Figure 2-3 Gaseous phase—random structure.

2-3 INTENSIVE PROPERTIES OF THERMODYNAMIC SUBSTANCES

An intensive property is one that is *independent* of the mass of the substance. Examples of intensive properties are

- Density, ρ.
- Pressure, p.
- Temperature, T.
- Specific volume, v.
- Specific weight, w.
- Specific gravity, s.g.
- Specific internal energy, u.

2-3.1 DENSITY

Density (ρ) is an intensive property that is often used in determining the mass of a system. It is defined as the mass per unit volume and thus is the inverse of specific volume, v. The following equations may be used for determining the density of homogeneous materials:

$$\rho = \frac{M}{V} \qquad [\text{kg/m}^3] \tag{2-2}$$

and

$$\rho = \frac{1}{v} \qquad [\text{kg/m}^3] \tag{2-3}$$

where

$$M = \text{mass} \qquad [\text{kg}]$$
$$V = \text{volume} \qquad [\text{m}^3]$$
$$v = \text{specific volume} \qquad [\text{m}^3/\text{kg}]$$

The unit associated with density is kilogram per cubic meter (kg/m^3).

Example Problem 2-5

Calculate the density of a homogeneous fluid such as water whose mass is 53.2 kg within a volume of 0.053 m^3.

Equation

$$\rho = \frac{M}{V} \qquad [\text{kg/m}^3] \qquad (\text{Eq. 2-2})$$

Parameters

$$M = 53.2\ \text{kg} \qquad (\text{given})$$
$$V = 0.053\ \text{m}^3 \qquad (\text{given})$$

Substitution

$$\rho = \frac{53.2}{0.053} \qquad \frac{[\text{kg}]}{[\text{m}^3]}$$

Answer

$$\boldsymbol{\rho = 1003.8\ \text{kg/m}^3}$$

2-3.2 PRESSURE

Pressure (p), an intensive property, is defined as force per unit area as shown by the following equation:

$$p = \frac{F}{A} \qquad [\text{N/m}^2 = \text{Pa}] \tag{2-4}$$

where

$$F = \text{force} \qquad [\text{N}]$$
$$A = \text{area} \qquad [\text{m}^2]$$

Pascal's law states that in liquids or gases, a pressure at a point in one direction produces that same pressure in all directions at the same point.

The unit associated with pressure is pascal [Pa], which is equivalent to newtons per square meter [N/m^2] as shown in Equation 2-4. For most problems in thermodynamics the pressure expressed in pascals would be an extremely small number; therefore, megapascals [MPa] will be the common term. The standard atmospheric pressure at sea level is equal to 0.1013 MPa = 0.76 m (or 760 mm) Hg.

From a consideration of potential energy, we can develop the principle of hydrostatic pressure as follows:

$$p = wh \qquad [\text{Pa}] \tag{2-5}$$

where

$$p = \text{hydrostatic pressure} \qquad [\text{Pa}]$$
$$w = \text{specific weight of fluid} \qquad [\text{N/m}^3]$$
$$h = \text{height (vertical distance)} \qquad [\text{m}]$$

In Equation 2-5 the sign convention for height is positive if the liquid surface is above (increased elevation) the point of reference; and is negative if the liquid surface is below the point of reference. Consequently, submerged objects, such as in the ocean, can be subjected to large hydrostatic pressures when the object is submerged at great depths.

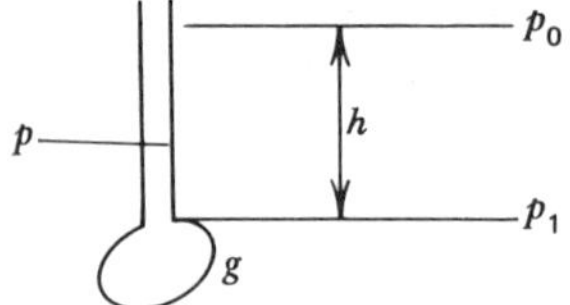

Figure 2-4 Sketch for Example Problem 2-6.

Example Problem 2-6

Determine the hydrostatic pressure of 2 meters of water in a 1-g gravitational field. (See Figure 2-4.)

Equation

$$p_1 = p_0 + (w)(h) \qquad [\text{Pa}]$$

Parameters

$$p_0 = 0.0 \qquad [\text{Pa}]$$

$$w = \rho g \qquad [\text{N/m}^3]$$

$$\rho = 1000 \text{ kg/m}^3 \qquad (\text{Table 2-5})$$

$$g = 9.81 \text{ m/s}^2 \qquad (\text{given})$$

$$w = (1000)(9.81) \qquad [\text{kg/m}^3][\text{m/s}^2] = [\text{kg}\cdot\text{m/s}^2][\text{m}^{-3}] = [\text{N/m}^3]$$

$$= 9810 \text{ N/m}^3$$

$$h = +2 \text{ m} \qquad (\text{given})$$

Substitution

$$p_1 = 0 + (9810)(+2) \qquad [\text{N/m}^3][\text{m}] = [\text{N/m}^2] = [\text{Pa}]$$

Answer

$$\mathbf{p_1 = +19620\ Pa \quad or \quad 0.0196\ MPa}$$

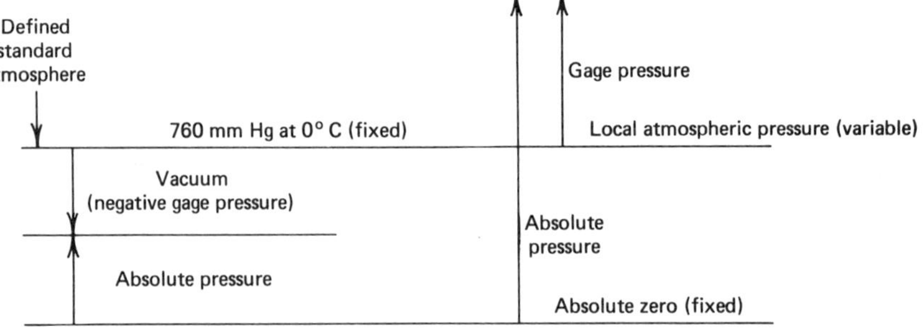

Figure 2-5 Relationship between absolute and gage pressure.

Most mechanical types of pressure gages measure the pressure above local atmospheric (barometric) pressure. This pressure is called gage pressure and is expressed in the SI unit pascal. The relationship between gage pressure and absolute pressure is shown in Figure 2-5.

Pressure values can be expressed in either absolute or gage, and are related (from Figure 2-5) by the following equation:

$$p_{\text{absolute}} = p_{\text{gage}} + p_{\text{atmospheric}} \qquad [\text{Pa}] \tag{2-6}$$

In Equation 2-6 the local atmospheric pressure is usually measured with a barometer. A negative value for gage pressure is called vacuum. Thus, rewriting Equation 2-6, we can obtain:

$$p_{\text{absolute}} = p_{\text{atmospheric}} - p_{\text{vacuum}} \qquad [\text{Pa}] \tag{2-7}$$

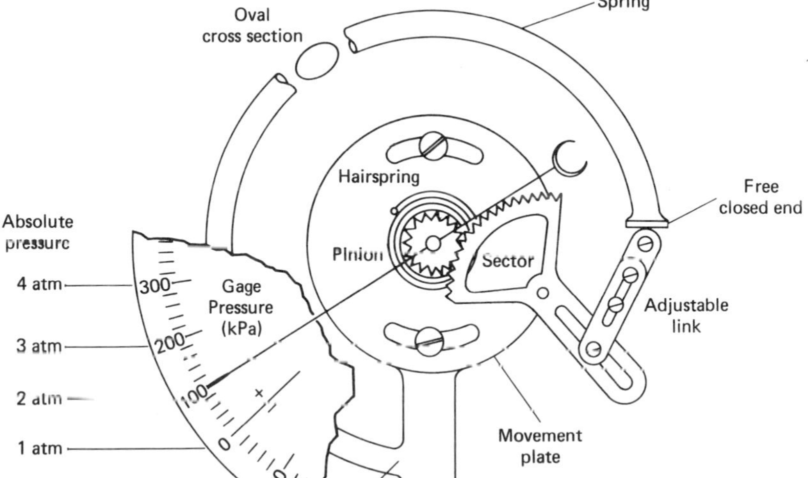

Figure 2-6 A Bourdon gage.

The most common mechanical pressure gage is called the Bourdon gage. This type of gage can be made to measure either absolute or gage pressure; however, most Bourdon tube gages are used for indicating gage pressure. A typical Bourdon gage is shown in Figure 2-6.

2-3.3 TEMPERATURE

Temperature (T), an intensive property, is used to indicate the amount of energy within the molecules of the substances. This molecular energy is called internal energy, and is discussed in the section on internal energy (page 31).

Often a student mistakenly equates temperature [°C] to heat [J]. Temperature is an indication of the amount of internal energy in a substance. Heat is energy in transit across a system boundary that, when entering a system, can become work or additional internal energy (see Section 1-4.3). When the internal energy level of a substance increases, its temperature increases.

The SI unit of temperature measurement is the Kelvin [K]. The Kelvin temperature scale is an absolute temperature scale because it starts from absolute zero. The Kelvin [K] is the preferred temperature scale to describe temperature levels, and the Kelvin temperature unit [ΔK] to describe differences or changes in temperature levels. However, in engineering and nonscientific areas wide use is still being made of the Celsius [°C] scale, which is based on 0°C as the temperature level at which ice and liquid water are in equilibrium at atmospheric pressure.

In thermodynamics there are three different types of temperature parameters regardless of the temperature scale used, namely:

- Temperature level [°C].
- Temperature difference [Δ°C].
- "Per degree" [Δ_1°C].

The temperature level is the normal temperature reading of a thermometer. It is a measure of how hot or cold a substance is.

The temperature difference is simply the difference between two temperature levels. For example, if a substance at a temperature (level) of 25°C were to be heated to a temperature (level) of 35°C, the temperature difference (in this case an increase) is 10 Δ°C. A temperature level [°C] can be increased (or decreased) by a temperature difference. In units [°C]+[Δ°C]=[°C].

The "per degree" temperature parameter is associated with values for various thermodynamic properties for a change in temperature (level) of one (1) degree. An example is specific heat where the value used is the amount of heat required to change the temperature of a unit mass of substance one (1) degree Kelvin. Common arrangement of [Δ°C] and

$[\Delta_1^\circ\mathrm{C}]$ units after equation substitutions with their reduction are as follows:

$$\frac{[\Delta^\circ\mathrm{C}]}{[\Delta_1^\circ\mathrm{C}]}=[-]$$

$$\frac{[-]}{[1/\Delta_1^\circ\mathrm{C}]}=[\Delta^\circ\mathrm{C}]$$

$\frac{[^\circ\mathrm{C}]}{[^\circ\mathrm{C}]}$ or $\frac{[^\circ\mathrm{C}]}{[\mathrm{K}]}$ or $\frac{[\mathrm{K}]}{[^\circ\mathrm{C}]}$ is not proper, and therefore does not cancel.

$$\frac{[\Delta K]}{[\Delta_1\mathrm{K}]}=[-]$$

$$\frac{[\mathrm{K}]}{[\mathrm{K}]}=[-]$$

$$\frac{[\mathrm{K}]}{[\Delta_1\mathrm{K}]}=[-]$$

Many authors use the temperature scale units such as °C for all three of the temperature parameters. Because this can confuse the beginning student, in this book the three different temperature parameters are given different units as indicated above.

Conversion of Temperature Level

°C to K. When there is a need to express a temperature level on the Celsius scale [°C] to the absolute temperature scale of Kelvin [K], the relationship equation is

$$T=T'+273.15 \qquad [\mathrm{K}] \tag{2-8}$$

where

T = the temperature level in K

T' = the temperature level in °C

°F to °C. The conversion of a temperature level on the English Fahrenheit scale [°F] to the corresponding temperature level on the Celsius scale [°C] can be made using the following equation:

$$T=\tfrac{5}{9}(T'-32) \qquad [^\circ\mathrm{C}] \tag{2-9}$$

where

T = the temperature level on the Celsius scale °C
T' = the temperature level on the Fahrenheit scale °F

Conversion of Temperature Units

Δ°C to ΔK. The temperature unit of Kelvin [ΔK] and Celsius [Δ°C] are the same. This is to say that a change of temperature level of one degree Celsius [Δ°C] equals a change of one Kelvin [ΔK]. The equation of equality is

$$[\Delta°\mathrm{C}] = [\Delta\mathrm{K}] \tag{2-10}$$

Δ°F to Δ°C. The equation of equality for a change in temperature between the Fahrenheit and Celsius temperature scales is

$$9\,[\Delta°\mathrm{F}] = 5\,[\Delta°\mathrm{C}] \tag{2-11}$$

From this equation of equality, the conversion factor F_c for converting Δ°F to Δ°C is

$$F_c = \frac{5}{9}\,\frac{[\Delta°\mathrm{C}]}{[\Delta°\mathrm{F}]} \tag{2-12}$$

Example Problem 2-7

A thermodynamic system consisting of a substance such as liquid helium is at −218.0°C. Convert this temperature to Kelvin.

Equation

$$T = T' + 273.15 \qquad [\mathrm{K}] \qquad (\text{Eq. 2-8})$$

Parameter

$$T' = -218.0°\mathrm{C} \qquad (\text{given})$$

Substitution

$$T = (-218.0) + (273.15) \qquad [°\mathrm{C}] + 273.15 = [\mathrm{K}]$$

Answer

$$\mathbf{T = +55.15\ K}$$

Example Problem 2-8

A thermodynamic substance undergoes a temperature change of 36°F. How many degrees Celsius would this be?

Equation

$$T = F_c T' \qquad [\Delta°C] \qquad \text{(Eq. 2-1)}$$

Parameters

$$F_c = \frac{5}{9}\frac{\Delta°C}{\Delta°F} \qquad \text{(Eq. 2-12)}$$

$$T' = 36\ \Delta°F \qquad \text{(given)}$$

Substitution

$$T = \left(\frac{5}{9}\right)(36) \qquad \left[\frac{\Delta°C}{\Delta°F}\right][\Delta°F] = [\Delta°C]$$

Answer

$$\mathbf{T = 20\ \Delta°C}$$

Temperature Scale Datum Points

In order to calibrate temperature-measuring instruments more accurately over a wide range of temperatures, an international agreement of datum points has been established as listed in Table 2-4. These temperature points are referred to as the International Temperature Scale.

Other secondary points have been established through the temperature range from 4.22 K (liquid helium) to 692.65 K (melting point of zinc).

2-3.4 SPECIFIC VOLUME

Specific volume (v) is an intensive property, since it is defined as the volume occupied by a unit of mass. If the substance or matter is homoge-

Table 2-4 International Temperature Scale—Basic Points

Element	Melting or Boiling Point at 1 atm	°C	K
Oxygen	Boiling	−182.97	90.18
Sulfur	Boiling	444.60	717.75
Antimony	Melting	630.50	903.65
Silver	Melting	960.8	1233.95
Gold	Melting	1063.0	1336.15
Water	Boiling	100.0	373.15
Water (Ice)	Melting	0.0	273.15
— (Absolute zero)	—	−273.15	0.0

neous, then the following equation can be used:

$$v = \frac{V}{M} \qquad [\mathrm{m^3/kg}] \tag{2-13}$$

The unit associated with the specific volume is cubic meters per kilogram $[\mathrm{m^3/kg}]$.

2-3.5 SPECIFIC WEIGHT

Specific weight (w) is an intensive property, which is defined as the weight per unit volume. It can be calculated for homogeneous materials by the following equation:

$$w = \frac{W}{V} \qquad [\mathrm{N/m^3}] \tag{2-14}$$

The unit associated with specific weight is newtons per cubic meter $[\mathrm{N/m^3}]$. Since mass and weight can be related using Equation 2-19, Equation 2-14 becomes:

$$w = \frac{Mg}{V} \qquad [\mathrm{N/m^3}] \tag{2-15}$$

where

g = the acceleration of gravity $[\mathrm{m/s^2}]$

Specific weight (w) can also be defined in terms of density as follows:

$$w = \rho g \qquad [\mathrm{N/m^3}] \tag{2-16}$$

2-3.6 SPECIFIC GRAVITY

Specific gravity (s.g.) is an intensive property that is often used to describe fluids. This property is the ratio of the density of any fluid to the density of water at 4°C for liquids. The density of water at this reference temperature has been determined to be 1000 $\mathrm{kg/m^3}$. Thus for liquids we can use the following equation:

$$\text{s.g.} = \frac{\rho}{1000\ \mathrm{kg/m^3}} \qquad [-] \tag{2-17}$$

Note that for density expressed in $[\mathrm{kg/m^3}]$, specific gravity is dimensionless or unitless. Table 2-5 lists values for density and specific gravity of some common liquids.

2-3.7 SPECIFIC INTERNAL ENERGY

Specific internal energy (u) is an intensive property. Internal energy is the energy within the molecules of the substance. This energy is in the form of

Table 2-5 Properties of Common Liquids (0.1 MPa and 25°C)

Substance	Density, ρ [kg/m^3]	Specific Gravity, s.g. (dimensionless)
Acetone	787	0.787
Alcohol, ethyl	787	0.787
Alcohol, methyl	789	0.789
Alcohol, propyl	802	0.802
Ammonia	826	0.826
Benzene	876	0.876
Carbon tetrachloride	1,590	1.590
Castor oil	960	0.960
Ethylene glycol	1,100	1.100
Fuel oil, heavy	906	0.906
Fuel oil, medium	852	0.852
Gasoline	721	0.721
Glycerine	1,263	1.263
Kerosene	823	0.823
Linseed oil	930	0.930
Mercury	13,633	13.633
Propane	495	0.495
Seawater	1,026	1.026
Turpentine	870	0.870
Water	1,000	1.000

vibrations and motions of the particles internal to the molecular structure as well as the motions of the molecule itself. In thermodynamics this internal energy u is not heat q, as heat (defined in Section 1-4.3) is "a form of energy in transit across a system boundary." The internal energy u increases as the temperature T of the substance increases. The magnitude of the internal energy at a particular temperature varies from substance to substance due to the differences in the molecular activity of the substances.

If the substance is homogeneous, the following equation can be used:

$$u = \frac{U}{M} \qquad [\text{kJ/kg}] \tag{2-18}$$

The unit usually associated with specific internal energy is energy per unit mass, expressed in kilojoules per kilogram [kJ/kg].

2-4 EXTENSIVE PROPERTIES OF THERMODYNAMIC SUBSTANCES

Extensive properties are those properties which are *dependent* on the magnitude of the mass of the substance in the thermodynamic system. Examples of extensive properties are

- Mass, M.
- Weight, W.

- Volume, V.
- Energy, E.

It should be noted that any intensive property can be converted to an extensive property by multiplying the intensive property by the mass within the system. For example,

$$V = vM \qquad [\mathrm{m}^3] \qquad \text{(Eq. 2-12)}$$

where

V = total volume, $[\mathrm{m}^3]$ an extensive property

v = specific volume, $[\mathrm{m}^3/\mathrm{kg}]$ and intensive property

M = total mass of the system $[\mathrm{kg}]$

2-4.1 MASS

In describing a thermodynamic system, one of the first properties that we must consider is *mass* (M). Mass is a property of matter to which a substance owes its inertia. In this book we will use the SI units, and therefore denote the basic unit of mass by the symbol M and express it in kilograms [kg].

2-4.2 WEIGHT

Weight (W) is a derived unit (not basic); thus the relationship between newtons, kilograms, meters, and seconds is given by the first law of motion ($F = Ma$). If a body or particle of matter at rest on the Earth's surface is released from the forces holding it at rest, it will experience the acceleration of free fall (acceleration of gravity, g). The force required to restrain it against free fall is commonly called weight. Thus by definition (or derivation) and substitution into the first law of motion:

$$W = F = Mg \qquad [\mathrm{N}] \qquad (2\text{-}19)$$

where

W = weight $[\mathrm{N}]$

F = force $[\mathrm{N}]$

M = mass $[\mathrm{kg}]$

g = acceleration $[\mathrm{m/s^2}]$

Example Problem 2-9

What is the weight of a system having a mass of 10 kg when the system is on the equator at an altitude of 3000 m?

Equation

$$W = F = Ma \qquad [\mathrm{N}] \qquad \text{(Eq. 2-19)}$$

Parameters

$M = 10$ kg (given)

$g = 9.7713\ \mathrm{m/s^2}$ (From Appendix A, Table A-2, on the equator at an elevation of 3000 m)

Substitution

$$W = (10)(9.7713) \qquad [\mathrm{kg}][\mathrm{m/s^2}] = [\mathrm{N}]$$

Answer

$$\mathbf{W = 97.713\ N}$$

2-4.3 VOLUME

Volume (V) is an extensive property and is defined as the amount of space within an identified boundary. The unit associated with volume is cubic meter [$\mathrm{m^3}$]. Although the SI unit for volume is officially the cubic meter, a commonly used unit for liquids is the liter [l]. The liter is equal to 1000 cubic centimeters or 1000 $\mathrm{cm^3} = 0.001\ \mathrm{m^3}$.

2-4.4 ENERGY

Energy (E) is an extensive property, which is defined as the capacity of a given body to produce physical effects external to that body. There are many forms of energy, but the primary ones that we will use in this book are potential (E_p), kinetic (E_k), and internal (U) energy. In the study of thermodynamics it is convenient to consider potential and kinetic energy as separate terms of mechanical energy (E_m).

$$E_m = E_p + E_k \qquad [\mathrm{J}] \tag{2-20}$$

The unit associated with energy is the joule (J), which is also the unit identified with work. A joule is therefore equal to a newton meter (N · m).

2-5 STATE OF THERMODYNAMIC SUBSTANCES

The term *state* is generally defined as the "mode or condition of being." The description of the state of a thermodynamic substance provides a key for the determination of the amount of energy in the thermodynamic substance (or vice versa) from thermodynamic tables.

In a thermodynamic system, work (Wk) output is produced or input required when the thermodynamic system substance changes from one state (state 1) to another state (state 2). Such work output or input is related to:

- The change of energy ($E_1 - E_2$) in the thermodynamic system substance
- The heat flow (Q) across the system boundary during the interval between state 1 and state 2

This relationship is expressed by the following equation:

$$Wk = Q + (E_1 - E_2) \qquad [\mathrm{J}] \tag{2-21}$$

The presence and magnitude of work (Wk) and/or heat flow Q during the interval between state 1 and state 2 affects the magnitude of the change in energy ($E_1 - E_2$) in the substance between the two states. However, the work Wk and the heat flow Q are not directly a part of the description of either state, as discussed in the following sections.

2-5.1 DESCRIPTION OF STATE—SINGLE PHASE

The *state* of a pure substance in one phase neglecting the mechanical energy E_m requires the evaluation of only two intensive properties in combination with one extensive property. These three properties must be independent, that is, not related to each other. Whether or not two intensive properties are independent can readily be determined by examining their units. Units of [m^3], [J], [K], and [Pa] are independent for single-phase substances, and [m^3], [J], [K] or [Pa] (but not both), and quality [—] are independent for two-phase substances, for example:

1. Specific volume [m^3/kg] and specific internal energy [J/kg] are independent.
2. Specific volume [m^3/kg] and density [kg/m^3] are *not* independent because both include the same units [m^3].

An example of a description of state is given in Table 2-6.

The authors recommend the use of the following properties:

Mass M as the extensive property because (a) it is the basic extensive property, and (b) processes between two states either have a constant

Table 2-6 Table of State

Type of Substance	Name or Identification
Mass M	Extensive property (independent)
Temperature T	Intensive property (independent)
Pressure p	Intensive property (independent)

mass or a change (addition or removal) of mass. The use of mass assists the student to better visualize the differences between two states.

Temperature T and pressure p as the two intensive properties because (a) thermodynamic tables for single phase substances are usually organized with temperature and pressure as the key properties for listing other data values, and (b) in an actual system temperature and pressure are usually the two properties that can most readily be measured.

With such a description of a unique state, the energy in the substance can be determined from thermodynamic tables, such as the steam tables discussed later on in this chapter.

The common intensive properties of a thermodynamic substance are listed below in two groups. The first group includes the three properties the authors recommend for the description of state, which we have called "primary," and the second group, which we have called "secondary."

Primary Intensive Properties of State		
Temperature	T	[°C] or [K]
Pressure	p	[Pa]
Quality	x	[—]

Secondary Intensive Properties of State		
Specific volume[a]	v	$[m^3/kg]$
Density[a]	ρ	$[kg/m^3]$
Specific weight[a]	w	$[N/m^3]=[kg/m^3][g]$
Specific gravity[a]	s.g.	$[-]=[kg/m^3][m^3/kg]$
Specific internal energy	u	[kJ/kg]
Specific enthalpy	h	[kJ/kg]
Specific entropy	s	$[kJ/kg \cdot \Delta_1 K]$

[a] Not independent: basic units [m/kg] that may be inverted or multiplied by a constant. For description of state these should be converted to specific volume.

To evaluate the two primary intensive properties of state: temperature and pressure or quality from values of secondary intensive properties different methods are used for different combinations of known intensive properties.

In general there are three different combinations of the intensive properties, which can be grouped as follows:

Group *1*

Primary Property—temperature (or pressure) given.
Secondary property—any one given.

Group 2

Primary Property—quality given.
Secondary Property—any one given.

Group 3

Primary Property—none given.

Secondary Property—any two independent properties given.

A method for obtaining the primary intensive properties for each of these groups is illustrated in the following discussion.

Group 1

Primary Property

Temperature (or pressure)

Secondary property

One of the following:

Specific volume (v)

Specific internal energy (u)

Specific enthalpy (h)

Specific entropy (s)

Step 1: using a saturated thermodynamic property table, go down the temperature (or pressure) column until the given value of the primary property is reached. Then move across that row to the appropriate column for the secondary property.

(a) If the value of the secondary property is larger than the value of the corresponding property for saturated vapor at the given temperature (or pressure) the *substance is single phase superheated vapor*. To complete the description of the state of the substance its *pressure* (*or temperature*) must be determined.

To determine its pressure (or temperature) turn to the tables for thermodynamic properties of superheated vapor.

When the substance temperature is given the values of the corresponding property for different pressures at the given temperature are reviewed to determine the two pressures that "bracket" the given value of the secondary property. The value of the pressure is then obtained by linear interpolation.

When the substance pressure is given, use the table for that pressure and determine the two temperatures that "bracket" the given value of the secondary property. The value of the temperature is then obtained by linear interpolation (see Appendix D-1).

(b) If the value of the secondary property equals or lies between the values for saturated vapor and saturated liquid of the given temperature (or pressure) the *substance is in the saturated region*, since the corresponding pressure (or temperature) is not an independent property. To complete the description of state of the substance its *quality* must be determined.

The quality is evaluated by using Equation 2-26x substituting the given value of the secondary property and the corresponding values for saturated vapor and liquid at the given temperature/pressure.

(c) If the value of the secondary property is smaller than the value of the corresponding property for saturated liquid at the given temperature (or pressure) the substance is *single phase subcooled liquid*. To complete the description of the state of the substance its *pressure* (*or temperature*) must be determined. This is done in a manner analogous to that described for superheated vapors.

Group 2

Primary Property

Quality

Secondary Property

One of the following:

Specific volume (v)

Specific internal energy (u)

Specific enthalpy (h)

Specific entropy (s)

The evaluation of the second primary property (temperature or pressure) is accomplished by an iterative process. A value of temperature (or pressure) is selected and the corresponding value of quality is calculated using Equation 2-26x. A second value is selected and the value of the temperature or pressure is established as discussed in Appendix D-2.

Group 3

Primary property

None given

Secondary property

Any two independent of the following:

Specific volume (v)

*Specific internal energy (u)

*Specific enthalpy (h)

Specific entropy (s)

*Specific internal energy and specific enthalpy are not independent (they have the same units).

The selection of the required primary properties to describe the state, and the subsequent iterative process to determine the values of the two primary properties are discussed in detail in Appendix D-2.

A convenient way to present the description of the state of a substance is illustrated in Table 2-7.

Example Problem 2-10

Describe the state of a thermodynamic substance in a system that contains 2 kg of superheated steam at 300°C at a pressure of 0.20 MPa.

Based upon this description of the state of the substance, various dependent properties can readily be obtained from Appendix A, Table A-3.3, such as specific internal energy u, specific volume v, and so forth. The values of temperature and

Table 2-7 Table of State

Type of Substance: H_2O		Superheated vapor (given) Single-Phase	
Mass M	[kg]	2	(given)
Temperature T	[°C]	300	(given)
Pressure p	[MPa]	0.20	(given)

pressure are the keys to obtaining the values of the other properties at this state.

In order to establish a description of state, it may be necessary at times to use other intensive properties as the independent properties defining the state. However, such properties must be capable of permitting the determination of the three basic properties listed and used in the illustrative example problem, and the three basic properties should be determined and listed. For instance, volume V may be directly available (instead of mass) as the extensive property. The mass can then be obtained by using the "dependent property" specific volume v.

$$M = \frac{V}{v} \quad [\text{kg}] \quad (\text{Eq. 2-13a})$$

When other intensive or extensive properties are involved in a description of state, the authors suggest using a description format beginning with the properties of mass, temperature, and pressure as illustrated in Example Problem 2-10, then drawing a demarkation line, followed by a listing of all pertinent parameters, such as shown in Table 2-8.

2-5.2 TABLES OF THERMODYNAMIC PROPERTIES

In order to solve thermodynamic problems, the values of the properties of the thermodynamic system substance must be available. The values of thermodynamic properties of substances commonly used in thermodynamic systems have been determined by various investigators and have been published, being available in various text books and reference literature. The data needed for the problems in this text are presented in Appendix A.

2-5.3 STEAM TABLES

Water always has been an important thermodynamic substance. Consequently, many studies have been conducted worldwide to determine accu-

Table 2-8 Table of State

Type of Substance: H_2O		Superheated Vapor (given) Single-Phase	
Mass M	[kg]	2.0	(calculated)
Temperature T	[°C]	300	(given)
Pressure p	[MPa]	0.20	(given)
Volume V	[m^3]	2.6324	(given)
Specific volume v	[m^3/kg]	1.3162	(steam tables)

rate values of the properties of water. These tables of property values are called "steam tables."

The most recent compilation of the most accurate values of the thermodynamic properties of water was published in 1969 by Keenan, Keyes, Hill, and Moore.[6] This latest version of the steam tables includes nine tables of thermodynamic properties that cover the phase regions of ice, water, and steam.

Van Wylen and Sonntag[7] have converted the principal tables into SI (metric) units. A sample of the Van Wylen and Sonntag steam tables is given in Tables 2-9 and 2-10. Both tables present columns of specific volume, specific internal energy, specific enthalpy and specific entropy, with rows of values at saturation for various temperatures in Table 2-9 and for various pressures in Table 2-10.

Two values of specific volume v [m^3/kg] are given in both tables: one for saturated liquid, one for saturated vapor.

Three values of specific internal energy u [kJ/kg] are also given. The first column is for saturated liquid; the last column is for saturated vapor. The middle column is the difference between the saturated vapor and saturated liquid columns, representing the internal energy of evaporation.

Both the properties of enthalpy and entropy are introduced later in the chapters and therefore will not be discussed at this time.

In Table 2-9 the second column is the saturation pressure associated with the corresponding temperature in the first column. In Table 2-10 the second column is the saturation temperature associated with the corresponding pressure in the first column.

Separate tables are given for single-phase superheated vapor such as in Table A-3.3 for H_2O. In these tables, for selected pressures columns of temperature T, specific volume v, specific internal energy u, specific enthalpy h, and specific entropy s are given.

Through the use of the thermodynamic information in the steam tables, the basic intensive properties of pressure and/or temperature to describe the state of a single-phase substance can be determined from other intensive properties such as specific volume v or specific internal energy u. For example, if a superheated (single-phase) vapor of H_2O were at a pressure of 1.0 MPa and had a value of internal energy u of 2957.3 kJ/kg, using the steam table (Appendix A-3.3), by moving down the specific internal energy column in the table for a pressure of 1.0 MPa, the value of 2957.3 kJ/kg would be found to be the value at 400°C. Hence the temperature of the superheated vapor would be 400°C.

Example Problem 2-11

If the specific volume v of a superheated water vapor at a pressure of 0.050 MPa were 5.746 m^3/kg, what would be the temperature?

Solution

First step. Go steam table (Appendix A 3.3) for $p = 0.050$ MPa; then go down the column for specific volume v. The value of v at 300°C is smaller and the value at

Table 2-9 Saturation Temperatures, Steam Tables—SI (Metric)

		m^3/kg		kJ/kg			kJ/kg			kJ/kg·K		
		Specific Volume		Internal Energy			Enthalpy			Entropy		
Temperature °C T	Pressure kPa P	Sat. Liquid v_f	Sat. Vapor v_v	Sat. Liquid u_f	Evap. u_{fv}	Sat. Vapor u_v	Sat. Liquid h_f	Evap. h_{fv}	Sat. Vapor h_g	Sat. Liquid s_f	Evap. s_{fv}	Sat. Vapor s_g
0.01	0.6113	0.001 000	206.14	.00	2375.3	2375.3	.01	2501.3	2501.4	.0000	9.1562	9.1562
5	0.8721	0.001 000	147.12	20.97	2361.3	2382.3	20.98	2489.6	2510.6	.0761	8.9496	9.0257
10	1.2276	0.001 000	106.38	42.00	2347.2	2389.2	42.01	2477.7	2519.8	.1510	8.7498	8.9008
15	1.7051	0.001 001	77.93	62.99	2333.1	2396.1	62.99	2465.9	2528.9	.2245	8.5569	8.7814
20	2.339	0.001 002	57.79	83.95	2319.0	2402.9	83.96	2454.1	2538.1	.2966	8.3706	8.6672
25	3.169	0.001 003	43.36	104.88	2304.9	2409.8	104.89	2442.3	2547.2	.3674	8.1905	8.5580
30	4.246	0.001 004	32.89	125.78	2290.8	2416.6	125.79	2430.5	2556.3	.4369	8.0164	8.4533
35	5.628	0.001 006	25.22	146.67	2276.7	2423.4	146.68	2418.6	2565.3	.5053	7.8478	8.3531
40	7.384	0.001 008	19.52	167.56	2262.6	2430.1	167.57	2406.7	2574.3	.5725	7.6845	8.2570
45	9.593	0.001 010	15.26	188.44	2248.4	2436.8	188.45	2394.8	2583.2	.6387	7.5261	8.1618
50	12.349	0.001 012	12.03	209.32	2234.2	2443.5	209.33	2382.7	2592.1	.7038	7.3725	8.0763
55	15.758	0.001 015	9.568	230.21	2219.9	2450.1	230.23	2370.7	2600.9	.7679	7.2234	7.9913
60	19.940	0.001 017	7.671	251.11	2205.5	2456.6	251.13	2358.5	2609.6	.8312	7.0784	7.9096
65	25.03	0.001 020	6.197	272.02	2191.1	2463.1	272.06	2346.2	2618.3	.8935	6.9375	7.8310
70	31.19	0.001 023	5.042	292.95	2176.6	2469.6	292.98	2333.8	2626.8	.9549	6.8004	7.7553
75	38.58	0.001 026	4.131	313.90	2162.0	2475.9	313.93	2321.4	2635.3	1.0155	6.6669	7.6824
80	47.39	0.001 029	3.407	334.86	2147.4	2482.2	334.91	2308.8	2643.7	1.0753	6.5369	7.6122
85	57.83	0.001 033	2.828	355.84	2132.6	2488.4	355.90	2296.0	2651.9	1.1343	6.4102	7.5445
90	70.14	0.001 036	2.361	376.85	2117.7	2494.5	376.92	2283.2	2660.1	1.1925	6.2866	7.4791
95	84.55	0.001 040	1.982	397.88	2102.7	2500.6	397.96	2270.2	2668.1	1.2500	6.1659	7.4159

Adapted from Joseph H. Keenan, Frederick G. Keyes, Philip G. Hill, and Joan G. Moore, *Steam Tables*, (New York: John Wiley & Sons, Inc., 1969).

Table 2-10 Saturation Pressure, Steam Tables—SI (Metric)

		m^3/kg		kJ/kg			kJ/kg			kJ/kg·K		
		Specific Volume		Internal Energy			Enthalpy			Entropy		
Pressure MPa	Temperature °C T	Sat. Liquid v_f	Sat. Vapor v_v	Sat. Liquid u_f	Evap. u_{fv}	Sat. Vapor u_v	Sat. Liquid h_f	Evap. h_{fv}	Sat. Vapor h_g	Sat. Liquid s_f	Evap. s_{fv}	Sat. Vapor s_g
0.40	143.63	0.001 084	0.4625	604.31	1949.3	2553.6	604.74	2133.8	2738.6	1.7766	5.1193	6.8959
0.45	147.93	0.001 088	0.4140	622.77	1934.9	2557.6	623.25	2120.7	2743.9	1.8207	5.0359	6.8565
0.50	151.86	0.001 093	0.3749	639.68	1921.6	2561.2	640.23	2108.5	2748.7	1.8607	4.9606	6.8213
0.55	155.48	0.001 097	0.3427	655.32	1909.2	2564.5	655.93	2097.0	2753.0	1.8973	4.8920	6.7893
0.60	158.85	0.001 101	0.3157	669.90	1897.5	2567.4	670.56	2086.3	2756.8	1.9312	4.8288	6.7600
0.65	162.01	0.001 104	0.2927	683.56	1886.5	2570.1	684.28	2076.0	2760.3	1.9627	4.7703	6.7331
0.70	164.97	0.001 108	0.2729	696.44	1876.1	2572.5	697.22	2066.3	2763.5	1.9922	4.7158	6.7080
0.75	167.78	0.001 112	0.2556	708.64	1866.1	2574.7	709.47	2057.0	2766.4	2.0200	4.6647	6.6847
0.80	170.43	0.001 115	0.2404	720.22	1856.6	2576.8	721.11	2048.0	2769.1	2.0462	4.6166	6.6628
0.85	172.96	0.001 118	0.2270	731.27	1847.4	2578.7	732.22	2039.4	2771.6	2.0710	4.5711	6.6421
0.90	175.38	0.001 121	0.2150	741.83	1838.6	2580.5	742.83	2031.1	2773.9	2.0946	4.5280	6.6226
0.95	177.69	0.001 124	0.2042	751.95	1830.2	2582.1	753.02	2023.1	2776.1	2.1172	4.4869	6.6041
1.00	179.91	0.001 127	0.194 44	761.68	1822.0	2583.6	762.81	2015.3	2778.1	2.1387	4.4478	6.5865
1.10	184.09	0.001 133	0.177 53	780.09	1806.3	2586.4	781.34	2000.4	2781.7	2.1792	4.3744	6.5536
1.20	187.99	0.001 139	0.163 33	797.29	1791.5	2588.8	798.65	1986.2	2784.8	2.2166	4.3067	6.5233
1.30	191.64	0.001 144	0.151 25	813.44	1777.5	2591.0	814.93	1972.7	2787.6	2.2515	4.2438	6.4953
1.40	195.07	0.001 149	0.140 84	828.70	1764.1	2592.8	830.30	1959.7	2790.0	2.2842	4.1850	6.4693
1.50	198.32	0.001 154	0.131 77	843.16	1751.3	2594.5	844.89	1947.3	2792.2	2.3150	4.1298	6.4448
1.75	205.76	0.001 166	0.113 49	876.46	1721.4	2597.8	878.50	1917.9	2796.4	2.3851	4.0044	6.3896
2.00	212.42	0.001 177	0.099 63	906.44	1693.8	2600.3	908.79	1890.7	2799.5	2.4474	3.8935	6.3409
2.25	218.45	0.001 187	0.088 75	933.83	1668.2	2602.0	936.49	1865.2	2801.7	2.5035	3.7937	6.2972
2.5	223.99	0.001 197	0.079 98	959.11	1644.0	2603.1	962.11	1841.0	2803.1	2.5547	3.7028	6.2575
3.0	233.90	0.001 217	0.066 68	1004.78	1599.3	2604.1	1008.42	1795.7	2804.2	2.6457	3.5412	6.1869

400°C is larger than the given value of 5.746 m^3/kg. Therefore, the temperature is between 300 and 400°C.

Second step. Interpolate between 300 and 400°C.

Equations

$$T=300+\Delta T\frac{\Delta v}{\Delta v'} \qquad [°C]$$

$$\Delta T=400-300 \qquad [°C]-[°C]=[\Delta°C]$$

$$\Delta T=100\Delta°C$$

$$\Delta v=v-v_{300} \qquad [m^3/kg]$$

$$=5.746-5.284 \qquad [m^3/kg]-[m^3/kg]=[m^3/kg]$$

$$\Delta v=0.462 \qquad [m^3/kg]$$

$$\Delta v'=v_{400}-v_{300} \qquad [m^3/kg]$$

$$=6.209-5.284 \qquad [m^3/kg]-[m^3/kg]=[m^3/kg]$$

$$\Delta v'=0.925 \qquad [m^3/kg]$$

Substitution

$$T=300+(100)\frac{0.462}{0.925} \qquad [°C]+[\Delta°C]\frac{m^3/kg}{m^3/kg}=[°C]$$

$$=300+(100)(0.50)=300+50$$

Answer

$$\mathbf{T=350°C}$$

2-5.4 SATURATION

When a system contains both liquid and vapor phases of a substance in equilibrium, both the liquid and vapor phases are saturated. In the saturated state, pressure and temperature are *not* independent properties. The pressure/temperature relationship for the saturated condition is unique for each substance. When the pressure is plotted against temperature, a single curve is obtained with the general form shown in Figure 2-7. This curve is called the saturation curve. The saturation curve ends at the critical point because the densities of the saturated liquid and vapor are equal and two phases do not exist. Critical temperatures and pressures for some common thermodynamic substances are listed in Appendix A, Table A-9.

The saturation curve separates the single-phase, superheated vapor region from the single-phase, subcooled liquid region. As the internal energy of superheated vapor is reduced at constant pressure, the temperature of the superheated vapor decreases until the saturation temperature is reached, at which point it becomes saturated vapor. A further reduction in

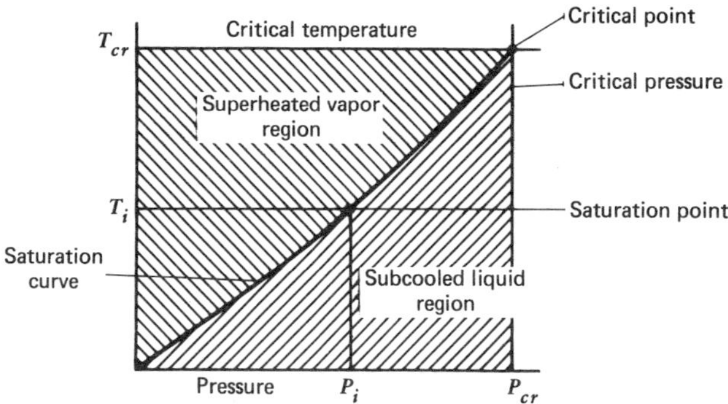

Figure 2-7 A typical saturation curve.

the internal energy effects no decrease in temperature until all of the saturated vapor has been condensed into saturated liquid. As soon as the substance has become a completely saturated liquid, any further decrease in internal energy will result in a decrease in temperature because the liquid is in the single-phase, subcooled liquid region. This is illustrated in Figure 2-8.

For example, if we consider superheated H_2O vapor at a constant pressure of 0.10 MPa, as shown in Figure 2-8, as the specific internal energy u is reduced from 4683.5 kJ/kg to 2675.5 kJ/kg, the temperature of

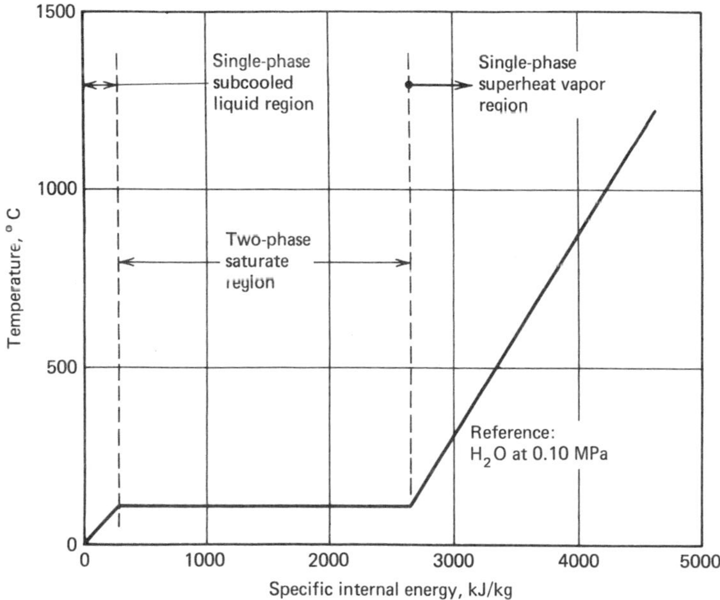

Figure 2-8 Relationship between temperature and specific internal energy of a substance at constant pressure.

the single-phase, superheated steam drops progressively from 1300°C to the saturation temperature (at 0.10 MPa) of 99.63°C. At this point, which is the lower boundary of the single-phase vapor region, the H_2O is 100 percent saturate vapor. As the specific internal energy *u* is further reduced (removing the latent heat of vaporization), the temperature remains constant at the saturation temperature (at 0.10 MPa) of 99.63°C. However, there is a progressive condensation of the saturated vapor into saturated liquid until the H_2O is 100 percent saturated liquid. This occurs when *u* is reduced to 417.36 kJ/kg, which is the upper boundary of the single-phase liquid region. Any further reduction in the specific internal energy *u* results in the liquid's being subcooled with a corresponding progressive reduction in temperature.

Specific values of the saturation temperature of H_2O for various pressures, and vice versa, are given in the pressure and temperature columns in the steam tables (Appendix A, Tables A-3.1 and A-3.2) portions of which have been duplicated as Table 2-9 and 2-10 for use in this chapter.

The saturation curve is important because it defines the temperature–pressure relationship for saturation of the substance involved. When the substance is saturated, its pressure and temperature are not independent intensive properties, being related in accordance with the saturation curve. In order to describe or define a saturated substance, another independent intensive property is required.

The saturation curves for several thermodynamic substances are shown in Figure 2-9.

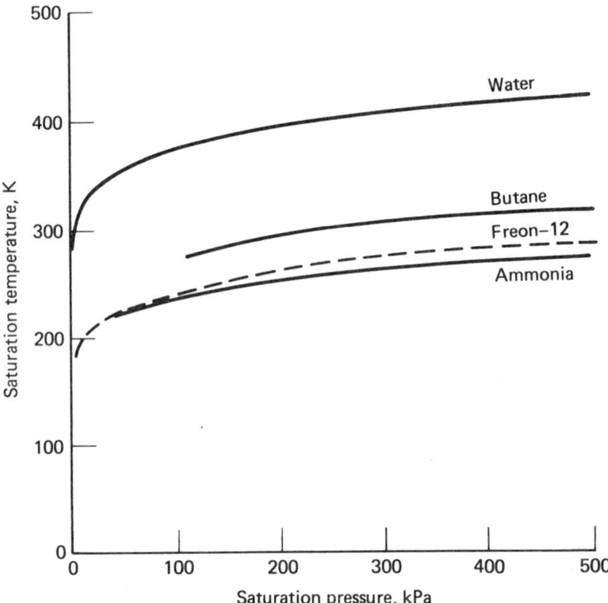

Figure 2-9 Saturation curves for several thermodynamic substances.

Example Problem 2-12

If the temperature of saturated steam is 82°C, what is its pressure?

Solution

First Step. Look up the saturation pressure in the steam tables. Note that, in Table 2-9, saturation pressures are listed for 80 and 85°C but not 82°C. This situation is commonly encountered in the solution of thermodynamic problems. To obtain the desired pressure, an interpolation must be made between the available values. The technique of interpolation is discussed in detail in Appendix D.

Second Step. Make an interpolation between the data for 80 and 85°C.

Equations

$$p_{82}^{0}=p_{80}^{0}+\Delta p \qquad [\mathrm{kPa}]$$

$$\Delta p=\Delta p^{1}\left(\frac{\Delta T}{\Delta T^{1}}\right) \qquad [\mathrm{kPa}]$$

$$\Delta p^{1}=p_{85}^{0}-p_{80}^{0} \qquad [\mathrm{kPa}]$$

Parameters

$$p_{80}^{0}=47.39\ \mathrm{kPa} \qquad \text{(Table 2-9)}$$

$$p_{85}^{0}=57.83\ \mathrm{kPa} \qquad \text{(Table 2-9)}$$

Substitution

$$\Delta p^{1}=57.83-47.39 \qquad [\mathrm{kPa}]-[\mathrm{kPa}]=[\mathrm{kPa}]$$

$$\Delta p^{1}=10.44\ \mathrm{kPa}$$

$$\Delta T^{1}=85-80 \qquad [°\mathrm{C}]-[°\mathrm{C}]=[\Delta°\mathrm{C}]$$

$$\Delta T^{1}=5\Delta°\mathrm{C}$$

$$\Delta T=82-80 \qquad [°\mathrm{C}]-[°\mathrm{C}]=[\Delta°\mathrm{C}]$$

$$p_{82}^{0}=(47.39)+(10.44)\frac{(2)}{(5)} \qquad [kPa]+[kPa]\frac{[\Delta°\mathrm{C}]}{[\Delta°\mathrm{C}]}=[\mathrm{kPa}]$$

$$=(47.39)+(4.18)\ \mathrm{kPa}$$

Answer

$$\mathbf{p_{82}^{0}=51.57\ kPa}$$

Saturated Liquid

Saturated liquid is liquid at the saturation temperature and pressure anywhere along the saturation curve.

Saturated Vapor

Saturated vapor is vapor at the saturation temperature and pressure anywhere along the saturation curve.

2-5.5 SUPERHEATED VAPOR

If the temperature of the substance is higher than the saturation curve, referring to Figure 2-7, the substance is in a single-phase superheat region. The properties are given in the superheat section of the steam tables in Appendix A-3. The properties are keyed to the pressure and temperature of the substance.

2-5.6 SUBCOOLED LIQUID

If the temperature of the substance is lower than the saturation temperature, referring to Figure 2-7, the substance is in a single-phase subcooled region. The properties are given in the subcooled liquid section of the steam tables in Appendix A-3. The properties are keyed to the pressure and temperature of the substance.

2-5.7 QUALITY

The fact that a substance is saturated (being at the state of T_i and p_i in Figure 2-7) indicates that if liquid or vapor or both are present, such liquid and vapor will be saturated. To indicate the amount of the substance in the vapor phase of a saturated mixture of liquid and vapor, the term quality (x) is used. *Quality* is defined as the ratio of the mass of the vapor phase to the total mass of the mixture, and is considered to be an intensive property. Quality is only applicable to the saturated state and may range in values from 0.00 (100 percent saturated liquid) to 1.00 (100 percent saturated vapor).

$$x = \frac{M\text{ vapor}}{M\text{ mixture}} \qquad [-] \tag{2-22}$$

Another term sometimes used to describe the quality region between 100 percent saturated liquid and 100 percent saturated vapor is *percent moisture*. A vapor–liquid mixture that has a quality of 0.90 has 10 percent moisture. In other words, percent moisture is 100 percent minus the quality in percent.

2-5.8 DESCRIPTION OF STATE—TWO-PHASE (SATURATION)

The state of a pure substance in two-phase (saturation) (neglecting mechanical energy E_m) requires the evaluation of two independent intensive properties in combination with one extensive property, in a similar manner to the single-phase description of state. However, in the case of two phases (saturation), temperature and pressure are not independent intensive properties.

The authors recommend the use of the following basic properties to describe the state:

- Mass M.
- Pressure, p, or temperature T.
- Quality x.

These properties either give direct access to listings in thermodynamic tables using p or T, or are properties used in the basic or common calculations, and hence must also be evaluated.

An example of a description of state is given in Table 2-11. Temperature T as the independent intensive property could be used instead of pressure.

The volume V of a two-phase system can be determined by adding the volume of the liquid phase and the vapor phase.

$$V_t = V_f + V_g \qquad [\mathrm{m}^3]$$

The volume of the liquid phase is

$$V_f = M_f v_f \qquad [\mathrm{m}^3]$$

In terms of the total mass M and quality x, the mass of the liquid phase M_f is

$$M_f = M(1-x) \qquad [\mathrm{kg}]$$

and

$$V_f = M(1-x)v_f \qquad [\mathrm{m}^3]$$

Also, the volume of the vapor phase is

$$V_g = M_g v_g \qquad [\mathrm{m}^3]$$

In terms of the total mass M and quality x, the mass of the vapor phase M_g is

$$M_g = Mx \qquad [\mathrm{kg}] \qquad \text{and} \qquad V_g = Mxv_g \qquad [\mathrm{m}^3]$$

Table 2-11 Table of State

Type of Substance	Name or Identification (saturated)
Mass M	Extensive property
Pressure p	Intensive property (independent)
Quality x	Intensive property (independent)

Adding the volumes of the vapor and liquid phases to obtain the total system volume V results in

$$V = M\left[xv_g + (1-x)v_f\right] \qquad [\mathrm{m^3}] \tag{2-23}$$

where

M = mass [kg]

x = quality [—]

v_g = specific volume of the vapor phase $[\mathrm{m^3/kg}]$

v_f = specific volume of the liquid phase $[\mathrm{m^3/kg}]$

In a similar manner the internal energy U can be determined by

$$U = M\left[xu_g + (1-x)u_f\right] \qquad [\mathrm{kJ}] \tag{2-24}$$

where

u_g = specific internal energy of the vapor phase $[\mathrm{kJ/kg}]$

u_f = specific internal energy of the liquid phase $[\mathrm{kJ/kg}]$

Quality can be defined by other properties. For example, the specific volume v might be given as the independent intensive property. Quality can be evaluated from the specific volume by using Equation 2-23, dividing both sides by M. This results in

$$\frac{V}{M} = v = xv_g + (1-x)v_f \qquad [\mathrm{m^3/kg}] \tag{2-25}$$

or

$$v = v_f + x(v_{fg}) \qquad [\mathrm{m^3/kg}] \tag{2-25a}$$

where

v = specific volume of the total substance $[\mathrm{m^3/kg}]$

x = quality [—]

v_g = specific volume of saturated vapor $[\mathrm{m^3/kg}]$

v_f = specific volume of saturated liquid $[\mathrm{m^3/kg}]$

v_{fg} = increase of specific volume between saturated vapor and liquid $[\mathrm{m^3/kg}]$

Solving Equation 2-24 for the quality x results in

$$x = \frac{v - v_f}{v_g - v_f}[-] \quad \text{or} \quad \frac{v - v_f}{v_{fg}} \quad [-] \qquad \text{(2-26)}$$

From this equation it is apparent that the quality of a system with a fixed specific volume will change with temperature as the values for the specific volumes of the vapor and liquid change with saturation temperature.

Example Problem 2-14

If the specific volume v of the saturated two-phase substance in a thermodynamic system is 2.0 [m^3/kg], and the specific volumes of the saturated liquid v_f and vapor v_g are 0.001 [m^3/kg] and 2.5 [m^3/kg] respectively, what is the quality x of the substance?

Equation

$$x = \frac{v - v_f}{v_g - v_f} \quad [-] \qquad \text{(Eq. 2-26)}$$

Parameters

$$v = 2.0 \text{ m}^3/\text{kg} \quad \text{(given)}$$

$$v_f = 0.001 \text{ m}^3/\text{kg} \quad \text{(given)}$$

$$v_g = 2.5 \text{ m}^3/\text{kg} \quad \text{(given)}$$

Substitution

$$x = \frac{(2.0) - (0.001)}{(2.5) - (0.001)} \qquad \frac{[\text{m}^3/\text{kg}] - [\text{m}^3/\text{kg}]}{[\text{m}^3/\text{kg}] - [\text{m}^3/\text{kg}]} = [-]$$

$$= \frac{1.999}{2.499}$$

Answer

$$\mathbf{x = 0.800} \quad [-]$$

If specific volume is not given, but volume V as an independent extensive property is given, the specific volume can readily be obtained as

$$v = \frac{V}{M} \quad [\text{m}^3/\text{kg}] \qquad \text{(Eq. 2-13)}$$

Example Problem 2-15

If a tank with a volume of 1 m^3 contains 10 kg of combined saturated water (liquid) and steam (vapor) at a temperature of 90°C, what is the quality x of the substance?

Equation

$$x = \frac{v - v_f}{v_g - v_f} \qquad [—] \qquad \text{(Eq. 2-26)}$$

Parameters

$$v = \frac{V}{M} \qquad [\text{m}^3/\text{kg}] \qquad \text{(Eq. 2-13)}$$

$$V = 1\ \text{m}^3 \qquad \text{(given)}$$

$$M = 10\ \text{kg} \qquad \text{(given)}$$

Therefore,

$$v = \tfrac{1}{10}[\text{m}^3]/[\text{kg}]$$

$$= 0.10\ \text{m}^3/\text{kg}$$

$v_f = 0.001036\ \text{m}^3/\text{kg}$
$v_g = 2.361\ \text{m}^3/\text{kg}$
(From the steam tables at 90°C, Table 2-9 in the text, or Appendix A, Table A-3.1)

Substitution

$$x = \frac{(0.10) - (0.001036)}{(2.361) - (0.001036)} \qquad \frac{[\text{m}^3/\text{kg}] - [\text{m}^3/\text{kg}]}{[\text{m}^3/\text{kg}] - [\text{m}^3/\text{kg}]} = [—]$$

$$= \frac{0.09896}{2.03600}$$

Answer

$$\mathbf{x = 0.042\ [—]}$$

Quality can also be obtained from specific internal energy u or from the total internal energy U in the same manner as by the use of specific volume v or total volume V. For a given state the quality will be the same value regardless of the intensive property used (i.e., specific volume v, specific internal energy u, specific enthalpy h, or specific entropy s).

Example Problem 2-16

If a tank containing a saturated mixture of water and steam at 95°C has an internal energy U of 20 MJ with a mass M of 10 kg, what will be the quality x of the substance?

Equation

$$x = \frac{u - u_f}{u_g - u_f} \qquad [—] \qquad \text{(Eq. 2-26)}$$

Parameters

$$u = \frac{U}{M} \quad [\text{kJ/kg}] \qquad (\text{Eq. 2-18})$$

$$U = 20\ [\text{MJ}] = 20{,}000 \quad [\text{kJ}] \qquad (\text{given})$$

$$M = 10\ \text{kg} \qquad (\text{given})$$

Therefore,

$$u = \frac{20{,}000}{10} \quad [\text{kJ}]/[\text{kg}]$$

$$= 2000\ \text{kJ/kg}$$

$u_f = 397.88$ kJ/kg
$u_g = 2500.6$ kJ/kg

(From the steam tables at 95°C, Table 2-9 in the text, or Appendix A, Table A-3.1)

Substitution

$$x = \frac{(2000) - (397.88)}{(2500.6) - (397.88)} \qquad \frac{[\text{kJ/kg}] - [\text{kJ/kg}]}{[\text{kJ/kg}] - [\text{kJ/kg}]} = [—]$$

$$= \frac{1602.12}{2102.72}$$

Answer

$$\mathbf{x = 0.762\ [—]}$$

In the case where neither the pressure nor quality is given, the state can be described if the specific volume v and the specific internal energy u are both given. As the quality of the state must be the same value whether it is calculated using the specific volume or specific internal energy, the following equation can be made:

$$\frac{v - v_f}{v_g - v_f} = \frac{u - u_f}{u_g - u_f} \qquad [—] \tag{2-27}$$

If we use Equation 2-27, the answer is obtained through a series of iterations, selecting a saturation pressure and substituting the values for the saturated liquid and vapor properties, repeating until a value of saturation pressure is found where the two parts of Equation 2-27 are equal. The value of the equation at this point is the quality x.

Example Problem 2-17

If a steam–water mixture in equilibrium has a specific volume v of 2.00 m^3/kg and a specific internal energy u of 2170 kJ/kg, what is the quality x of the substance and the temperature?

Plan

Use Equation 2-27 and assume trial temperature, iterating until Equation 2-27 is satisfied.

Equation

$$\frac{v-v_f}{v_g-v_f}=\frac{u-u_f}{u_g-u_f} \qquad [—] \qquad \text{(Eq. 2-27)}$$

Parameters

$$v=2.00\ \text{m}^3/\text{kg} \qquad \text{(given)}$$
$$u=2170\ \text{kJ/kg} \qquad \text{(given)}$$

First Iteration

Assume $T=80°\text{C}$.

$$\left.\begin{aligned} v_f&=0.001029\ \text{m}^3/\text{kg}\\ v_g&=3.407\ \text{m}^3/\text{kg}\\ u_f&=334.86\ \text{kJ/kg}\\ u_g&=2482.2\ \text{kJ/kg}\end{aligned}\right\} \quad \text{Table 2-9 in text or Appendix A, Table A-3.1}$$

Quality by Specific Volume

Substitution

$$x_v=\frac{2.00-0.001029}{3.407-0.001029}=\frac{1.99897}{3.40597}=0.587 \qquad [—]$$

Quality by Specific Internal Energy

$$x_u=\frac{2170-334.86}{2482.2-334.86}=\frac{1835.14}{2147.34}=0.855 \qquad [—]$$
$$x=x_u\neq x_v$$

Since $0.855\neq 0.587$, the quality by specific volume is not equal to the quality by specific internal energy; therefore, assume another temperature.

Second Iteration

Assume that $T=90°\text{C}$.

$$\left.\begin{aligned} v_f&=0.001036\ \text{m}^3/\text{kg}\\ v_g&=2.361\ \text{m}^3/\text{kg}\\ u_f&=376.85\ \text{kJ/kg}\\ u_g&=249h.5\ \text{kJ/kg}\end{aligned}\right\} \quad \text{Table 2-9 in text or Appendix A, Table A-3.1}$$

Quality by Specific Volume

$$x_v=\frac{2.00-0.001036}{2.361-0.001036}=\frac{1.9990}{2.3600}=0.847\ [—]$$

Table 2-12 Table of State

Type of Substance	Water–Steam	
Mass M	2 kg	(given)
Temperature T	90°C	(given Problem 2-16 solution)
Quality x	0.847 [—]	(given Problem 2-16 solution)

Quality by Specific Internal Energy

$$x_u = \frac{2170 - 376.85}{2494.5 - 376.85} = \frac{1793.15}{2117.65} = 0.847\ [—]$$

Answer

$$\boldsymbol{x = 0.847}$$

Thus

$$x_u = x_v (0.847 = 0.847)$$

Therefore, the assumed temperature of the second iteration is correct.
Answer

$$\boldsymbol{T = 90°\mathrm{C}}$$

Example Problem 2-18

If the mass of the substance in Example Problem 2-16 is 2 kg, give a description of state.

Plan

Make a table of the key independent descriptive properties as shown in Table 2-12.

2-5.9 CHANGE OF STATE—TWO-PHASE (SATURATION)

In thermodynamics the change of state from an initial state (state 1) to another state (state 2) by the addition (or removal) of mass, and so forth is important. The change may occur at constant volume or constant pressure, or under some other conditions. In order to describe state 2 in terms of mass M, pressure p (or temperature T), and quality x, the common properties between state 1 and state 2 and the changed property values are listed in the table of state. When sufficient intensive and extensive property values have been identified to establish state 2, the property values for mass, pressure (or temperature), and quality can be established. This is illustrated by the following example problem.

Example Problem 2-19

A butane tank has a volume of 1 m^3. It is one-quarter full of liquid butane at 21.1°C. A mass of 400 kg of liquid butane saturated at 18.3°C is added. What is

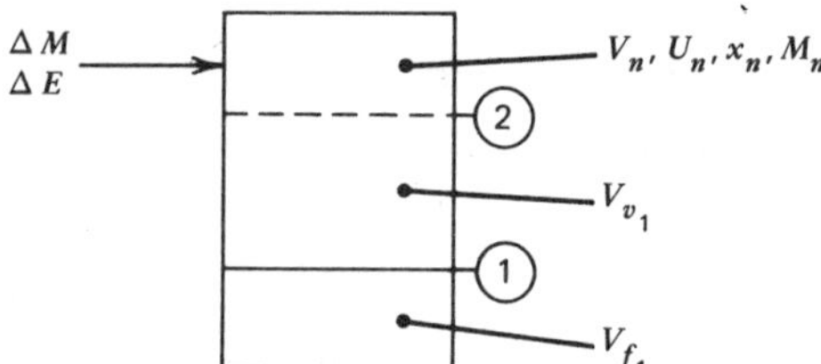

Figure 2-10 Sketch for Example Problem 2-19.

the state of the butane in the tank at equilibrium after adding the butane? (See Figure 2-10.)

Plan

Make a table of state listing state 1 and state 2. (See Table 2-13.) The volume remains constant: $V_1 = V_2$. Then proceed as follows.

(C-1) Determine the mass (M) for state 1. $M_2 = M_1 + \Delta M$, which is given. Verify that state 2 is two-phase (the volume liquid less than the tank volume)

Identify the intensive properties that can be used with M_2 to define state 2. (They are v_2 and u_2.)

(C-2) v_2 is determined from V_2 and M_2.

(C-3) u_2 determination requires evaluating U_2

(C-4) U_2 is the sum of $U_1 + \Delta E$, where ΔE is the energy added to the system by the injection of the additional mass of butane.

(C-5) U_1 calculated from the description of state 1.

(C-6) ΔE is the sum of the internal energy in the injected liquid plus the flow work of injection across the system boundary (into the tank).

$$\Delta E = \Delta M(u_\Delta + p_1 v_\Delta)$$

(C-7) Determine the quality from u_1 and v_2 and p_2.

(C-1) Equation

$$M_2 = M_1 + \Delta M \qquad [\text{kg}]$$

Table 2-13 Table of State

Type of Substance	Butane	1 Two-phase sat. (g)	2 Two-phase sat. (C–1)
Mass M	kg	148.7 (C–1)	548.7 (C-1)
Pressure p	kPa		212 (C-7)
Quality x	[—]	217.9 (st)	0.0005 (C-7)
Volume V	m^3	1 (g)	1 (g)
Specific volume v	m^3/kg		0.001822 (C-2)
Internal energy U	kJ	49,981 (C–5)	177,873 (C-4)
Specific internal energy u	kJ/kg		324.17 (C-3)
Temperature T	°C	21.1 (g)	

Parameters

M_1— Determine from the given total volume V_1 and the relative volumes of the liquid and vapor phases of state 1

$$M_1 = M_{f_1} + M_{g_1} \qquad [\text{kg}]$$

$$M_{f_1} = \frac{V_{f_1}}{v_{f_1}} \qquad [\text{kg}] \qquad (\text{Eq. 2-13}x)$$

$$M_{g_1} = \frac{V_{g_1}}{v_{g_1}} \qquad [\text{kg}] \qquad (\text{Eq. 2-13}x)$$

$$V_{g_1} = 3\ V_{f_1} \qquad (\text{given})$$

Therefore,

$$M_{g_1} = \frac{3\ V_{f_1}}{v_{v_1}} \qquad [\text{kg}]$$

Substituting the above terms for M_{f_1} and M_{g_1} in the M_1 equation, we obtain

$$M_1 = \frac{V_{f_1}}{v_{f_1}} + \frac{3\ V_{f_1}}{v_{g_1}} \qquad [\text{kg}]$$

$$V_{f_1} = 0.25\ \text{m}^3 \qquad (\text{given}) \left(\tfrac{1}{4} \text{ of a 1-m}^3 \text{ tank}\right)$$

$$v_{f_1} = 0.73 \times 10^{-3}\ \text{m}^3/\text{kg (st) (saturated liquid at 21.1°C)}$$

$$v_{g_1} = 0.180\ \text{m}^3/\text{kg (st) (saturated vapor at 21.1°C)}$$

$$M_1 = \frac{0.25}{1.73 \times 10^{-3}} + \frac{(3)(0.25)}{0.180} \quad \frac{[\text{m}^3]}{[\text{m}^3/\text{kg}]} + \frac{[\text{m}^3]}{[\text{m}^3/\text{kg}]} = [\text{kg}]$$

$$M_1 = 144.5 + 4.2$$

$$M_1 = 148.7 \text{ kg}$$

$$\Delta M = 400 \text{ kg} \qquad (\text{given})$$

Substitution

$$M_2 = 148.7 + 400 \qquad [\text{kg}] + [\text{kg}] = [\text{kg}]$$

Answer

$$\mathbf{M_2 = 548.7\ [kg]}$$

The following procedure verifies that state 2 is two-phase.

Test Equation

$$M_2 v_{f_1} < V_1$$

Parameters

$M_2 = 548.7$ [kg] (C-1)

$v_{f_1} = 1.73 \times 10^{-3}$ m^3/kg (st) (larger specific volume)

$V_1 = 1$ m^3 (given)

Substitution

$$(548.7)(1.73 \times 10^{-3}) < 1$$
$$0.949 < 1$$

Therefore, state 2 is two-phase.

(C-2) Equation

$$v_2 = \frac{V_2}{M_2} \quad \text{m}^3/\text{kg} \qquad \text{(Eq. 2-13)}$$

Parameters

$V_2 = 1$ m^3 (given)

$M_2 = 548.7$ kg (C-1)

Substitution

$$v_2 = \frac{1}{548.7} \qquad \frac{[\text{m}^3]}{[\text{kg}]} = [\text{m}^3/\text{kg}]$$

Answer

$$\mathbf{v_2 = 0.001822} \qquad [\mathbf{m^3/kg}]$$

(C-5) Equation

$$U_1 = M_1 x u_{g_1} + M_1(1-x)u_{f_1} \qquad \text{[kJ]} \qquad \text{(Eq. 2-24)}$$

Parameters

$M_1 = 148.7$ kg (C-1)

$$x = \frac{M_{v_1}}{M_1} \qquad [\text{—}]$$

$$M_{g_1} = \frac{V_{g_1}}{v_{g_1}} \qquad \text{[kg]}$$

$V_{g_1} = 0.75$ m^3 (given)

$v_{g_1} = 0.180$ m^3/kg (st) (saturated vapor at 21.1°C)

$$M_{g_1}=\frac{0.75}{0.180} \qquad \frac{[\mathrm{m}^3]}{[\mathrm{m}^3/\mathrm{kg}]}=[\mathrm{kg}]$$

$$=4.17\ \mathrm{kg}$$

$$x=\frac{4.17}{148.7} \qquad \frac{[\mathrm{kg}]}{[\mathrm{kg}]}$$

$$=0.028$$

$$u_{g_1}=691.95\ [\mathrm{kJ/kg}]\ (\mathrm{st})\ (\text{saturated vapor at } 21.1^\circ\mathrm{C})$$

$$u_{f_1}=325.87\ [\mathrm{kJ/kg}]\ (\mathrm{st})\ (\text{saturated liquid at } 21.1^\circ\mathrm{C})$$

$$U_1=(148.7)(0.028)(691.95)+(148.7)(1-0.028)(325.87)$$

$$[\mathrm{kg}][—][\mathrm{kJ/kg}]+[\mathrm{kg}][—][\mathrm{kJ/kg}]=[\mathrm{kJ}]$$

$$=2881+47{,}100$$

Answer

$$\mathbf{U_1=49{,}981\ kJ}$$

(C-6). Equation

$$\Delta E=\Delta M(u_\Delta+p_1 v_\Delta) \qquad [\mathrm{kJ}]$$

Parameters

$$\Delta M=400\ \mathrm{kg} \qquad (\text{given})$$

$$u_\Delta=319.36\ \mathrm{kJ/kg}\ (\mathrm{st})\ (\text{saturated liquid at } 18.3^\circ\mathrm{C})$$

$$p_1=217.9\ \mathrm{kPa}\ (\mathrm{st})\ (\text{saturated at } 21.1^\circ\mathrm{C})$$

$$v_\Delta=1.72\times10^{-3}\ \mathrm{m^3/kg}\ (\mathrm{st})\ (\text{saturated liquid at } 18.3^\circ\mathrm{C})$$

Substitution

$$\Delta E=(400)(319.36)+(217.9)(1.72\times10^{-3})$$

$$\cdot[\mathrm{kg}][\mathrm{kJ/kg}]+[\mathrm{kPa}][\mathrm{m^3/kg}]$$

$$[\mathrm{kPa}][\mathrm{m^3/kg}]=\left[\frac{\mathrm{kN}}{\mathrm{m}^2}\right]\left[\frac{\mathrm{m}^3}{\mathrm{kg}}\right]=\left[\frac{\mathrm{kNm}}{\mathrm{kg}}\right]=\left[\frac{\mathrm{kJ}}{\mathrm{kg}}\right]$$

$$[kg]\left[\frac{\mathrm{kJ}}{\mathrm{kg}}+\frac{\mathrm{kJ}}{\mathrm{kg}}\right]=[\mathrm{kJ}]+[\mathrm{kJ}]=[\mathrm{kJ}]$$

$$\Delta\mathrm{E}=(400)(319.36+0.37)$$

$$=(400)(319.73)$$

Answer

$$\mathbf{\Delta E=127{,}892\ kJ}$$

(C-4). Equation

$$U_2=U_1+\Delta E \qquad [\mathrm{kJ}]$$

Parameters

$$U_1=49{,}981 \quad [\text{kJ}] \quad (\text{C-5})$$

$$\Delta E=127{,}892 \quad [\text{kJ}] \quad (\text{C-6})$$

Substitution

$$U_2=49{,}981+127{,}892 \quad [\text{kJ}]+[\text{kJ}]=[\text{kJ}]$$

Answer

$$\mathbf{U_2=177{,}873} \quad [\mathbf{kJ}]$$

(C-3). Equation

$$u_2=\frac{U_2}{M_2} \quad [\text{kJ/kg}] \quad (\text{Eq. 2-18})$$

Substitution

$$u_2=\frac{177873}{548.7} \quad \frac{[\text{kJ}]}{[\text{kg}]}=[\text{kJ/kg}]$$

Answer

$$\mathbf{u_2=324.17} \quad [\mathbf{kJ/kg}]$$

(C-7). Equation

$$\frac{v_2-v_{f_2}}{v_{g_2}-v_{f_2}}=\frac{u_2-u_{f_2}}{u_{g_2}-u_{f_2}}=x \quad (\text{Eq. 2-27})$$

Solve by assuming a saturation pressure, to evaluate v_{f_2}, v_{g_2}, u_{f_1}, and u_{f_2}, from the steam tables; when the two sides of the equation are equal, the saturation pressure and quality have been established.

First Iteration
Assume that $p_2=210$ kPa.
Parameters
(by interpolation)

$$v_f=v_{f_{199.3}}+\Delta v_f$$

$$v_{f_{199.3}}=0.00172 \quad [\text{m}^3/\text{kg}]$$

$$\Delta v_f=\simeq 0 \quad \text{by inspection}$$

$$v_f=0.00172$$

$$v_g=v_{g_{199.3}}+\Delta v_g \quad [\text{m}^3/\text{kg}]$$

$$v_{g_{199.3}}=0.195 \quad [\text{m}^2/\text{kg}]$$

Interpolation data are given in Table 2-14.

Table 2-14 Interpolation Data

p	v_g
199.3	0.195
210.0	v_g
217.9	0.180

$$\Delta v_g = \left(\frac{210.0-199.3}{217.9-199.3}\right)(0.180-0.195)$$

$$= \left(\frac{10.7}{18.6}\right)(-0.015) = (0.575)(-0.015)$$

$$\Delta v_g = -0.009 \qquad [\mathrm{m^3/kg}]$$

$$v_g = 0.195 - 0.009$$

$$v_g = 0.186 \qquad [\mathrm{m^3/kg}]$$

$$u_f = u_{f_{199.3}} + \Delta u_f \qquad [\mathrm{kJ/kg}]$$

$$u_{f_{199.3}} = 319.36 \qquad [\mathrm{kJ/kg}]$$

Interpolation data are given in Table 2-15.

Table 2-15 Interpolation Data

p	u_f
199.3	319.36
210.0	u_f
217.9	325.87

$$\Delta u_f = \left(\frac{210.0-199.3}{217.9-199.3}\right)(325.87-319.36)$$

$$= (0.575)(6.51)$$

$$= 3.74$$

$$u_f = 319.36 + 3.74$$

$$= 323.10\ \mathrm{kJ/kg}$$

$$u_g = u_{g_{199.3}} + \Delta u_g \qquad [\mathrm{kJ/kg}]$$

$$u_{g_{199.3}} = 688.22 \qquad [\mathrm{kJ/kg}]$$

Interpolation data are given in Table 2-16.

Table 2-16 Interpolation Data

p	u_g
199.3	688.22
210.0	u_g
217.9	691.95

$$\Delta u_g = \left(\frac{210.0-199.3}{217.9-199.3}\right)(691.95-688.22)$$

$$= (0.575)(3.73)$$

$$\Delta u_g = 2.14 \qquad [\mathrm{kJ/kg}]$$

$$u_g = 688.22 + 2.14$$

$$= 670.36 \qquad [\mathrm{kJ/kg}]$$

Substitution

$$\frac{0.001822-0.00172}{0.186-0.00172} \qquad \frac{324.17-323.10}{670.36-323.10} \qquad \frac{0.00010}{0.184} \; \frac{1.07}{347.26}$$

$$0.00054 \neq 0.00308$$

By inspection we find that there is little change in the specific volumes, mainly in internal energy; u_f must be higher.

Second Iteration

Try $p = 213$ kPa.

$$v_f = 0.00173 \qquad [\mathrm{m^3/kg}]$$

$$v_g = 0.195 + \Delta v_g$$

Interpolation data are given in Table 2-17.

Table 2-17 Interpolation Data

p	v_g
199.3	0.195
213.0	v_g
217.9	0.180

$$\Delta v_g = \left(\frac{213.0-199.3}{217.9-199.3}\right)(0.180-0.195)$$

$$= \left(\frac{13.7}{18.6}\right)(-0.015) = (0.7366)(-0.015)$$

$$= -0.011$$

$$v_g = 0.195 - 0.011$$

$$= 0.184 \qquad [\mathrm{m^3/kg}]$$

$$u_f = 319.36 + \Delta u_f \qquad [\mathrm{kJ/kg}]$$

Interpolation data are given in Table 2-18.

Table 2-18 Interpolation Data

p	u_f
199.3	319.36
213.0	u_f
217.9	325.87

$$\Delta u_f = \left(\frac{213.0-199.3}{217.9-199.3}\right)(325.87-319.36)$$
$$= (0.7366)(6.51)$$
$$= 4.80$$
$$u_f = 319.36 + 4.80$$
$$= 324.16 \qquad [\text{kJ/kg}]$$
$$u_g = 688.22 + \Delta u_g \qquad [\text{kJ/kg}]$$

Interpolation data are given in Table 2-19.

Table 2-19 Interpolation Data

p	u_g
199.3	688.22
213.0	u_g
217.9	691.95

$$\Delta u_g = \left(\frac{213.0-199.3}{217.9-199.3}\right)(691.95-688.22)$$
$$= (0.7366)(3.73)$$
$$\Delta u_g = 2.75 \qquad [\text{kJ/kg}]$$
$$u_g = 688.22 + 2.75$$
$$= 690.97 \qquad [\text{kJ/kg}]$$

Substitution

$$\frac{0.001822-0.00173}{0.184-0.00173} \qquad \frac{324.17-324.16}{690.97-324.16}$$
$$\frac{0.00009}{0.182} \neq \frac{0.01}{366.81}$$
$$0.00049 \neq 0.000027$$

The pressure has been bracketed. It is a little less than the second iteration assumption. The available data do not have sufficient figures to obtain an accurate answer.

Final Answer

pressure $p_2 = 212$ kPa

and

quality $x = 0.0005$

2-6 PROBLEMS

2-1. Define a dimension and a unit. How are they related?

2-2. Give the basic SI (metric) units by listing:
(a) The dimensions by the physical parameter and symbol.
(b) The associated units by name and abbreviation.

2-3. Give the secondary SI (metric) units used in thermodynamics by listing:
(a) The dimensions by the physical parameter and symbol.
(b) The associated units by name and abbreviation.
(c) The equivalence in terms of the basic SI units.

2-4. Convert the following to SI (metric units):
(a) 3 [Btu] to joules [J].
(b) 10 [Btu/s] to watts [W].
(c) A temperature change of 9Δ°F to a temperature change in Δ°C; a temperature change in Kelvin [ΔK].
(d) 212°F to °C.
(e) 500 [ft·lbs] to joules [J].
(f) 2 [HP] to watts [W].
(g) Convert 80°C to Kelvin [K].
(h) 10 joules [J] to [BTU].
(i) 5 watts [W] to [BTU/hr].
(j) A temperature change of 15°C to a temperature change in Δ°F.
(k) 68°F to °C.
(l) 100 Newton meters [Nm] to joules [J].
(m) 5 [kW] to Horsepower [HP].
(n) Convert 293 K to °C.

2-5. Name four common intensive properties of a substance, and give the SI unit associated with each one.

2-6. Convert a gage pressure of 200 kPa to an absolute pressure.

2-7. What is the absolute pressure of a gas having a 150-kPa gage pressure at a barometric pressure of 80 kPa?

2-8. Convert an absolute pressure of 2.0 MPa to a gage pressure when the barometer is 100 kPa.

2-9. The absolute pressure of a gas in a tank is 350 kPa. If the ambient barometric pressure is 90 kPa, what would the gage pressure be?

2-10. What is the difference between intensive and corresponding extensive properties of a substance?

2-11. What intensive properties have no extensive property counterpart?

2-12. When an intensive property has an extensive property counterpart, what is the relationship between the two? Describe by an equation.

2-13. Name three common extensive properties of a substance and give the SI units associated with each one.

2-14. Why is the description of the *state* of a thermodynamic substance important?

2-15. Name a set of two independent intensive properties that would be needed to describe the state of a single phase pure substance.

2-16. Would the set of specific internal energy u and specific volume v meet the requirements for intensive properties needed to describe the state of a single phase pure substance? Give the reasoning for your answer.

2-17. Would the set of specific volume v and mass M describe the state of a single-phase pure substance? Give the reasoning for your answer.

2-18. Name the extensive property that would, in conjunction with the intensive properties cited in Problem 2-15, complete the description of the state of a single-phase pure substance.

2-19. (a) Describe the state of a thermodynamic system that contains 2 kg of liquid water (saturated) at 85°C.
(b) What is the volume of the substance in the system?

2-20. Describe the state of 2 kg saturated liquid water having an internal energy U of 1208.62 kJ.

2-21. (a) Describe the state of a thermodynamic system with 2 kg of steam (saturated) at 85°C.
(b) What is the volume of the substance in the system?

2-22. Describe the state of steam in terms of the primary properties of mass M, pressure p, and temperature T for the following conditions:
(a) A pressure of 0.30 MPa, and a specific volume v of 0.7964 m^3/kg.
(b) A specific volume v of 2.172 m^3/kg at a temperature of 200°C.
(c) A specific volume v of 0.2358 m^3/kg at a temperature of 300°C.

2-23. Steam is at a pressure of 3.0 MPa and a temperature of 500°C. How much volume will 10 kg of the steam occupy?

2-24. What is the saturation curve? Why is it important?

2-25. Define:
(a) Saturated liquid.
(b) Saturated vapor.
(c) Superheated vapor.
(d) Subcooled liquid.
(e) Quality (with respect to a two-phase substance).

2-26. Name a set of two independent intensive properties that would be needed to describe the state of a two phase pure substance.

2-27. Would the set of temperature T and pressure p meet the requirements for the intensive properties needed to describe the state of a two-phase pure substance? Give the reasoning for your answer.

2-28. Would the set of quality x and specific volume v meet the requirements for the intensive properties needed to describe the state of a two-phase pure substance? Give the reasoning for your answer. How would you establish the state of the substance from these properties?

2-29. Would the set of specific internal energy u and specific volume v meet the requirements for the intensive properties needed to describe the state of a two phase pure substance? Give the reasoning for your answer. How would you establish the state of the substance from these properties?

2-30. Name the extensive property that would, in conjunction with the intensive properties cited in Problem 2-29, complete the description of the state of a two-phase pure substance.

2-31. Describe the state of a thermodynamic system in equilibrium that contains 2 kg of liquid water and 2 kg of vapor at 100°C.

2-32. What is the volume V of the system described in Problem 2-31?

2-33. What is the total internal energy U in the system described in Problem 2-31?

2-34. What is the specific volume v of the system described in Problem 2-31?

2-35. What is the specific internal energy u in the system described in Problem 2-31?

2-36. Describe the state of steam in terms of the primary properties of mass M, pressure p, temperature T, and quality x, as required, for the following:

(a) Steam with a specific internal energy of 2482 kJ/kg at a temperature of 300°C.

(b) Steam exhausting from a turbine at 0.1013 MPa pressure with an enthalpy h of 2552 kJ/kg.

2-37. If heat were added to the system described in Problem 2-31 raising the temperature to 120°C, but maintaining the volume constant, what would be the quality? How does this compare to the quality in Problem 2-31?

2-38. If 1 kg of water vapor at a pressure of 200 kPa and a temperature of 200°C were injected into the system described in Problem 2-31 with no volume change, describe the final state. Consider only the internal energy of the injected steam.

2-39. Describe the state of a thermodynamic system in equilibrium that contains 2 kg of liquid water, a quality x of 0.2 [—]; and a specific volume v of 0.68222 m^3/kg.

2-40. What is the volume V of the system described in Problem 2-39?

2-41. What is the specific internal energy u in the system described in Problem 2-39?

2-42. If a thermodynamic system has 2 kg of mass M, a specific volume v of 0.1541 m^3/kg and a specific internal energy u of 1621.63 kJ/kg, describe its state in terms of mass, temperature and quality.

2-43. A tank half full by volume of normal saturated butane is at 26.7°C.

(a) Describe the state of the closed system for 1 kg of substance (butane saturated liquid and saturated vapor combined).

(b) What is the internal energy per unit mass of this substance?

2-44. The tank in Problem 2-43 is cooled down to 21.1°C (the total volume of the saturated liquid and vapor remains the same).

(a) What is the pressure in the tank?

(b) What is the internal energy per unit mass of this substance?

(c) How much heat (kJ/kg) would have to be added to the 21.1°C butane liquid-vapor mixture to bring it back to the 26.7°C condition?

2-45. If the butane tank in Problem 2-43 were equipped with a "safety" burst diaphragm that would rupture at 1.0 MPa gage pressure, how much heat could the butane in the tank absorb before the burst diaphragm would rupture?

2-46. If the butane tank in Problem 2-45 were $\frac{7}{8}$ full at 26.7°C how much heat could the butane in the tank absorb before the burst diaphragm would rupture? Note that when the liquid volume equals the tank volume, any increase in temperature will cause a rupture of the burst diaphragm.

2-47. Assume that 2 kg of water (generic term) are inside a tank of 1 m^3 volume at 115°C. Then 1.1 kg of superheated steam with 3313.1 kJ of energy is injected into the tank. Describe the state of the substance in the tank after equilibrium has been established.

2-48. A tank half full of liquid ammonia at 24°C is left in the sunshine and becomes heated to 40°C. What will be the pressure in the tank?

2-49. What is the volume of the liquid ammonia in the tank of Problem 2-48 when it has been heated to 40°C? Give the answer in terms of volume (m^3) per kilogram of ammonia. What volumetric fraction of the tank is this?

2-6 - $P_{ABS} = P_{gage} + P_{ATM}$

CHAPTER 3

EQUIVALENCE OF WORK AND HEAT

The physical principle known as the conservation of energy is the basis for the first law of thermodynamics. The equivalence of work and heat was first proposed by Isaac Newton, and this concept led to the study of thermodynamics as a discipline. The first law of thermodynamics equates the various forms of energy and relates the transformation from one form of energy to another.

3-1 FIRST LAW OF THERMODYNAMICS

The principle of conservation of energy is the first law of thermodynamics. The energy involved in a thermodynamic system can be divided into three forms, namely:

- Work Wk, output or input.
- Heat Q, input or output.
- Total energy E in the system substance.

One way of stating the first law of thermodynamics is: "The sum of the energies going out of the system must equal the sum of the energies going into the system." This statement can be put into the form of an equation in the following manner:

$$\text{energy going out} = \text{energy going in}$$

$$E_0 + Q_0 + Wk_o = E_i + Q_i + Wk_i$$

$$(Wk_o - Wk_i) = (Q_i - Q_o) + (E_i = E_o)$$

since

$$(Wk_o - Wk_i) = Wk_{\text{net}}$$

$$(Q_i - Q_o) = Q_{\text{net}}$$

$$(E_i - E_o) = \Delta E$$

$$Wk_{\text{net}} = Q_{\text{net}} + \Delta E \qquad [\text{J}] \tag{3-1}$$

Equation 3-1 equates the net work output of a system (Wk_{net}) to any change in the total energy (ΔE) of the system and the net amount of heat transferred (Q_{net}) into (or out of) the system. Equation 3-1 also indicates that

1. If *no heat transfer is involved*, $\Delta Q = 0$.

The work output comes from a reduction in the total energy of the system substance

$$Wk = +\Delta E \qquad [\mathrm{J}] \tag{3-1a}$$

2. If *no work output or input is involved*, $Wk = 0$.

The total energy of the system substance increases or decreases with heat transfer into or out of the system

$$\Delta E = \Delta Q \qquad [\mathrm{J}] \tag{3-1b}$$

3-2 FORMS OF ENERGY

The energy in the system substance may be in the form of:

- Kinetic energy (E_k).
- Potential energy (E_p).
- Internal energy (U)—for closed systems.
- Internal energy plus the flow work entering or leaving the system ($U+pV$)—for open systems

Each of these forms of energy is discussed in the following sections. The total energy (E) in the system substance is the sum of the kinetic (E_k), potential (E_p), and the internal energy (U).

$$E = E_k + E_p + U \qquad [\mathrm{J}]\text{—closed systems} \tag{3-2}$$

$$= E_k + E_p + (U + pV) \qquad [\mathrm{J}]\text{—open systems} \tag{3-3}$$

Sometimes the sum of the kinetic energy E_k and the potential energy E_p is called the total mechanical energy E_m of the system. Thus

$$E_m = E_k + E_p \qquad [\mathrm{J}] \tag{3-4}$$

Substituting E_m for $(E_k + E_p)$, we find that Equation 3-2 becomes

$$E = E_m + U \qquad [\mathrm{J}] \tag{3.2a}$$

and Equation 3-3 becomes

$$E = E_m + (U + pV) \qquad [\mathrm{J}] \tag{3-3a}$$

3-2.1 KINETIC ENERGY

The formula for kinetic energy E_k follows; the derivation from Newtonian mechanics can be found in a physics text.

$$E_k = \frac{M\mathbf{V}^2}{2} \qquad [\mathrm{J}] \tag{3-5}$$

where

$$M = \text{mass} \qquad [\mathrm{kg}]$$
$$\mathbf{V} = \text{velocity} \qquad [\mathrm{m/s}]$$

Example Problem 3-1

Determine the kinetic energy of a jet of steam whose mass is 2 kg traveling at a speed of 1000 meters per second.

Equation

$$E_k = \tfrac{1}{2}M\mathbf{V}^2 \qquad [\mathrm{J}] \qquad (\text{Eq. 3-5})$$

Parameters

$$M = 2\ \mathrm{kg} \qquad (\text{given})$$
$$\mathbf{V} = 1000\ \mathrm{m/s} \qquad (\text{given})$$

Substitution

$$\tfrac{1}{2}(2)(1000)^2 \qquad [\mathrm{kg}][\mathrm{m/s}]^2$$

$$[\mathrm{kg}][\mathrm{m}^2]/[\mathrm{s}^2] = \frac{[\mathrm{kg}][\mathrm{m}]}{[\mathrm{s}^2]}[\mathrm{m}] = [\mathrm{N}][\mathrm{m}] = [\mathrm{J}]$$

Answer

$$\mathbf{E_k = 1{,}000{,}000\ J\ or\ 1.0\ MJ}$$

3-2.2 POTENTIAL ENERGY

Equation 3-6 is the formula for potential energy E_p. The elevation Z is based upon any selected datum, which must be defined for the system being considered.

$$E_p = MgZ \qquad [\mathrm{J}] \tag{3-6}$$

where:

$$M = \text{mass} \qquad [\mathrm{kg}]$$
$$Z = \text{elevation above datum} \qquad [\mathrm{m}]$$
$$g = \text{local acceleration of gravity} \qquad [\mathrm{m/s^2}]$$

Example Problem 3-2

Determine the potential energy of 2 kg of steam at an elevation of 5 m above a datum (reference plane).

Equation

$$E_p = MgZ \qquad [\mathrm{J}] \qquad (\text{Eq. 3-6})$$

Parameters

$M = 2\ \mathrm{kg}$ (given)

$Z = 5\ m$ (given)

$g = 9.8\ \mathrm{m/s^2}$ (nominal Earth's gravitational) (constant from Appendix A-2)

Substitution

$$E_p = (2)(9.8)(5) \qquad [\mathrm{kg}][\mathrm{m/s^2}][\mathrm{m}] = [\mathrm{kg \cdot m^2/s^2}] = [\mathrm{J}]$$

Answer

$$\mathbf{E_p = 98\ J}$$

3-2.3 TOTAL MECHANICAL ENERGY

The total mechanical energy E_m of a system is the sum of the kinetic and potential energy. This relationship can be written as follows:

$$E_m - E_k + E_p \qquad [\mathrm{J}] \qquad (\text{Eq. 3-4})$$

Example Problem 3-3

Determine the total mechanical energy of the steam described in the two previous example problems.

Equation

$$E_m = E_k + E_p \qquad [\mathrm{J}] \qquad (\text{Eq. 3-4})$$

Parameters

$$E_k = 1{,}000{,}000 \qquad [\mathrm{J}] \qquad (\text{given})$$

$$E_p = 98 \qquad [\mathrm{J}] \qquad (\text{given})$$

Substitution

$$E_m = (1{,}000{,}000) + (98) \qquad [\mathrm{J}] + [\mathrm{J}] = [\mathrm{J}]$$

Answer

$$E_m = 1{,}000{,}098 \text{ J} \qquad \text{or} \qquad 1.0 \text{ MJ}$$

3-2.4 INTERNAL ENERGY

Internal energy U is often described as the property that reflects the mechanical energy of the atoms or molecules of the substance. In general, the various components of internal energy are

- Translational or kinetic energy of the atoms or molecules in random motion.
- Vibrational energy of the individual molecules due to straining of the atomic bonds.
- Rotational energy of those molecules which spin about their axis.
- Chemical energy.

There are other forms of energy such as strain energy, electromagnetic energy and nuclear energy that must be included if applicable in the overall analysis of a thermodynamic problem; however, these forms of energy are of lesser importance to us in this introductory book and therefore will be omitted.

In solving thermodynamic problems wherein internal energy is involved, the magnitude of the change in internal energy during the thermodynamic process is the important value.

$$\Delta U = U_2 - U_1 \qquad [\text{J}] \tag{3-7}$$

where

ΔU = change in internal energy between states 1 and 2 [J]

U_1 = internal energy at state 1 [J]

U_2 = internal energy at state 2 [J]

As there is no simple definition for evaluating an "absolute" value of internal energy level (in contrast to the evaluation of kinetic and potential energy), values of the amount of increase or change U_I in internal energy from a convenient datum D are used. In the analysis of a thermodynamic process when the internal energy value of the final state is subtracted from the internal energy of the initial state, the net change in the internal energy ΔU during the process is obtained and is independent of the datum upon which the internal energy value was based (as long as both values are based on the same datum), as shown by the following equations.

$$U_1 = D + U_{I_1}$$
$$U_2 = D + U_{I_2}$$
$$\Delta U = U_2 - U_1$$
$$= (D + U_{I_2}) - (D + U_{I_1})$$
$$= U_{I_2} - U_{I_1} \qquad \text{Eq. 3-7}$$

The amount of internal energy increase or decrease from a selected datum is obtained empirically (by experiment). Such values for internal energies for substances at temperature and pressure conditions commonly encountered in engineering problems can be found in reference books. Tables for water (steam), Freon, ammonia, and butane are given in Appendix A of this text.

The parameter $U + pV$ for open systems is treated in a manner similar to U as discussed in Section 3-5.2.

3-3 ENERGY SIGN CONVENTIONS

The usual sign conventions for the three forms of energy in the thermodynamic equations are

$+Wk$ = work is done *by* the system on something in the surroundings.
$-Wk$ = work is done *on* the system by something in the surroundings.
$+Q$ = heat is transferred *into* the system.
$-Q$ = heat is transferred *out* of the system.
$+U$ = energy level higher than tabulated datum.
$-U$ = energy level lower than tabulated datum.

Since signs can be a source of confusion, absolute values often will be used to define net heat transferred. Thus

$$Q_{\text{net}} = |Q_{\text{in}}| - |Q_{\text{out}}|$$

3-4 FIRST LAW APPLIED TO CLOSED SYSTEMS

When considering the first law of thermodynamics as it applies to a closed (nonflow) system, the kinetic and potential energy of the system usually do not change or are negligible (zero). Consequently, only the net heat flow

Q_{net}, net work Wk_{net} and the change in the internal energy of the substance ΔU are involved. The change of state of the substance in the closed system occurs during a time period wherein the input/output of work Wk or heat Q may be involved with an associated change in the internal energy ΔU of the substance. The energy equation for the closed system based upon the first law of thermodynamics is

$$\Delta U = Q_{net} - Wk_{net} \qquad [\mathrm{J}]$$

If one is primarily interested in the net work Wk_{net}, this equation can be rearranged as follows:

$$Wk_{net} = Q_{net} - \Delta U \qquad [\mathrm{J}] \tag{3-8}$$

or

$$wk_{net} = q_{net} - \Delta u \qquad [\mathrm{kJ/kg}] \tag{3-8a}$$

but

$$\Delta U = U_2 - U_1$$

$$\Delta u = u_2 - u_1$$

Substituting for Δu in Equation 3-8a, we obtain

$$wk_{net} = q_{net} - (u_2 - u_1) \qquad [\mathrm{kJ/kg}] \tag{3-8b}$$

where subscript 1 refers to the initial state and subscript 2 refers to the final state.

Example Problem 3-4

During the compression stroke of an internal combustion engine, the heat rejected is 50 kJ/kg and the work input is 100 kJ/kg. Calculate the change in internal energy of the working fluid, stating whether it is an increase or a decrease. (See Figure 3-1.)

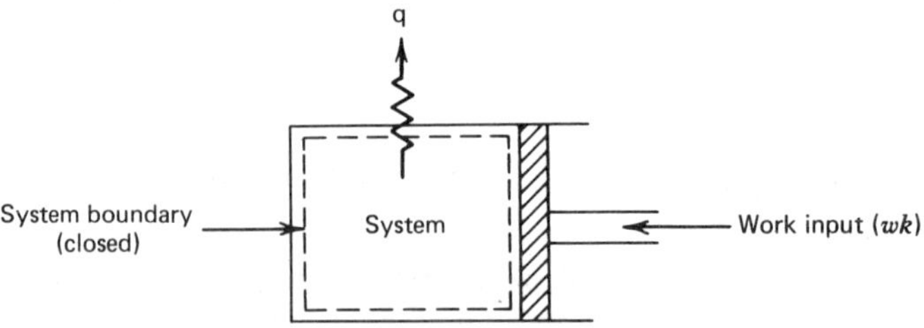

Figure 3-1 Sketch for Example Problem 3-4.

Equation

$$\Delta u = q - wk \qquad [\text{kJ/kg}] \qquad (\text{Eq. 3-8x})$$

Parameters

$$q = -50 \text{ kJ/kg} \qquad (\text{given})$$
$$wk = -100 \text{ kJ/kg} \qquad (\text{given})$$

Substitution

$$\Delta u = (-50) - (-100) \qquad [\text{kJ/kg}] - [\text{kJ/kg}] = [\text{kJ/kg}]$$

Answer

$$\boldsymbol{\Delta u = +50 \text{ kJ/kg}}$$

Since Δu is positive, there is an increase in the internal energy of this system during compression.

3-4.1 ADIABATIC PROCESS

In a closed thermodynamic process or a system where no heat flow is involved (adiabatic process or system) Equation 3-8 becomes:

$$Wk_{\text{net}} = -\Delta U \qquad [\text{J}] \tag{3-9}$$

$$Wk_{\text{net}} = U_1 - U_2 \qquad [\text{J}] \tag{3-9a}$$

$$wk_{\text{net}} = u_1 - u_2 \qquad [\text{kJ/kg}] \tag{3-9b}$$

Example Problem 3-5

In a Diesel engine air is compressed adiabatically by a piston moving in a closed cylinder. To effect the compression stroke, an input of 150 J of energy is required. Determine the change in the internal energy of the air between the start and completion of the compression stroke. (See Figure 3-2.)

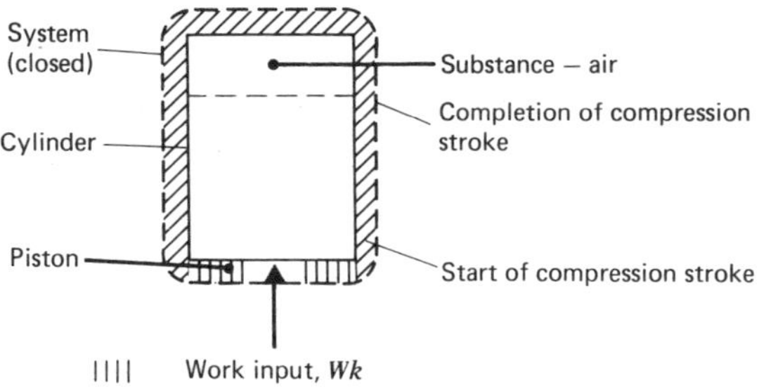

Figure 3-2 Sketch for Example Problem 3-5.

Equation

$$\Delta U = -Wk_{\text{net}} \qquad [\text{J}] \qquad (\text{Eq. 3-9})$$

Parameters

$$Wk_{\text{net}} = -150 \qquad [\text{J}] \qquad (\text{given})$$

Substitution

$$\Delta U = -(-150) \qquad [\text{J}]$$

Answer

$$\mathbf{\Delta U = +150\ J}$$

Thus an increase in internal energy in the air results from the work input (energy added).

3-4.2 HEAT FLOW ONLY

In a closed thermodynamic process or system where no work is involved, Equation 3-7 becomes

$$\Delta U = Q_{\text{net}} \qquad [\text{J}] \qquad (3\text{-}10)$$

Example Problem 3-6

A tank of 0.4654 m^3 volume is filled with steam at a pressure of 1.00 MPa and a temperature of 250°C. If 631.6 kJ of heat is added,

(a) How much would the internal energy of the steam increase?
(b) Describe the final state of the steam.

Solution (a):
(See Figure 3-3.)

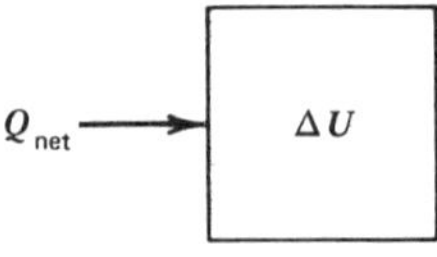

Figure 3-3 Sketch for Example Problem 3-6a.

Equation

$$\Delta U = Q_{\text{net}} \qquad [\text{J}] \qquad (\text{Eq. 3-10})$$

Parameters

$$Q_{\text{net}} = 631.6\ \text{kJ} \qquad (\text{given})$$

Substitution

$$\Delta U = 631.6 \text{ kJ}$$

Answer

$$\mathbf{\Delta U = 631.6 \text{ kJ}}$$

Solution (b)
(See Figure 3-4 and Table 3-1.)

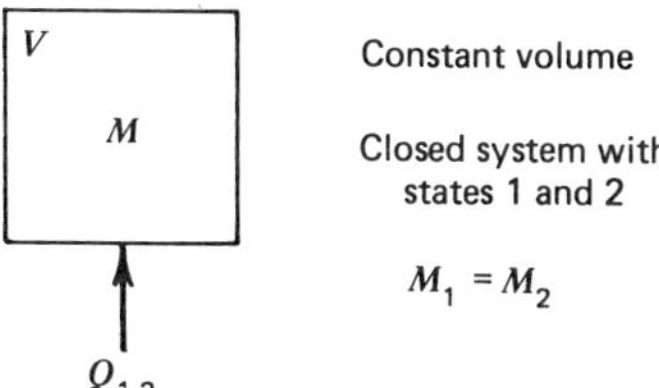

Figure 3-4 Sketch for Example Problem 3-6b.

The required parameters are

- Condition of the substance to establish the properties needed to define state 2.
- Mass M_2, which is the same as M_1.
- Temperature T_2.
- Pressure p_2 if in superheat, or quality x if in a two-phase saturate condition.

These parameters can be determined in the following manner:

1. Establish the condition of the H_2O in state 1. This is done by referring to the steam tables using p_1 and T_1, which were given. The H_2O is found to be in the superheat region.
2. Establish the condition of the H_2O in state 2. As heat Q is added at constant mass and volume, state 2 is also in the superheat region.

Table 3-1 Description of State

	1		2	
Substance H_2O	Superheated steam	(st)	Superheated steam	(st)
Mass M	2 kg	(C−1)	2 kg	
Pressure p	1.00 MPa	(g)	1.40 MPa	(C-6)
Temperature T	250°C	(g)	443.4°C	(C-5)

Symbols. (g) given; (st) from steam tables, Appendix A-3.3; and (C−) calculated.

Therefore, to define state 2, mass M_2, temperature T_2, and pressure p_2 are all that is required.

3. Establish T_2 through u_2. This requires the determination of the mass (M) in the system.
4. Establish p_2 by referring to the superheated steam tables to find the pressure p_2, which corresponds to T_2 and u_2.

Determination of M_2. Calculation C-1

$$M_2 = M_1 = M \qquad [\text{kg}]$$

$$M = \frac{V}{v_1} \qquad [\text{kg}] \qquad (\text{Eq. 2-13a})$$

$$V = 0.4654\ \text{m}^3 \qquad (\text{given})$$

$$v_1 = 0.2327\ \text{m}^3/\text{kg} \qquad (\text{Appendix, A-3.3 at 1.0 MPa, 250°C})$$

$$M = \frac{0.4654}{0.2327} \qquad \frac{[\text{m}^3]}{[\text{m}^3/\text{kg}]} = [\text{kg}]$$

$$M = 2.0\ \text{kg} \qquad (\text{To table of state and C-2})$$

Determination of u_2. Calculation C-2

$$u_2 = \frac{U_2}{M} \qquad [\text{kJ/kg}] \qquad (\text{Eq. 2-18})$$

$$U_2 = 6051.4\ \text{kJ} \qquad (\text{C-3})$$

$$\text{M} = 2.0\ \text{kg} \qquad (\text{C-1})$$

$$u_2 = \frac{6051.4}{2.0} \qquad \frac{[\text{kJ}]}{[\text{kg}]} = [\text{kJ/kg}]$$

$$u_2 = 3025.7\ \text{kJ/kg}$$

Determination of U_2. Calculation C-3

$$U_2 = U_1 + \Delta U \qquad [\text{kJ}] \qquad (\text{Eq. 3-7a})$$

$$U_1 = 5419.8\ \text{kJ} \qquad (\text{C-4})$$

$$\Delta U = 631.6\ \text{kJ} \qquad (\text{given})$$

$$U_2 = 5419.8 + 631.6 \qquad [\text{kJ}] + [\text{kJ}] = [\text{kJ}]$$

$$U_2 = 6051.4\ \text{kJ} \qquad (\text{to C-2})$$

Determination of U_1. Calculation C-4

$U_1 = Mu_1$ [kJ] (Eq. 2-18)

$M = 2.0$ kg (C-1)

$u_1 = 2709.9$ kJ/kg (Appendix, A-3.3 at 1.0 MPa and 250°C)

$U_1 = (2.0)(2709.9)$ [kg][kJ/kg] = [kJ]

$U_1 = 5419.8$ kJ (to C-3)

Determination of T_2. Calculation C-5

Since the specific internal energy of a substance is a function of temperature, the steam tables can be used to obtain the corresponding temperature. It can be seen from the steam tables that for superheated steam there is a slight increase in the specific internal energy with an increase in pressure as real superheated steam characteristics deviate slightly from the theoretical values.

For the calculated value of u_2 (3025.7 kJ/kg), the temperature T_2 will lie between 400 and 500°C throughout the pressure range of 1.00 to 10.0 MPa. The specific volume v_2 (0.2327 m^3/kg) lies in the 400 to 500°C range around a pressure of 1.40 MPa. Therefore, the temperature T_2 will be determined by interpolating between 400 and 500°C using the steam table u values at 1.40 MPa.

$$T_2 = 400 + \Delta T °C$$

$$\Delta T = \Delta T^1 \frac{\Delta u}{\Delta u^1} \Delta °C$$

The interpolation data are listed in Table 3-2.

Table 3-2 Interpolation Data

T[°C]	u[kJ/kg]
400	2952.5
T_2	3025.7
500	3121.1

$$\Delta T^1 = 500 - 400 = 100\ \Delta °C$$

$$\Delta u = 3025.7 - 2952.5$$

$$\Delta u = 73.2 \text{ kJ/kg}$$

$$\Delta u^1 = 3121.1 - 2952.5$$

$$\Delta u^1 = 168.6 \text{ kJ/kg}$$

$$\Delta T=(100)\frac{(73.2)}{(168.6)}=[\Delta°C]\frac{[\text{kJ/kg}]}{[\text{kJ/kg}]}=[\Delta°\text{C}]$$
$$=43.4\ \Delta°\text{C}$$
$$T_2=400+43.4 \qquad [°\text{C}]+[\Delta°\text{C}]=[°\text{C}]$$
$$T_2=443.4°\text{C}$$

Determination of p_2. Calculation C-6

The pressure p is determined by the use of the specific volume v_2. The pressure may lie between 1.20 and 1.40 MPa, or between 1.40 and 1.60 MPa.

The first step is to calculate the specific volume at a pressure of 1.40 MPa and a temperature of 443.4°C.

$$v_{443.4}=v_{400}+\Delta v \qquad [\text{m}^3/\text{kg}]$$
$$\Delta v=\Delta v^1\left(\frac{\Delta u}{\Delta u^1}\right) \qquad [\text{m}^3/\text{kg}]$$

The interpolation data are listed in Table 3-3.

$$\Delta v^1=0.2521-0.2178=0.0343$$
$$\frac{\Delta u}{\Delta u^1}=0.434[—] \qquad (\text{C-5})$$
$$\Delta v=(0.0343)(0.434) \qquad [\text{m}^3/\text{kg}][—]$$
$$=0.0149\ \text{m}^3/\text{kg}$$
$$v_{443.4}=0.2178+0.0149$$
$$=0.2327\ \text{m}^3/\text{kg}$$

If the value of v at 1.40 MPa and 443.4°C (0.2327 [m^3/kg]} were smaller than v_2, the pressure (1.40 MPa) would be higher than p_2. To determine p_2, an interpolation between v_2 and the specific volumes at 443.4°C at 1.20 and 1.40 MPa would be necessary. If the value of v at 1.40 MPa and 443.4°C were larger than v_2, an interpolation between v_2 and the specific volumes at 1.40 and 1.60 MPa would be necessary.

It turns out that v at 1.40 MPa and 443.4°C equals v_2, so p_2 is 1.40 MPa. Therefore,

$$p_2=1.40\ \text{MPa}$$

Table 3-3 Interpolation Data

T[°C]	v[m^3/kg] at 1.40 MPa	
400	0.2178	Table A-3.3
443.4	v_2	
500	0.2521	Table A-3.3

3-4.3 CLOSED SYSTEM WITH MULTIPLE COMPONENTS

A closed cycle system can encompass several processes and/or components. Heat may flow into the system during one process or into one component, and out during another process or from another component. This can also be true of work.

An example of such a closed system is the steam powerplant shown in Figure 3-5.

In the analysis of such a system, only the net heat flow Q_{net} and the net work Wk_{net} are considered. As the system is at steady state, there is no change in U, and Equation 3-7 becomes

$$Wk_{net} = Q_{net} \qquad [\mathrm{J}] \tag{3-11}$$

or

$$wk_{net} = q_{net} \qquad [\mathrm{J/kg}] \tag{3-11a}$$

Here Wk_{net} is the sum of all work crossing the system boundary. According to the sign conventions, Wk leaving the system is positive (+) and work entering the system is negative (−). Numerically, Wk_{net} is the sum of all work leaving the system minus the sum of all work entering the system.

$$Wk_{net} = \left|\sum Wk_{out}\right| - \left|\sum Wk_{in}\right|$$

In a similar manner Q_{net} is the sum of all heat entering and leaving the system. The heat entering is positive (+) and the heat leaving is negative (−).

$$Q_{net} = \left|\sum Q_{in}\right| - \left|\sum Q_{out}\right|$$

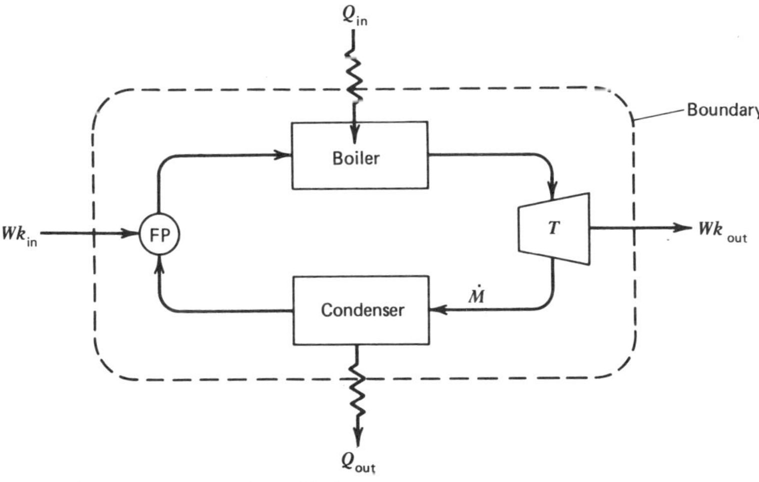

Figure 3-5 Closed system with heat and work crossing the boundary.

In the analysis of thermodynamic systems, producing-power Equation 3-11 becomes

$$P_{net} = \dot{Q}_{net} \qquad [J/s] = [W] \tag{3-12}$$

Example Problem 3-7

In a small electrical power generating plant the steam turbine develops 2500 kW. The heat supplied to the steam in the boiler is 3000 kJ/kg. The heat rejected by the system to the cooling water in the condenser is 2000 kJ/kg. The power required by the feed pump to return the condensate to the boiler is 10 kW. Determine the mass flow rate of steam for this cycle. (Refer to Figure 3-5.)

Equation

$$\dot{M} q_{net} = \dot{Q}_{net} = p_{net} \qquad [kW]$$

Therefore,

$$\dot{M} = \frac{p_{net}}{q_{net}} \qquad [kg/s] \tag{3-12a}$$

Parameters

$$P_{net} = p_{out} + p_{in} \qquad [kW]$$

$$P_{out} = +2500 \text{ kW} \qquad \text{(given)}$$

$$P_{in} = -10 \text{ kW} \qquad \text{(given)}$$

$$P_{net} = (+2500) + (-10) \qquad [kW] + [kW] = [kW]$$

$$= +2490 \text{ kW}$$

$$q_{net} = q_{in} + q_{out} \qquad [kW/kg]$$

$$q_{in} = +3000 \text{ kJ/kg} \qquad \text{(given)}$$

$$q_{out} = 2000 \text{ kJ/kg} \qquad \text{(given)}$$

$$q_{net} = (+3000) + (-2000) \qquad [kJ/kg] + [kJ/kg] = [kJ/kg]$$

$$= +1000 \text{ kJ/kg}$$

Substitution

$$\dot{M} = \frac{+2490}{+1000} \qquad \frac{[kW]}{[kJ/kg]} = \frac{[kJ/s]}{[kJ/kg]} = [kg/s]$$

Answer

$$\dot{M} = \mathbf{2.49\ kg/s}$$

3-5 FIRST LAW APPLIED TO OPEN SYSTEMS

In the open system, mass passes across the system boundary in addition to heat and work as described in Chapter 1. As the mass enters the system, it brings energy; when the mass leaves the system it takes out energy. In the study of thermodynamics where fluid flows are involved, there are several important considerations. These include the following:

- The principle of flow continuity (conservation of mass).
- Fluid flow mechanical energy losses.

The principle of flow continuity is discussed in the following section. Fluid flow mechanical energy losses are discussed in Chapter 13.

3-5.1 FLUID FLOW CONTINUITY (CONSERVATION OF MASS)

The principle of conservation of mass underlies the continuity equation. Many thermodynamic problems are concerned with those systems that have a given amount of mass flowing per unit time.

The mass flow rate $\dot{M}$ in a duct is

$$\dot{M}=\rho\dot{V}=\rho A\mathbf{V} \tag{3-13}$$

where

$\dot{M}$ = mass flow rate [kg/s]

$\dot{V}$ = volumetric flow rate [m^3/s]

ρ = fluid density [kg/m^3]

A = flow cross-sectional area [m^2]

$\mathbf{V}$ = fluid flow velocity (average) [m/s]

Equation 3-13 is commonly referred to as the continuity equation.

A steady state mass flow system is one wherein the net mass flowing into the system is equal to the net mass flowing out of the system at any instant of time. Such a system is illustrated in Figure 3-6. In this system water flows through a length ΔX of a uniform pipe, with no work or heat crossing the system boundary. The volume of water contained in the system at any given instant of time is

$$V=A(\Delta X) \tag{3-14}$$

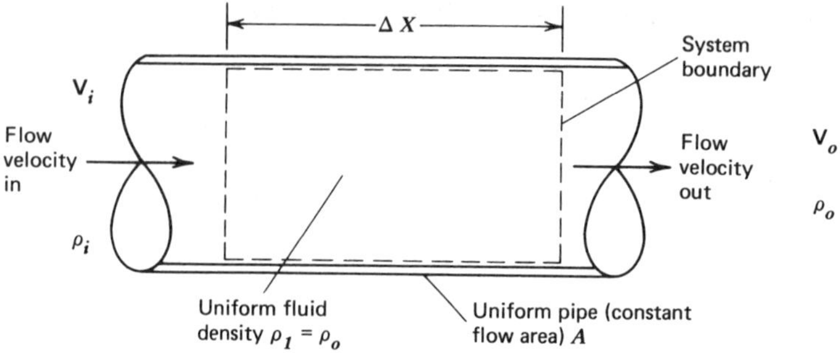

Figure 3-6 A steady mass flow system.

where

$$V = \text{volume} \qquad [\text{m}^3]$$

$$A = \text{flow cross-sectional area} \qquad [\text{m}^2]$$

$$\Delta X = \text{length of pipe} \qquad [\text{m}]$$

and the mass M of water contained in the system is

$$M = \rho V = \rho A(\Delta X)$$

where

$$M = \text{mass of fluid} \qquad [\text{kg}]$$

$$\rho = \text{fluid density} \qquad [\text{kg/m}^3]$$

The mass flow rate $\dot{M}_i$ entering the system is

$$\dot{M}_i = \rho_i A_i \mathbf{V}_i \qquad [\text{kg/s}]$$

The mass flow rate leaving the system is

$$\dot{M}_o = \rho_o A_o \mathbf{V}_o \qquad [\text{kg/s}]$$

For a steady mass flow system the two flow rates are equivalent. Thus,

$$\dot{M}_i = \dot{M}_o$$

so that

$$\rho_i A_i \mathbf{V}_i = \rho_o A_o \mathbf{V}_o$$

In the steady state system illustrated in Figure 3-6,

$\rho_i = \rho_o$ as there is no work done or heat transferred, (temperature remains constant)

$A_i = A_o$ by definition (uniform pipe)

Therefore,

$$\mathbf{V}_i = \mathbf{V}_o$$

In a closed thermodynamic system, the mass of the fluid in the system with respect to time is constant (does not change). Therefore,

$$M = \text{constant}$$

Thus the initial and final values for the mass in any closed system can be equated. This can often simplify the solution of this type of problem.

For an open system, the statement for continuity (conservation of mass) is as follows:

$$M_{\text{in}} - M_{\text{out}} = \Delta M_{\text{system}} \tag{3-15}$$

where

M_{in} = mass entering the system

M_{out} = mass leaving the system

ΔM_{system} = change in mass of the system

If the term ΔM_{system} is positive, the system is increasing in mass (i.e., more mass enters than leaves the system). If the term ΔM_{system} is negative, the mass of the system is decreasing. This can be expressed in rate terms as:

$$\dot{M}_{\text{in}} - \dot{M}_{\text{out}} = \Delta \dot{M}_{\text{system}} \tag{3-16}$$

where:

$\dot{M}_{\text{in}}$ = mass flow rate entering the system [kg/s]

$\dot{M}_{\text{out}}$ = mass flow rate leaving the system [kg/s]

$\Delta \dot{M}_{\text{system}}$ = mass change rate of the system [kg/s]

In the case of a steady mass flow (with respect to time), then

$$\dot{M}_{\text{in}} - \dot{M}_{\text{out}}$$

If the volumetric flow rate is required, the continuity equation can be expressed as follows:

$$\dot{V} = A\mathbf{V} \qquad [\mathrm{m^3/s}] \qquad (3\text{-}17)$$

where

$\dot{V}$ = volumetric flow rate of fluid $[\mathrm{m^3/s}]$

A = cross-sectional area of flow $[\mathrm{m^2}]$

$\mathbf{V}$ = average velocity of fluid flowing $[\mathrm{m/s}]$

Example Problem 3-8

A 0.02-m-diameter pipe has a steady flow of water at a velocity of 25 m/s. Determine the mass flow rate in this pipe. Assume that the density of water is 998.3 kg/m³. (See Figure 3-7.)

Equation

$$\dot{M} = \rho A\mathbf{V} \qquad [\mathrm{kg/s}] \qquad (\text{Eq. } 3\text{-}13)$$

Parameters

$$\rho = 998.3\ \mathrm{kg/m^3} \qquad \text{(given)}$$

$$A = \frac{\pi d^2}{4}\ \mathrm{m^2}$$

$$d = 0.02\ \mathrm{m} \qquad \text{(given)}$$

$$A = \frac{\pi (0.02)^2}{4} \qquad [\mathrm{m^2}]$$

$$= 3.14 \times 10^{-4} \qquad [\mathrm{m^2}]$$

$$\mathbf{V} = 25[\mathrm{m/s}] \qquad \text{(given)}$$

Substitution

$$\dot{M} = (998.3)(3.14 \times 10^{-4})(25) \qquad [\mathrm{kg/m^3}][\mathrm{m^2}][\mathrm{m/s}] = [\mathrm{kg/s}]$$

Answer

7.84 kg/s

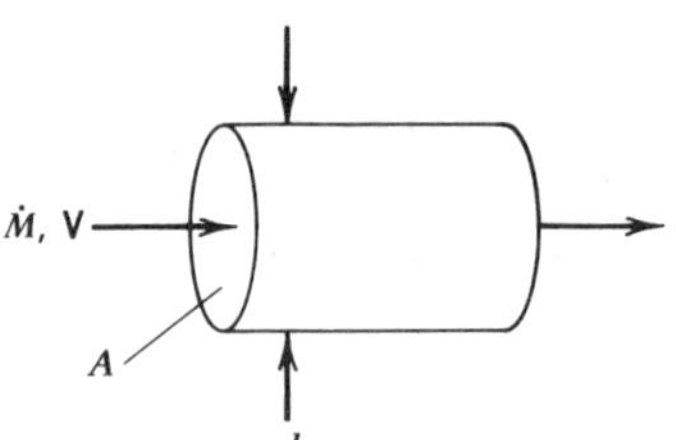

Figure 3-7 Sketch for Example Problem 3-8.

3-5.2 OPEN STEADY STATE SYSTEMS

The steady state open system is treated differently from the closed system. In the closed system the mass of the substance does not change, but its state does change during the time increment when work and heat cross the system boundary. In the open system there is a continuous steady flow of mass entering and leaving the system, as contrasted with the fixed body of mass in the closed system. Consequently, in the application of the first law of thermodynamics to the open system, the rates of energy flow (power) into and out of the system are the parameters of concern. These power terms are not identified with any specific body of mass because they are applicable to the mass flow rate.

To simplify the introduction of the student to the open system by giving a direct comparison with the equations and units of the closed system, the initial treatment of the open system in this text will be in terms of energy units [J] instead of the actual power units [J/s or W].

As mass passes across the system boundary in an open system, the energy associated with the mass entering and with the mass leaving the system must be considered in addition to the net work Wk_{net} and/or net heat Q_{net} input or output to the substance while it is within the open system.

The forms of energy that can enter or leave an open thermodynamic system with the mass flow are

- Mechanical potential energy $E_p = (Mgz)$ [J].
- Mechanical kinetic energy $E_k = (MV^2/2)$ [J].
- Internal energy U [J].
- Flow work (flow energy) pV [J].

An open thermodynamic system is diagrammed in Figure 3-8. The energy terms in the figure, with the exception of the flow work pV, have been defined previously. The term *flow work* is the product of the pressure and volume of the substance (mass) entering and leaving the system. This product is the work input to the system when the substance (mass) enters

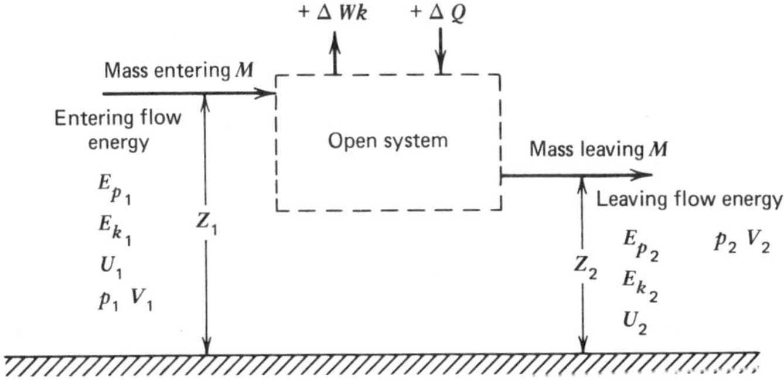

Figure 3-8 Diagram of an open thermodynamic system.

and the work output from the system when the substance (mass) leaves the system.

By applying the first law equation to the open system:

$$Wk_{\text{net}} = Q_{\text{net}} + \Delta E \qquad [\text{J}] \qquad (\text{Eq. 3-1})$$

where

$$\Delta E = \Delta E_p + \Delta E_k + \Delta U + (p_1 V_1 - p_2 V_2) \qquad [\text{J}]$$

Equation 3-1 can be rewritten as follows:

$$Wk_{\text{net}} = Q_{\text{net}} - \left[\Delta E_p + \Delta E_k + \Delta U + (p_1 V_1 - p_2 V_2)\right] \qquad [\text{J}] \qquad (3\text{-}18)$$

Substituting flow parameters for ΔE_p and ΔE_k, we obtain

$$Wk_{\text{net}} = Q_{\text{net}} - \left[Mg(Z_2 - Z_1) + \frac{M}{2}(\mathbf{V}_2^2 - \mathbf{V}_1^2) + (U_2 - U_1) + M(P_2 V_2 - p_1 V_1) \right] \qquad [\text{J}] \qquad (3\text{-}19)$$

Equation 3-19 can be rewritten as follows:

$$Wk_{\text{net}} = Q_{\text{net}} + M\left[g(Z_1 - Z_2) + \frac{\mathbf{V}_1^2 - \mathbf{V}_2^2}{2} + (U_1 + p_1 V_1) - (U_2 + p_2 V_2) \right] \qquad [\text{J}] \qquad (3\text{-}20)$$

or

$$wk_{\text{net}} = q_{\text{net}} + \left[g(Z_1 - Z_2) + \frac{\mathbf{V}_1^2 - \mathbf{V}_2^2}{2} + (u_1 + p_1 v_1) - (u_2 + p_2 v_2) \right] \qquad [\text{J/kg}] \qquad (3\text{-}20\text{a})$$

For the case when the change in the potential and kinetic energies of the mass flow is zero, as in the case of the closed system, Equation 3-20a reduces to

$$wk_{\text{net}} = q_{\text{net}} + (u_1 + p_1 v_1) - (u_2 + p_2 v_2) \qquad [\text{kJ/kg}] \qquad (3\text{-}21)$$

and for adiabatic systems (where $q_{\text{net}} = 0$), Equation 3-21 becomes

$$wk_{\text{net}} = (u_1 + p_1 v_1) - (u_2 + p_2 v_2) \qquad [\text{kJ/kg}] \qquad (3\text{-}22)$$

These open system equations (3-21 and 3-22) have comparable form with the closed system equations (3-8b and 3-9b), but the property $u+pv$ appears in place of the property u in the closed system equations. This property $u+pv$ is present in all open system equations and is called enthalpy or heat content. A discussion of enthalpy follows.

3-5.3 ENTHALPY

In an open system, as shown in Figure 3-8, the energy entering or leaving the system with the mass flow is the sum of (1) the mechanical energy E_k+E_p, (2) the internal energy U, and (3) a pV term (which is called "flow work" by some authors). In open system equations the energy term combination of $U+pV$ or $u+pv$ occurs to such an extent that it is convenient to replace this combination with a single term, which is called enthalpy or sometimes heat content. Enthalpy can be used either as an extensive property H or as an intensive property h. Values of h are usually listed in thermodynamic tables in the same manner as internal energy u values.

The relationship between the intensive value of enthalpy h and the extensive value H is defined by the following equation:

$$h=\frac{H}{M} \qquad [\text{kJ/kg}] \tag{3-23}$$

where

h = specific enthalpy of the substance [kJ/kg]
H = total enthalpy of the substance [J]
M = mass of substance in the system [kg]

Also, by the definition the specific enthalpy is

$$h=u+pv \qquad [\text{kJ/kg}] \tag{3-24}$$

where

h = specific enthalpy [kJ/kg]
u = specific internal energy [kJ/kg]
p = fluid pressure [Pa]
v = specific volume [m^3/kg]

and enthalpy H is

$$H=U+pV \qquad [\text{kJ}] \tag{3-24a}$$

where

$$U = \text{internal energy} \quad [\text{kJ}]$$
$$p = \text{fluid pressure} \quad [\text{Pa}]$$
$$V = \text{volume} \quad [\text{m}^3]$$

Since enthalpy is a property (like its components: internal energy, pressure, and specific volume), it can be used in solving thermodynamic problems; its values are listed in the steam tables. Values of enthalpy are based on an arbitrary temperature datum. Because the change in enthalpy is usually involved in thermodynamic problems, the temperature level of the arbitrary datum for the enthalpy values is not important, as long as the two values used to obtain the change in enthalpy are based on the same datum. For example, enthalpy values in the steam tables (Appendix A, Table A-3) are based on an enthalpy value of zero (0.000) for saturated water at 0.01°C.

Example Problem 3-9

Calculate the specific enthalpy of saturated steam vapor having a specific internal energy value of 2506.1 kJ/kg, a pressure of 100 kPa, and a specific volume of 1.6940 m^3/kg.

Equation

$$h = u + pv \qquad [\text{kJ/kg}] \qquad (\text{Eq. 3-24})$$

Parameters

$$u = 2506.1 \text{ kJ/kg} \qquad (\text{given})$$

$$p = 100 \text{ kPa} \qquad (\text{given})$$

$$v = 1.6940 \text{ m}^3/\text{kg} \qquad (\text{given})$$

Substitution

$$h = (2506.1) + (100)(1.6940) \qquad [\text{kJ/kg}] + [\text{kPa}][\text{m}^3/\text{kg}]$$

$$[\text{kPa}][\text{m}^3/\text{kg}] = \left[\frac{\text{kN}}{\text{m}^2}\right]\left[\frac{\text{m}^3}{\text{kg}}\right] = \left[\frac{\text{kNm}}{\text{kg}}\right] = [\text{kJ/kg}]$$

$$= (2506.1) + (169.40) \qquad [\text{kJ/kg}] + [\text{kJ/kg}] = [\text{kJ/kg}]$$

Answer

$$\mathbf{h = 2675.5 \text{ kJ/kg}}$$

Compare this answer with the value of h in the steam tables for saturated steam vapor at 100 kPa pressure.

3-5.4 OPEN SYSTEM EQUATIONS USING ENTHALPY

The general open system equation (Equation 3-22) can be written using enthalpy as follows:

$$Wk_{\text{net}} = Q_{\text{net}} + (H_1 - H_2) + M\left[g(Z_1 - Z_2) + \frac{(\mathbf{V}_1^2 - \mathbf{V}_2^2)}{2}\right] \quad [\text{J}] \tag{3-25}$$

or

$$wk_{\text{net}} = q_{\text{net}} + (h_1 - h_2) + \frac{g(Z_1 - Z_2)}{1000} + \left(\frac{\mathbf{V}_1^2 - \mathbf{V}_2^2}{2000}\right) \quad [\text{kJ/kg}] \tag{3-25a}$$

In the steady flow, general energy equation (Equation 3-19) the mechanical potential energy term *MgZ*, while being of importance in hydraulics, is usually of negligible value in thermodynamic systems. Neglecting this mechanical potential energy term, we find that the steady flow energy equation for the open system is

$$Wk_{\text{net}} = Q_{\text{net}} + (H_1 - H_2) + M\left(\frac{\mathbf{V}_1^2 - \mathbf{V}_2^2}{2}\right) \quad [\text{J}] \tag{3-26}$$

or

$$wk_{\text{net}} = q_{\text{net}} + (h_1 - h_2) + \left(\frac{\mathbf{V}_1^2 - \mathbf{V}_2^2}{2000}\right) \quad [\text{kJ/kg}] \tag{3-26a}$$

If the change in kinetic energy $[\mathbf{V}_1^2 - \mathbf{V}_2^2)/2]$ is negligible, Equation 3-26 becomes

$$Wk_{\text{net}} = Q_{\text{net}} + (H_1 - H_2) \quad [\text{J}] \tag{3-27}$$

and

$$wk_{\text{net}} = q_{\text{net}} + (h_1 - h_2) \quad [\text{kJ/kg}] \tag{3-27a}$$

For processes with no net work crossing the system boundary, such as a boiler or condenser, where $Wk_{\text{net}} = 0$, Equation 3-27 becomes

$$H_2 - H_1 = \Delta Q \quad [\text{J}] \tag{3-28}$$

or

$$h_2 - h_1 = \Delta_q \quad [\text{kJ/kg}] \tag{3-28a}$$

If the process or system, in addition, is adiabatic Q_{net} or $q_{net}=0$, Equation 3-28 becomes

$$Wk_{net}=(H_1-H_2) \qquad [J] \tag{3-29}$$

or

$$wk_{net}=(h_1-h_2) \qquad [kJ/kg] \tag{3-29a}$$

Applying the first law Equation 3-26a to a nozzle where wk_{net} and q_{net} are zero, we obtain

$$h_1-h_2=\frac{\mathbf{V}_2^2-\mathbf{V}_1^2}{2000} \qquad [kJ/kg] \tag{3-30}$$

As discussed at the beginning of Section 3-5.2, the open systems produce power [kJ/s]=[kW] instead of the work units [kJ/kg] of the above equations. Power is

$$p=(\dot{M})(wk_{net}) \qquad [kg/s][kJ/kg]=[kJ/s]=[kW]$$

The previous net work equations convert to power equations as follows.
Equation 3-25a

$$P=\dot{M}\left[q_{net}+(h_1-h_2)+g(Z_1-Z_2)+\left(\frac{\mathbf{V}_1^2-\mathbf{V}_2^2}{2000}\right)\right] \qquad [kW] \tag{3-31}$$

Equation 3-26a

$$P=\dot{M}\left[q_{net}+(h_1-h_2)+\left(\frac{\mathbf{V}_1^2-\mathbf{V}_2^2}{2000}\right)\right] \qquad [kW] \tag{3-32}$$

Equation 3-27a

$$P=\dot{M}[q_{net}+(h_1-h_2)] \qquad [kW] \tag{3-33}$$

Equation 3-28a

$$P=\dot{M}(h_2-h_1)=\dot{M}(\Delta q) \qquad [kW] \tag{3-34}$$

Equation 3-29a

$$P=\dot{M}(h_1-h_2) \qquad [kW] \tag{3-35}$$

Example Problem 3-10

Superheated steam at 1.4 MPa and 350°C is expanded through a turbine, exhausting at 10 kPa. If the turbine produces 910 kJ/kg of net work, what is the state of

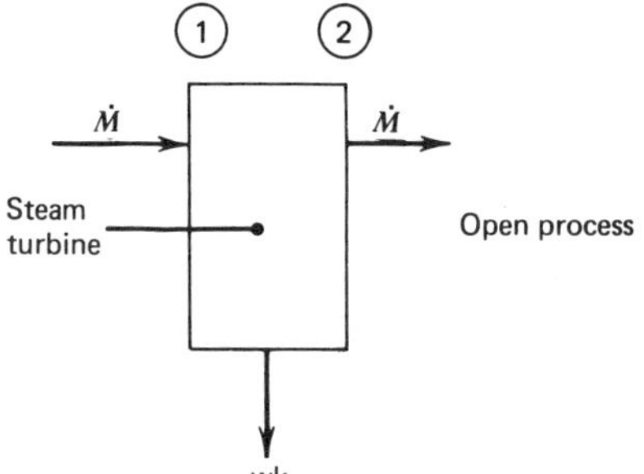

Figure 3-9 Sketch for Example Problem 3-10.

the exhaust steam? Assume an adiabatic expansion with negligible mechanical energy in the flow. (See Figure 3-9 and Table 3-4.)

Solution Sequence

1. Determine h_2 and compare with h_{g_2} to determine if state 2 is in saturation or superheat.
2. If in saturation, calculate the quality x and obtain the saturation temperature from the steam tables. If in superheat, use the superheat steam table for p_2 and determine the temperature corresponding to h_2. Quality is not applicable (NA).

Determine h_2 (C-1)

$$h_2 = h_1 - wk \qquad [\text{kJ/kg}] \qquad (\text{Eq. 3-29a})$$

$$h_1 = 3149.5 \text{ kJ/kg} \qquad (\text{Appendix A, Table A-3.3})$$

$$wk = 910 \text{ kJ/kg} \qquad (\text{given})$$

$$h_2 = 3149.5 - 910 \qquad [\text{kJ/kg}] - [\text{kJ/kg}] = [\text{kJ/kg}]$$

$$h_2 = 2239.5 \text{ kJ/kg}$$

Table 3-4 Table of State

Parameter	1		2	
Substance H_2O	sh vapor		sat 2ϕ	(C-2)
Mass (rate) $\dot{M}$	1 kg/s	(a)	1 kg/s	(a)
Temperature T	350°C	(g)	45.8	(C-4)
Pressure p	1.4 MPa	(g)	10 kPa	(g)
Quality x	[—] NA		0.856	(C-3)

Note: (g) given; (C-) calculated; and (a) assumed.

Determine if state 2 is Saturation or Superheat (C-2)

Test for saturation (two-phase): $h_{2_g} > h_2 > h_{2_f}$

$$h_{g_2} = 2584.7 \text{ kJ/kg} \qquad \text{(Appendix A, Table A-3.2)}$$

$$h_{f_2} = 191.8 \text{ kJ/kg} \qquad \text{(Appendix A, Table A-3.2)}$$

$$2584.7 > 2239.5 > 191.8$$

Therefore, h_2 is in saturation (two-phase).

Determine the quality x_2(C-3)

$$x_2 = \frac{h_2 - h_{f2}}{h_{g2} - h_{f2}} \qquad [-] \qquad \text{(Eq. 2-26)}$$

All the parameters were previously listed.

$$x_2 = \frac{2239.5 - 191.8}{2584.7 - 191.8} \qquad \frac{[\text{kJ/kg}] - [\text{kJ/kg}]}{[\text{kJ/kg}] - [\text{kJ/kg}]} = [-]$$

$$= \frac{2047.7}{2392.9}$$

$$x_2 = 0.856[-]$$

Determine the temperature T_2(C-4)

$$T_2 = \text{saturation temperature at 10 kPa}$$

$$T_2 = 45.8°\text{C} \qquad \text{(Appendix A, Table A-3.2)}$$

Example Problem 3-11

A gas turbine has a mass flow rate $\dot{M}$ of 35 kg/s, the specific enthalpy h_1 of the inlet gas is 1225 kJ/kg, while the exhaust gas has a specific enthalpy h_2 of 350 kJ/kg. The velocity of the inlet $\mathbf{V}_1$ and exhaust gases $\mathbf{V}_2$ are 150 m/s and 60 m/s, respectively, and a heat loss $\dot{Q}$ is 1000 kW. Calculate the power output from this turbine. (See Figure 3-10 and Table 3-5 on page 93.)

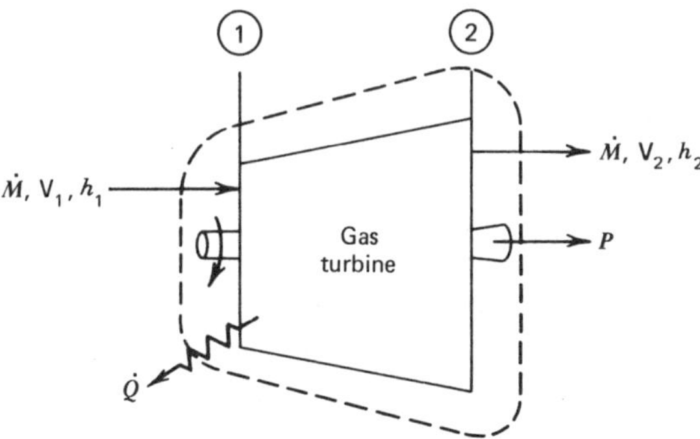

Figure 3-10 Sketch for Example Problem 3-11.

Table 3-5 Table of State

Parameter	Units	1	2
Substance		Gas	Gas
Mass flow rate $\dot{M}$	kg/s	35 (g)	35 (g)
Temperature T	°C	—	—
Pressure p	MPa	—	—
Enthalpy h	kJ/kg	1225 (g)	350 (g)
Velocity **V**	m/s	150 (g)	60 (g)

Note: (g) means given.

Equation

$$P=\dot{Q}+\dot{M}(h_1-h_2)+\dot{M}\left(\frac{\mathbf{V}_1^2-\mathbf{V}_2^2}{2000}\right) \quad [\text{kW}] \qquad (\text{Eq. 3-32})$$

Parameters

$$\dot{Q}=-1000\text{ kW} \qquad (\text{given})$$

The balance is listed under the Table of State.

Substitution

$$P=+(-1000)+(35)(1225-350)+\frac{(35)(150^2-60)^2}{2000}$$

$$\cdot[\text{kW}]+[\text{kg/s}][\text{kJ/kg}-\text{kJ/kg}]+\frac{[\text{kg/s}]\left[(\text{m/s})^2-(\text{m/s})^2\right]}{[\text{W/kW}]}$$

$$=[\text{kW}]+[\text{kJ/s}]+\frac{[\text{N}\cdot\text{m/s}]}{[\text{W/kW}]}$$

$$=[\text{kW}]+[\text{kW}]+[\text{kW}]=[\text{kW}]$$

$$P=(-1000)+(30{,}625)+(331)$$

Answer

$$\boldsymbol{P=29{,}956\text{ kW}} \qquad \textbf{Power output}$$

3-7 PROBLEMS

3-1. Explain the first law of thermodynamics. Why is it important?

3-2. List the three forms of energy involved in a thermodynamic system.

3-3. What forms can the energy in the system substance take?

3-4. Steam is discharged from the high pressure nozzle of a steam turbine at 1000 m/s. What is the kinetic energy of 1 kg of steam in joules?

3-5. Steam enters a deLaval nozzle at 30 m/s and leaves the nozzle at 1000 m/s. What is the increase in the kinetic energy of 1 kg of steam passing through the nozzle? Give the answer in joules [J].

3-6. Gas is flowing in a pipe with a velocity of 40 m/s. What is the kinetic energy of the gas per kg of gas?

3-7. A clock is powered by a 1.2-kg mass hung on a chain passing over and engaging a 5-cm-diameter pulley in the clock mechanism. How much energy has been put into the clock mechanism when the weight drops 1.2 m?

3-8. A boiler feed pumps water from a sump at 1-m elevation above a reference elevation datum into a boiler drum at 10-m elevation above the reference elevation datum. What is the change in the potential energy of 1 kg of water between the sump and the boiler drum? Give the answer in joules [J].

3-9. What is the specific internal energy u of saturated water vapor at 100°C (based on the internal energy of saturated liquid water at 0.01°C being 0.0 kJ/kg)?

3-10. What is the change in specific internal energy u between saturated liquid at a pressure of 20 kPa and saturated vapor at a pressure of 1.0 Mpa?

3-11. What is the change in the specific internal energy of water under the following conditions?

(a) Saturated vapor at 0.10 MPa pressure becomes saturated liquid.

(b) Saturated vapor at 0.10 MPa pressure becomes superheated vapor at 0.10 MPa and 300°C.

3-12. Compare the specific internal energies of steam at 400°C at pressures of 0.01, 0.1, and 1.0 MPa. Why are these values of specific internal energy similar?

3-13. If 2 kg of steam is changed from 200 kPa and 150°C to 0.80 MPa and 250°C by a closed-cycle adiabatic process, determine the work input required in joules (J).

3-14. If 1 kg of steam is changed from 1 MPa and 500°C to vapor at 20 kPa and 60°C by a closed-cycle adiabatic process, what is the work output in kJ/kg?

3-15. Superheated steam at 1.0 MPa and 500°C is expanded through a turbine, exhausting it at 0.10 MPa. If the turbine does work equivalent to 603.2 kJ/kg of inlet steam, what is the state of the exhaust steam? Assume that the kinetic energies of the inlet and exhaust flow velocities are negligible and the process is adiabatic.

3-16. For the steam inlet conditions in Problem 3-15, what amount of work would be produced by the turbine per unit mass of steam when the exhaust steam was 100 percent saturated vapor at a pressure of 0.1013 MPa?

3-17. To increase the amount of work produced by the turbine in Problem 3-16, various inlet and exhaust conditions are tried. What would be the quantitative effect of the following methods?

(a) Raising the inlet steam temperature to 600°C with a pressure of 1.0 MPa while maintaining the exhaust pressure at 0.10 MPa with 100 percent saturated vapor.

(b) Lowering the exhaust pressure to 10 kPa with 100 percent saturated vapor while maintaining the turbine inlet temperature at 500°C with a pressure of 1.0 MPa.

(c) Raising the inlet steam pressure to 1.20 MPa while maintaining the inlet temperature at 500°C and the exhaust pressure at 0.10 MPa with 100 percent saturated vapor.

3-18. A small, steam-driven, electrical power generating plant produces 1500 kW (net) when the steam flow rate is 2 kg/s. If the heat supplied to the steam is

3200 kJ/kg, how much heat ($\dot{Q}_{out}$) in [kJ/s] must be rejected by the condenser?

3-19. Calculate the specific enthalpy h of steam at 1.00 MPa and 500°C using the values of specific internal energy u and specific volume v from the steam tables. Compare with the value given in the steam tables for specific enthalpy h.

3-20. Treating liquid water as an incompressible fluid, determine the specific work (wk/kg) to pump water (saturated liquid) at 20 kPa pressure into a boiler at 3.0 MPa pressure.

3-21. An open system uses an incompressible working fluid with a specific volume of 0.0025 m^3/kg. If the initial pressure is 150 kPa, determine the specific work required to obtain a final pressure of 750 kPa.

3-22. If liquid water is assumed to be incompressible, how much work is required to raise the pressure from 20 kPa to 3.0 MPa of 5 kg of water?

3-23. Steam is expanded through a deLaval nozzle. The inlet steam pressure is 1.20 MPa; the temperature is 500°C; and the steam velocity to the nozzle entrance is 30 m/s. The expansion is adiabatic to saturate vapor at 20 kPa pressure. What is the discharge steam velocity?

3-24. Steam is expanded in a deLaval nozzle from 1.40 MPa to saturate vapor at 20 kPa pressure. In order to obtain an exhaust velocity of 1011.3 m/s what nozzle inlet steam temperature is necessary? Assume that the inlet steam velocity is zero.

3-25. If the inlet steam temperature were reduced to 400°C, what would the nozzle exhaust velocity be?

3-4 – $K.E. = \frac{1}{2} m(V)^2$

3-5 – $E_{ent.} = \frac{1}{2} m v^2$ } INCREASE = $E_{LV} - E_{ENT}$
$E_{LEAV.} = \frac{1}{2} m v^2$

3-8 – $P.E. = mg\Delta h$; $\Delta h = 9m$

3-15 – a.) $W_{NET} = h_1 - h_2$; $h_2 = h_1 - W_{NT}$
b.) A3.2 – h_g @ 0.1 MPa = 2675
since $h_2 > h_g$, CONDITION 2 IS ALSO S.H.
c.) A3.3 P = 0.1 MPa & h = 2875.3 → 200°C

3-20 – a) 3.0 MPa → 3000 kPa
b.) $W_k = \Delta P(v_f)$ → 20 kPa - 3000 kPa $(v_f) = W_k$
A3.2 $v_f = .001017$

CHAPTER 4

INTRODUCTION TO HEAT ENGINES

Energy is obtained from, or absorbed by, a thermodynamic system through a change of state of the substance in the system. Such a change of state of a substance in the thermodynamic system is termed a thermodynamic process. A thermodynamic process can occur in a number of ways, as discussed in Chapter 6, and may involve work input or output, and/or heat transfer across the boundary of the thermodynamic system during the thermodynamic process. The magnitude of such work input or output or heat transfer is the difference between the energy at the state of the substance before and after the thermodynamic process.

4-1 THE THERMODYNAMIC CYCLE

A thermodynamic cycle consists of a combination of two or more thermodynamic processes, which when completed returns the system to its initial state. The thermodynamic cycles of greatest interest include the following:

- Power-producing cycles.
- Heat-pumping cycles.
- Thrust (force)-producing cycles.

Heat engines based upon these cycles are discussed in the following sections. The power-producing engine produces work output from heat input. The heat pump engine pumps heat from a lower to a higher temperature by the input of work. The thrust heat engine produces force from an input of heat.

4-2 POWER-PRODUCING HEAT ENGINES

The power-producing heat engine includes automobile engines, electrical power plants, aircraft engines, and the like. The work-producing heat engine is a thermodynamic system (open or closed) operating in a complete thermodynamic cycle. Heat enters the system at a higher temperature; net work output is produced; and a smaller amount of heat is rejected, leaving the system at a lower temperature. The second law of thermodynamics, which will be discussed in Chapter 5, relates the heat input, heat rejection, and the net work output.

A diagram of the basic elements of the work (power)-producing heat engine thermodynamic system is shown in Figure 4-1.

The heat supplied from the high temperature source is Q_H, and the heat rejected to the lower temperature sink is Q_L. The net work produced is Wk.

Applying the *first law of thermodynamics* to the cycle, we find that

$$\text{net heat supplied} = \text{net work done}$$

$$[Q_H - Q_L] = Wk_{\text{net}} \qquad [\text{J}] \tag{4-1}$$

Applying the *second law of thermodynamics* (discussed in Chapter 5) to the cycle, we find that the gross heat supplied must be greater than the net work done, and therefore,

$$Q_H > Wk_{\text{net}}$$

The thermal efficiency, η_{th} (Greek letter eta), of a heat engine is defined as the ratio of the net work done in the cycle to the gross heat supplied in the cycle. It is usually expressed as a percentage rather than a ratio. Thus

$$\text{thermal efficiency} = \eta_{\text{th}} = \frac{Wk}{Q_H} \qquad [-] \tag{4-2}$$

Substituting Equation 4-1 into 4-2, we obtain

$$\eta_{\text{th}} = \frac{(Q_H - Q_L)}{Q_H} = \left[1 - \left(\frac{Q_L}{Q_H}\right)\right] \qquad [-] \tag{4-3}$$

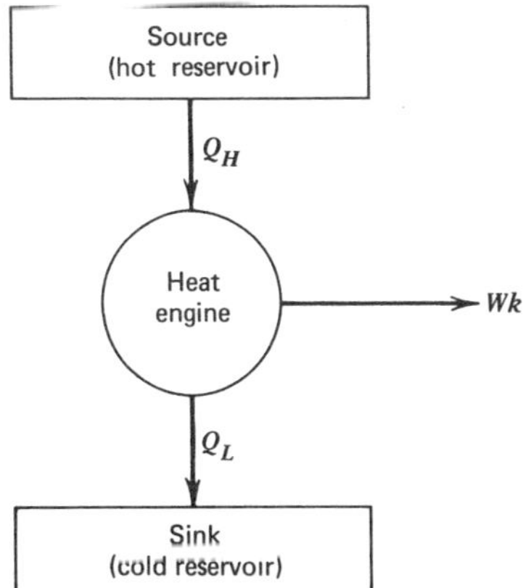

Figure 4-1 Diagram of a work-producing heat engine thermodynamic system.

It can be seen from this definition that due to constraints of the second law, the thermal efficiency of a heat engine is always less than 100 percent.

4-3 HEAT-PUMPING HEAT ENGINES

The heat-pumping engine is commonly called the reverse cycle heat engine. Its applications include refrigeration and air conditioning systems, as well as the heat pump used for heating buildings. Like the work-producing heat engine, the heat-pumping engine is a thermodynamic system (usually closed) operating in a complete thermodynamic cycle. Heat enters the system at a lower temperature; net work is put into the system; and a larger amount of heat is rejected at a higher temperature.

A diagram of the basic elements of the heat-pumping engine thermodynamic system is shown in Figure 4-2. In the heat pump (or refrigerator) cycle an amount of heat Q_L is supplied from the cold reservoir, and an amount of heat Q_H is rejected to the hot reservoir. The net work input to the system is Wk. Applying the first law, we obtain

$$Q_H = Q_L + Wk \qquad [\mathrm{J}] \tag{4-4}$$

There are three main performance measures for reverse cycle engines, namely:

- Coefficient of performance.
- Coefficient of refrigeration.
- Capacity.

4-3.1 THE COEFFICIENT OF PERFORMANCE

The coefficient of performance is a measure of the heating effectiveness of a heat pump, being concerned with the relationship between the amount of

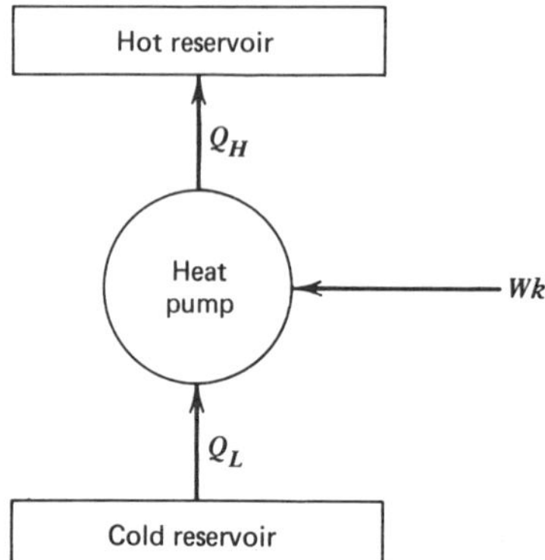

Figure 4-2 Diagram of a heat pump (or refrigerator) cycle.

heat output (rejected) and the amount of work input to the cycle. The equation is

$$COP = \left[\frac{Q_{out}}{Wk_{in}} \right] \qquad [—] \qquad (4\text{-}5)$$

where:

Q_{out} = net heat output (rejected) by heat pump [J]

Wk_{in} = work input required to drive the heat pump [J]

Example Problem 4-1

A heat pump rejects 1200 W of heat energy and requires an input work of 500 W to the compressor. Determine the coefficient of performance for this cycle. (See Figure 4-3.)

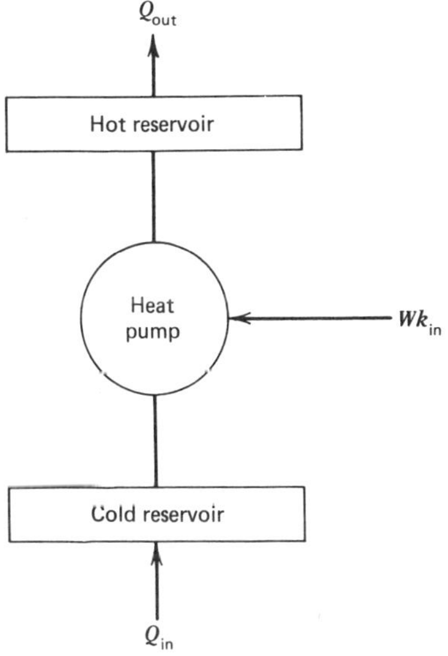

Figure 4-3 Sketch for Example Problem 4-1.

Equation

$$COP = \frac{Q_{out}}{Wk_{in}} \qquad [—] \qquad (\text{Eq. } 4\text{-}5)$$

Parameters

$$\dot{Q}_{out} = -1200 \text{ W} \qquad \text{(given)}$$

$$Wk_{in} = -500 \text{ W} \qquad \text{(given)}$$

Substitution

$$\text{COP} = \frac{(-1200)}{(-500)} \qquad \frac{[\text{W}]}{[\text{W}]} = [—]$$

Answer

$$\textbf{COP} = \textbf{2.4 [—]}$$

4-3.2 THE COEFFICIENT OF REFRIGERATION

The coefficient of refrigeration is a measure of the cooling effectiveness of a refrigerator, being concerned with the relationship between the amount of heat withdrawn from the refrigerated space and the amount of work input to the cycle. The equation is

$$\text{COR} = \frac{-Q_{\text{in}}}{Wk_{\text{in}}} \qquad [—] \tag{4-6}$$

where

Q_{in} = amount of heat withdrawn from the refrigerator [J]

Wk_{in} = actual work required to drive the refrigerator [J]

The numerical values for COP and COR are positive and always greater than 1.0.

Example Problem 4-2

A refrigeration cycle requires an input of 750 W to the compressor. If the amount of heat withdrawn from the refrigerated space is 1250 W, calculate the coefficient of refrigeration for this cycle. (See Figure 4-4.)

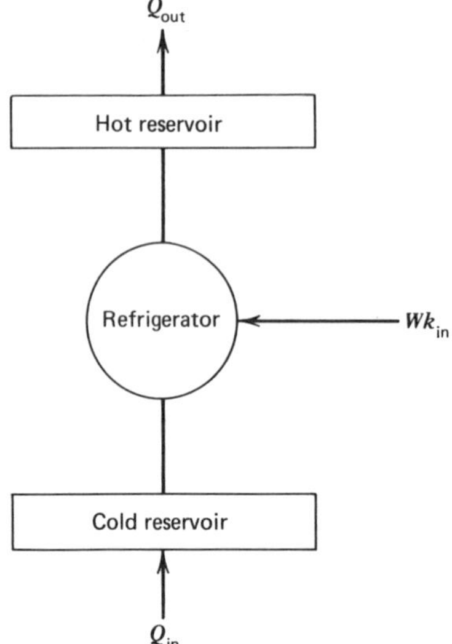

Figure 4-4 Sketch for Example Problem 4-2.

Equation

$$\text{COR} = \left[\frac{-Q_{\text{in}}}{Wk_{\text{in}}}\right] \quad [\text{—}] \qquad \text{(Eq. 4-6)}$$

Parameters

$$\dot{Q}_{\text{out}} = -1200\ \text{W} \qquad \text{(given)}$$

$$Wk_{\text{in}} = -500\ \text{W} \qquad \text{(given)}$$

Substitution

$$\text{COR} = \frac{-(1250)}{(-750)} \qquad \frac{[\text{W}]}{[\text{W}]} = [\text{—}]$$

Answer

$$\mathbf{COR = 1.67\ [—]}$$

4-3.3 THE CAPACITY

The capacity C_r is the measure of the rate at which a refrigerator removes heat or a heat pump reduces heat. Consequently, it is a measure of the size of the system from an energy standpoint. The equation for this parameter is

$$C_r = \dot{M}Q_r \qquad \text{(for heat pump)} \qquad [\text{W}] \qquad (4\text{-}7)$$

or

$$C_r = \dot{M}Q_a \qquad \text{(for refrigerator)} \qquad [\text{W}] \qquad (4\text{-}8)$$

where

C_r = the capacity [W]

$\dot{M}$ = refrigerant flow rate [kg/s]

Q_r = heat rejected by heat pump [kJ/kg]

Q_a = heat added from refrigerator [kJ/kg]

4-4 THRUST-PRODUCING CYCLES

The most common thrust-producing heat engines are the rocket engine and aircraft jet engine. They differ from power-producing heat engine cycles because the desired output is a force (thrust). The thrust is produced by a change in momentum of the propellant. In the case of rocket engines, the total propellant (solid or liquid) is carried in tanks that are part of the vehicle. The aircraft jet engine, however, only carries the fuel, since the atmospheric air is used as the primary propellant (mass). In both types of thrust producers, an exothermic (heat-releasing) reaction takes place in the combustion chamber to provide the thermal energy necessary for a high velocity jet exhaust.

A more detailed explanation of these thrust-producing cycles is given in Chapter 11.

4-5 PROBLEMS

4-1. Define a thermodynamic process.

4-2. Define a thermodynamic cycle. How does the thermodynamic cycle differ from a thermodynamic process?

4-3. For what useful purposes can thermodynamic cycles be applied? Name three categories and give a common example of each. How many of these thermodynamic cycles do you directly use or benefit from every day?

4-4. Describe in general terms the key components of a power producing heat engine by using a sketch to indicate what crosses the system boundary.

4-5. Determine the thermal efficiency of a heat engine that has 150 kJ of heat supplied from a high temperature source and 100 kJ of heat rejected to a low temperature sink.

4-6. Determine the thermal efficiency of a heat engine which rejects 200 kJ of heat to a low temperature sink and has a work output of 80 kJ.

4-7. Determine the amount of heat required from a high temperature source to obtain a thermal efficiency of 25 percent for a heat engine that has an output of 1.755 J.

4-8. Describe in general terms the key components of a heat pump by using a sketch to indicate what crosses the system boundary.

4-9. Define the parameter called the coefficient of performance (COP).

4-10. Calculate the COP for a heat pump that requires 125 kJ work input in order to provide 350 kJ net heat rejected.

4-11. A heat pump draws 1500 W of electrical power and supplies 4.5 kJ/s of heat to a heating system. What is the COP of the heat pump?

4-12. A heat pump with a COP of 2.5 supplies 5 kJ/s of heat. What is the power input required by the heat pump?

4-13. Describe in general terms the key components of a refrigerator by using a sketch to indicate what crosses the system boundary.

4-14. How does the thermodynamic cycle for a refrigerator differ from that of a heat pump?

4-15. Define the parameter called the coefficient of refrigeration (COR).

4-16. Calculate the COR for a refrigerator which requires an input of 350 W to the compressor while removing 775 W from a refrigerated space.

4-17. A refrigerator with a COR of 2.5 removes 5 kJ/s of heat from a refrigerated space. How much electrical power will the refrigeration system draw in kilowatts?

4-18. Define the parameter called the capacity.

4-19. Determine the capacity of the refrigeration system when the refrigerant flow rate is 5 kg/s and the heat removed is 155 kJ/kg of refrigerant flow.

4-20. The capacity of a refrigeration system is 10 kW. If the heat removed is 150 kJ/kg, what is the refrigerant flow rate in kg/s?

4-6 - TH. EFF. $= \eta_{TH} = \frac{W_K}{Q_H}$; $Q_H = W_K + Q_L$ $\eta = \frac{80 \text{ kJ}}{280 \text{ kJ}} = 28.57\%$

4-16 - COR $= \frac{-Q_{in}}{W_{K\,in}} = \frac{-775 \text{ W}}{-(350 \text{ W})} = 2.214$

4-19 - $C_r = \dot{M} Q_a$; C_r = CAPACITY OF REFR. SYSTEM.
$\dot{M}$ = REF. FLOW RATE
Q_a = HEAT REMOVED

4-20 - $\dot{M} = \frac{C_r}{Q_a}$

CHAPTER 5 -

5-4 - c) $x = \frac{s - s_f}{s_g - s_f} \Rightarrow s = [(x)(s_g - s_f)] + s_f$

5-7 ① $x_2 = \frac{s_2 - s_f}{s_g - s_f}$

② $h_2 = [(x_2)(h_g - h_f)] + h_f$

③ $W_K = h_1 - h_2$

5-13 - $\eta_{TH} = \frac{T_H - T_C}{T_H}$

CHAPTER 5

AVAILABILITY OF ENERGY

Although the net heat supplied to a thermodynamic system is equal to the net work done by the system (as discussed in Chapter 3), the gross energy supplied to the system must be greater than the net work done by the system. Not all of the input heat is available for producing output work because some heat must always be rejected by the system. The second law of thermodynamics is concerned with the relationship between the amount of energy supplied to a system and the amount of work that can be obtained from the system. Consequently, it is applicable only to thermodynamic systems that have work input or output. The second law of thermodynamics is concerned with the availability of energy from a thermodynamic cycle.

5-1 SECOND LAW OF THERMODYNAMICS

The second law of thermodynamics is difficult to define in a simple sentence and can best be presented by well-known statements or axioms that have been formulated by recognized investigators such as the following:

KELVIN: A system cannot operate cyclically and produce a net work output while exchanging heat at one fixed temperature.

PLANCK: It is impossible to construct an engine that will work in a complete cycle and produce no effect other than the raising of a weight and the cooling of a heat reservoir.

CLAUSIUS: Heat cannot, of itself, pass from a lower to a higher temperature reservoir.

It is important to note a fundamental fact when statements of the second law are considered: it is impossible to convert continuously a supply of heat completely into mechanical work; however, it is possible to convert mechanical work completely into heat. For example, when the brakes are applied in a car, bringing it to rest, the kinetic energy of that car is converted completely into heat at the brake drums.

Related to the second law statements are the concepts of availability of energy, entropy, process reversibility and thermal efficiency. These concepts are discussed in the following sections.

The concept of *availability of energy* is of essential importance in thermodynamics. The availability of energy refers to the maximum amount of energy in a given process that may be transformed into useful work. The processes of transformation are either (1) reversible or (2) irreversible. In all reversible processes (discussed in Section 5-3) there is no change in the availability of the energy evolved in the process. In irreversible processes the available portion of the total energy always decreases due to such effects as friction, turbulence and temperature differences.

Due to this concept of availability of energy, the following statements can be made:

- Only a portion of heat energy may be converted into work.
- Energy that is in the mechanical stored forms of kinetic or potential mechanical energy may by a reversible process (frictionless with no energy loss) be converted wholly into mechanical work.
- Only a part of the internal energy of a gas under pressure may be converted into work.
- The internal chemical energy of a fuel may, by the process of combustion, be first converted into heat and then only a portion of this heat may be converted into work.

5-2 ENTROPY

Entropy S is an abstract thermodynamic property of a substance that can be evaluated only by calculation. As T. E. Lawrence explains entropy in *The Seven Pillars of Wisdom*, entropy is a measure of the microscopic disorder of a substance. The higher the value of the entropy, the greater the microscopic disorder of the substance. For example, ice has a lower entropy value than water at the same temperature, and steam has a still higher value of entropy. The entropy parameter explains why water freezes and boils under a certain set of conditions, and why it cannot under other conditions.

The use of entropy in this text is primarily limited to the fact that reversible adiabatic expansions or compressions (discussed in Sections 5-3, 7-4, and 8-3) are constant entropy (isentropic) processes. As the end state of such a thermodynamic process has the same value of entropy as the initial state, the work (output or input) of the process is a dependent parameter dictated by the entropy value of the initial state of the substance and the pressure of the final state.

If the adiabatic expansion or compression is not reversible, the value of the entropy at the final state of the process will be higher than the value of the entropy at the initial state of the process. The greater the deviation from a reversible process, the larger will be the increase in the entropy between the initial and final states of the process.

Since the first law of thermodynamics is concerned only with the transformation of the various forms of energy, the work available from a thermodynamic process was discussed without any limitation or restriction on the end state of the process. The second law of thermodynamics places a restriction on the end state of a process as the law states that "it is impossible to convert continuously a supply of heat completely into mechanical work." To apply the second law, a means to establish the amount of work that can be available from a thermodynamic process is needed. To fulfill this need an abstract property of thermodynamic substances was devised. This property, called entropy, can be either intensive or extensive, and usually is given the symbol (S or s), has dimensions of Q/T, and units of [J/K] or [kJ/kg·K]. Since the values of entropy cannot be measured, being derived by calculation only, the student will find that entropy cannot be visualized in the usual physical sense. Numerical values of entropy s are usually given in thermodynamic tables in the same manner as internal energy u and enthalpy h. The way in which entropy controls thermodynamic processes is discussed in the following paragraphs.

When thermodynamic processes, forming a power producing cycle, are plotted on coordinates of absolute temperature T and entropy Q/T, as illustrated in Figure 5-1, the area of the plot enclosed by the processes is a measure of the available work from the cycle.

The diagram of a work producing heat engine operating between a set of heat source and heat sink temperatures is illustrated in Figure 4-1. The ideal power producing cycle (obtaining the maximum amount of work

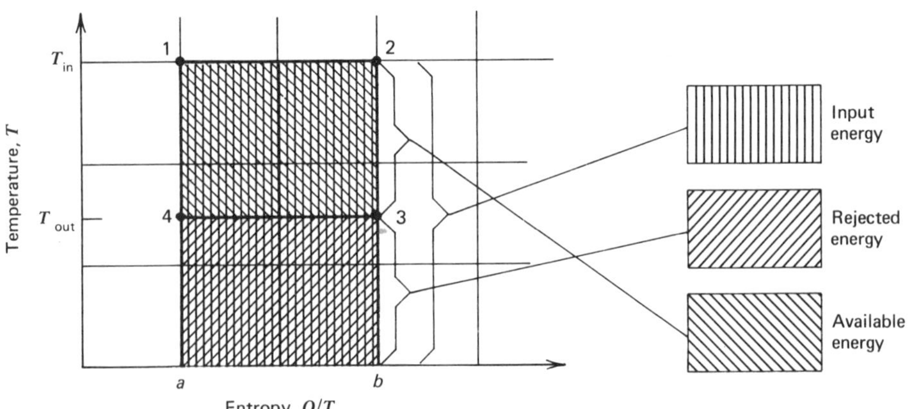

Figure 5-1 Temperature–entropy diagram for an ideal power producing thermodynamic cycle.

from a given set of heat source and heat sink temperatures) for this heat engine is plotted on temperature-entropy coordinates is shown in Figure 5-1. The cycle comprises four processes, and operates between the heat source temperature (T_i) and the heat sink temperature (T_o).

In following the thermodynamic cycle from point 1 back to point 1, the first process (1–2) is an increase in entropy at constant temperature. The next process between points 2 and 3 is a decrease in temperature at constant entropy. The next process between points 3 and 4 is a decrease in entropy at constant temperature. The last process, completing the thermodynamic cycle, between points 4 and 1 is an increase in temperature at constant entropy.

In the temperature-entropy diagram for the thermodynamic cycle presented in Figure 5-1, the input energy, rejected energy, and available energy are determined as follows:

- Input energy is represented by the area bounded by *a*–1–2–*b*–*a*.
- Rejected (wasted) energy is represented by the area bounded by *a*–4–3–*b*–*a*.
- Available energy is represented by the area bounded by 1–2–3–4–1.

Kelvin's statement of the second law can readily be illustrated by the temperature–entropy diagram shown in Figure 5-2. When the inlet temperature T_i equals the outlet or rejection temperature T_o, the area representing the available energy is zero.

Real power-producing cycles deviate from the ideal cycle illustrated in Figure 5-1, since the cycle processes do not form a perfect rectangle. In general, a power-producing cycle such as the simple steam powerplant, can be divided into four sections, as illustrated in Figure 5-3, (page 108), namely:

1–2. Heat input—boiler.
2–3. Work output—turbine (expansion).
3–4. Heat rejection—condenser.
4–1. Work input–boiler feed pump (compression).

The primary control entropy has of the real cycle is in the expansion and compression processes. Efficient modern expansion and compression

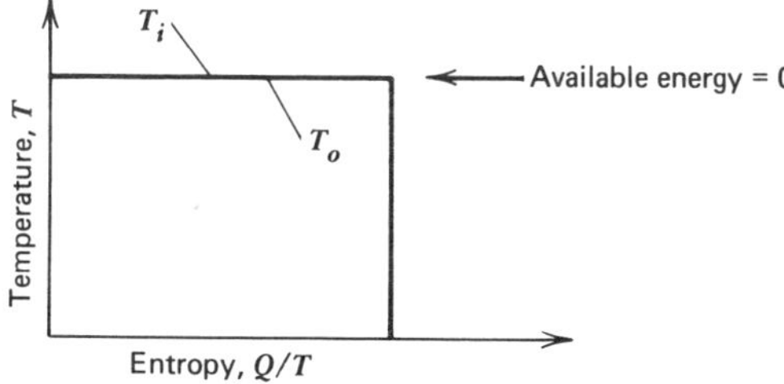

Figure 5-2 Temperature–entropy diagram for the thermodynamic cycle in Kelvin's second law statement.

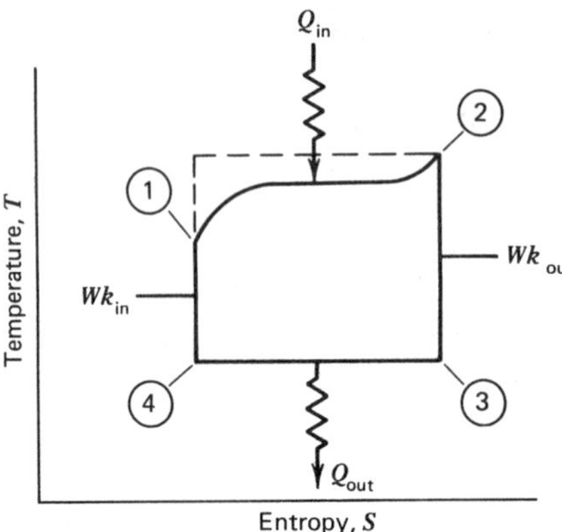

Figure 5-3 Temperature–entropy diagram for a real cycle.

machinery operates close to a constant entropy (isentropic) process, and a constant entropy process is usually assumed. Consequently, for example, the end state 3 of the expansion process (from state 2) is uniquely defined by (a) the final pressure or volume at the end of the expansion to state 3, and (b) the entropy value of state 3, which is the same as the value for state 2. These two independent intensive properties (p_3 or v_3 and s_3) define state 3. With both states 2 and 3 defined (through the use of entropy as described above), the amount of work output (or input) from the process 2–3 can be determined by obtaining the values of internal energy or enthalpy for states 2 and 3 from thermodynamic tables for substitution into the appropriate work equation as discussed in Chapter 3.

Example Problem 5-1

A steam turbine is receiving steam at 3.5 MPa and 600°C, and exhausting to 0.030 MPa. How much work does the turbine produce per kilogram of steam? (See Figure 5-4 and Table 5-1, page 109.)

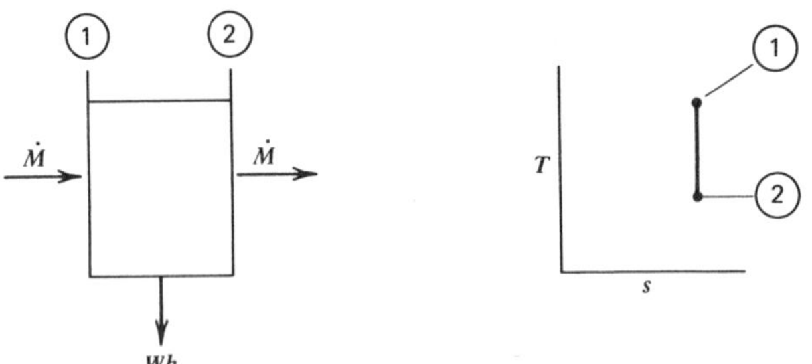

Figure 5-4 Open system isentropic expansion sketch for Example Problem 5-1.

Solution

1. Equation for "unknown" $wk = h_1 - h_2$ [kJ/kg] (Eq. 3-29a).
2. As state 1 is defined, h_1 can be obtained from steam tables A-3.3.

3. To evaluate h_2, state 2 must be defined. This can be done through the intensive properties of pressure p and specific entropy s. Based on the assumption of a constant entropy (isentropic) expansion, as discussed on page 108, $s_2 = s_1$.
4. Evaluate h_2.

Table 5-1 Table of State

Substance H_2O	sh①(g)	② sat. 2ϕ (C-1)
Mass M [kg]	1(a)	1(a)
Temperature T [°C]	600(g)	NA(C-1)
Pressure p [MPa]	3.5(g)	0.030(g)
Quality x [—]	NA	0.951(C-2)
Entropy s [kJ/kg·K]	7.4339(st)	7.4339

Equation

$$wk = h_1 - h_2 \quad [\text{kJ/kg}] \quad (\text{Eq. 3-29a})$$

$$h_1 = 3678.4 \quad [\text{kJ/kg}] \quad (\text{Table A-3.3 at 3.5 MPa and 600°C})$$

C-1. Establish State 2

$$s_2 = s_1 \quad [\text{kJ/kg·K}] \quad (\text{isentropic expansion})$$

$$s_1 = 7.4339 \quad [\text{kJ/kg·K}] \quad (\text{Table A-3.3 at 3.5 MPa and 600°C})$$

Therefore,

$$s_2 = 7.4339 \quad [\text{kJ/kg·K}]$$

$$s_g = 7.7686 \quad [\text{kJ/kg·K}] \quad (\text{Table A-3.2 at 0.03 MPa})$$

State 2 is in quality $s_2 < s_g$.

C-2. Calculate the Quality for State 2

$$x = \frac{s_2 - s_f}{s_g - s_f} \quad [—] \quad (\text{Eq. 2-26x})$$

$$s_2 = 7.4339 \quad [\text{kJ/kg·K}] \quad (\text{C-1})$$

$$s_f = 0.9439 \quad [\text{kJ/kg·K}] \quad (\text{Table A-3.2, saturated liquid at 30 kPa})$$

$$s_g = 7.7686 \quad [\text{kJ/kg·K}] \quad (\text{Table A-3.2, saturated vapor at 30 kPa})$$

$$x = \frac{7.4339 - 0.9439}{7.7686 - 0.9439} \quad \frac{[\text{kJ/kg·K}] - [\text{KJ/kg·K}]}{[\text{kJ/kg·K}] - [\text{kJ/kg·K}]} = [—]$$

$$= \frac{6.4900}{6.8247}$$

$$= 0.951 \ [—]$$

C-3. Calculate h_2

$$h_2=xh_g(1-x)h_f \quad [\text{kJ/kg}] \quad (\text{Eq. 2-25x})$$

$x=0.951$ [—] (C-2)

$h_g=2625.3$ [kJ/kg] (Table A-3.2 saturated vapor at 30 kPa)

$h_f=289.2$ [kJ/kg] (Table A-3.2 saturated liquid at 30 kPa)

$$h_2=(0.951)(2625.3)+(1-0.951)(289.2)$$

$$\cdot[—][\text{kJ/kg}]+[—][\text{kJ/kg}]$$

$$=2496.7+(0.049)(289.2)=2496.7+14.1 \quad [\text{kJ/kg}]$$

$$=2510.8 \quad [\text{kJ/kg}]$$

C-4. Calculate *wk*

$$wk=h_1-k_2 \quad [\text{kJ/kg}]$$

$$=3678.4-2510.8 \quad [\text{kJ/kg}]-[\text{kJ/kg}]=[\text{kJ/kg}]$$

Answer

$$\mathbf{wk=1167.6\ kJ/kg}$$

5-2.1 THE ENTROPY PARAMETER (*S*)

An important characteristic of entropy is that there is only one unique value of entropy for a particular state of a substance, irrespective of how the substance arrived at that particular state. The equation defining entropy is

$$\Delta s=\frac{du}{T}+\frac{pdv}{T} \qquad [\text{kJ/kg}\cdot\Delta_1\text{K}] \qquad (5\text{-}1)$$

The numerical value of entropy for a particular substance and state is a calculated value because entropy cannot be measured. The higher the value of the entropy parameter for a particular set of conditions, the less available the energy in the system for work output. An increase in temperature will cause an increase in entropy; a decrease in temperature will cause a decrease in entropy.

Numerical values of specific entropy s (an intensive property) are readily available in reference books, usually in thermodynamic tables of properties of various substances such as the steam tables (for water). The units for specific entropy are $\text{kJ/kg}\cdot\Delta_1\text{K}$.

If the total entropy of the state of the system is desired, the intensive property specific entropy is simply multiplied by the mass of the system, thereby obtaining the extensive property of total entropy S.

In the analysis of thermodynamic processes, as discussed in Chapter 6, and thermodynamic cycles, as discussed in Chapters 7, 8, and 9, the change in entropy (Δs) is the important parameter.

Temperature–entropy (T–s) diagrams of a thermodynamic process graphically show the typical changes in those properties during the process. The important *change in entropy* of the process depends only on the initial and final (end) states of the process and not on the intermediate states of the working fluid in the process.

5-2.2 THE THIRD LAW OF THERMODYNAMICS

The third law of thermodynamics provides a limitation through the following statement:

> Entropy tends to a minimum constant value as the temperature tends to absolute zero. For a pure element this minimum value is zero, but for all other substances it is not less than zero, but possibly more.

This statement of the third law was a result of research in the temperature regime near absolute zero and has not been violated. From a practical standpoint it means that it is impossible to attain a temperature of absolute zero by other than a reversible (ideal) process.

5-3 PROCESS REVERSIBILITY

A classical definition of reversibility is "When a fluid undergoes a reversible process, both the fluid and its surroundings can always be restored to their original state." All other processes are irreversible.

The basic criteria for reversibility are

1. The process must be frictionless.
2. The difference in pressure between the fluid and its surroundings during the process must be infinitesimally small.
3. The difference in temperature between the fluid and its surroundings during the process must be infinitesimally small.

In a *reversible process*, the thermodynamic system changes state in such a way that the substance passes through a continuous series of equilibrium states. At any instant during the reversible process the state point can be located on thermodynamic property diagrams. Therefore, the reversible process can be drawn as a solid line from one point to another such as that shown between point 1 and point 2 in Figure 5-5.

Actually, no real thermodynamic process is a truly reversible process. Many actual processes approximate reversible processes and from the engineering point of view may be treated as reversible processes. For example, an adiabatic reversible process is known as an isentropic process, since there is no change in the value of entropy (as discussed in Chapter 6).

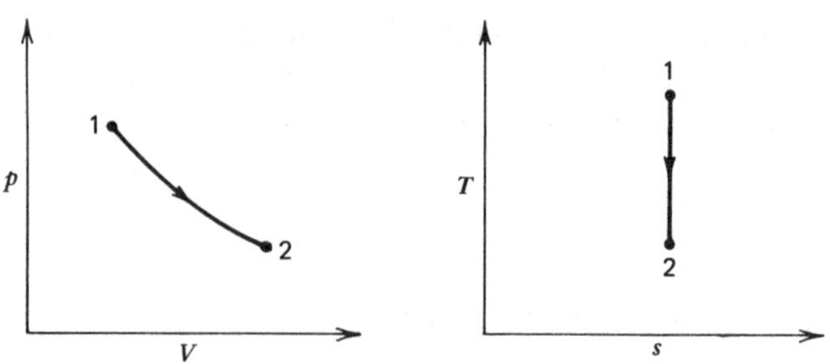

Figure 5-5 Thermodynamic property diagrams of a reversible adiabatic expansion process.

Example Problem 5-2

One kilogram of steam expands isentropically from 500 to 100 kPa. If the initial temperature is 500°C, determine the final temperature of the steam. (See Figure 5-6.)

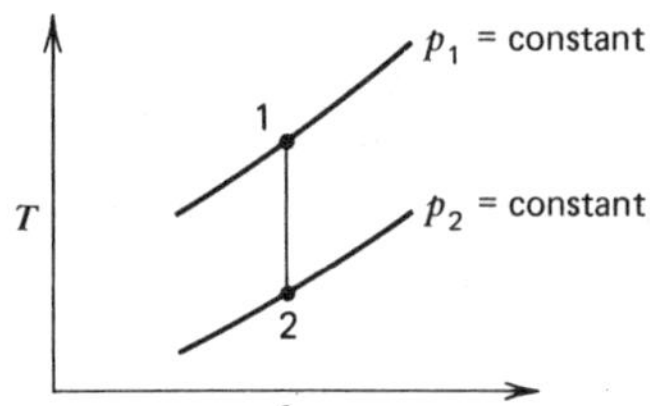

Figure 5-6 Sketch for Example Problem 5-2.

Equation

$$s_1 = s_2 = \text{constant (definition of isentropic)}$$

Parameters

$$T_1 = 500°\text{C} \qquad \text{(given)}$$

$$p_1 = 500 \text{ kPa} = 0.5 \text{ MPa} \qquad \text{(given)}$$

$$s_1 = 8.0873 \text{ kJ/kg}\cdot\Delta_1\text{K} \qquad \text{(Appendix A, Table A-3.3)}$$

$$p_2 = 100 \text{ kPa} = 0.1 \text{ MPa} \qquad \text{(given)}$$

Substitution

$$s_2 = 8.0873 \text{ kJ/kg}\cdot\Delta_1\text{K}$$

Linear interpolation from the steam tables at p_2 is given in Table 5-2.

Table 5-2 Interpolation Data

T (°C)	s[kJ/kg·Δ_1K]
250	8.0333
T	8.0873
300	8.2158

$$T_2 = 250 + \Delta T \qquad [°\text{C}]$$

$$\Delta T = \Delta T' \frac{\Delta s}{\Delta s'} \qquad [\Delta°\text{C}]$$

$$\Delta T' = 3000 - 250 = 50 \qquad [\Delta°\text{C}]$$

$$\Delta s = 8.0873 - 8.0333 = 0.054 \qquad [\text{kJ/kg}\cdot\Delta_1\text{K}]$$

$$\Delta s' = 8.2158 - 8.0333 = 0.1825 \qquad [\text{kJ/kg}\cdot\Delta_1\text{K}]$$

$$\Delta T = (50)\frac{0.054}{0.1825} \qquad [\Delta°C]\left[\frac{\text{kJ/kg}\cdot\Delta_1\text{K}}{\text{kJ/kg}\cdot\Delta_1\text{K}}\right] = [\Delta°\text{C}]$$

$$\Delta T = 14.8 \qquad [\Delta°\text{C}]$$

$$T_2 = 250 + 14.8 \qquad [°\text{C}] + [\Delta°\text{C}] = [°\text{C}]$$

Answer

$$\mathbf{T_2 = 264.8°C}$$

When the substance undergoing a process is not kept in equilibrium throughout the intermediate states, a continuous path (line) on the T–s property diagram cannot be shown. The process is then irreversible. Irreversible processes are usually represented by a dotted line connecting the end states to indicate that the intermediate points are indeterminate. An example of such a process is shown in Figure 5-7.

When an irreversible process is analyzed, the final state is always different from that of a reversible process. The irreversible process ends with a higher value for specific entropy and a higher temperature. Consequently, there is less available energy from an irreversible process.

5-4 THE CARNOT CYCLE

The simple Carnot cycle is often used as a standard against which the performance characteristics of other heat engine cycles are measured. The Carnot cycle is the most efficient cycle possible operating between two given temperature levels. The Carnot cycle is a reversible four-process cycle as indicated by the temperature-entropy diagram in Figure 5-8, page 114.

The four processes forming the Carnot cycle are in the following sequence, referring to the reference numbers in the T–s diagram:

1–2. Reversible adiabatic compression from a low temperature T_L to a higher temperature T_H.
2–3. Reversible isothermal expansion at temperature T_H.
3–4. Reversible adiabatic compression from the temperature T_H to T_L.

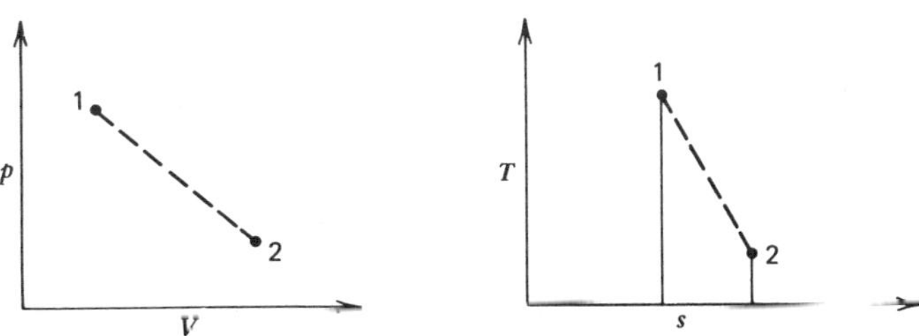

Figure 5-7 Thermodynamic property diagrams of an irreversible process.

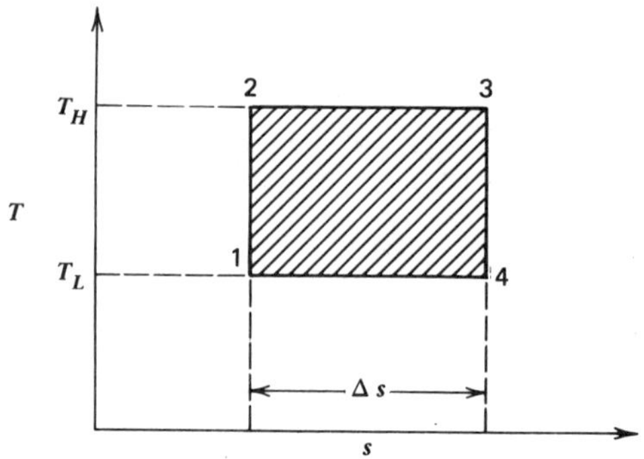

Figure 5-8 Carnot cycle temperature–entropy diagram.

4–1. Reversible isothermal expansion at temperature T_L to the initial state.

The thermal efficiency of the Carnot cycle is

$$\eta_{c_{th}} = \frac{(T_H - T_L)}{T_H} = \left[1 - \left(\frac{T_L}{T_H}\right)\right] \quad [-] \tag{5-2}$$

where

$\eta_{c_{th}}$ = Carnot thermal efficiency [—]
T_H = high temperature source [K]
T_L = low temperature sink [K]

Often the thermal efficiency of a heat engine cycle is compared with the thermal efficiency of a Carnot cycle at the same high and low cycle temperatures. The ratio between these two thermal efficiencies is called the "Carnot cycle efficiency." This parameter indicates how close the cycle thermal efficiency is to the highest possible thermal efficiency.

Example Problem 5-3

A Carnot engine operates with a high temperature source at 1400°C and a cold temperature sink at 50°C. Determine the thermodynamic efficiency of this engine. (See Figure 5-9, page 115).

Equation

$$\eta_{c_{th}} = 1 - \frac{T_L}{T_H} \quad [-] \quad \text{(Eq. 5-2)}$$

Parameters

$T_L = 50°C = 50 + 273 = 323$ K
$T_H = 1400°C = 1400 + 273 = 1673$ K

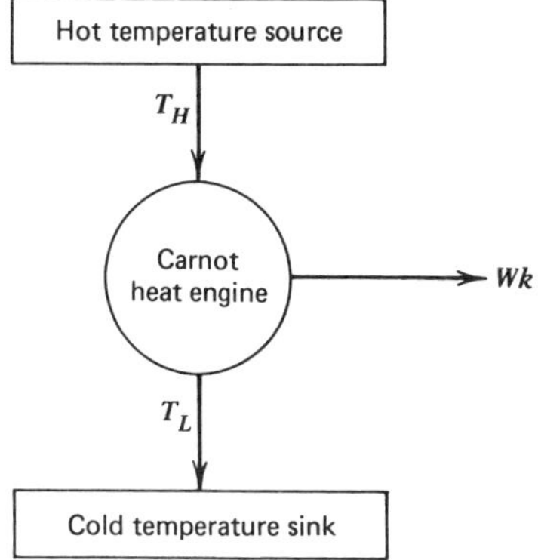

Figure 5-9 Sketch for Example Problem 5-3.

Substitution

$$\eta_{c_{th}} = 1 - \frac{323}{1673} = [—] - \frac{[K]}{[K]} = [—]$$

$$= 1 - 0.193$$

Answer

$$\boldsymbol{\eta_{c_{th}} = 0.807\ [—]}$$

5-5 PROBLEMS

5-1. How would you explain the second law of thermodynamics?

5-2. How does the second law of thermodynamics interact with, or relate to, the first law of thermodynamics?

5-3. What is entropy and how would you define it?

5-4. What is the specific entropy (s) for steam in the following states:
(a) 1.2 MPa 400°C.
(b) 0.4 MPa saturated vapor.
(c) 4 kPa $x=0.90$.
(d) 3 MPa 500°C.
(e) 0.5 MPa saturated liquid.
(f) 10 kPa $x=0.80$.

5-5. How does entropy control the amount of available energy from an isentropic expansion or the energy required for an isentropic compression?

5-6. A steam turbine receives steam at 1.0 MPa and 600°C, and exhausts at 0.10 MPa. Assuming an isentropic expansion,
(a) What is the state of the exhaust steam?
(b) How much work does the turbine produce per kilogram of steam?

5-7. A steam turbine is receiving steam at 3.50 MPa and 600°C, and exhausting to 0.030 MPa. Assuming an isentropic expansion,
(a) What is the state of the exhaust steam?
(b) How much work does the turbine produce per kilogram of steam?

5-8. If the inlet steam temperature in Problem 5-7 dropped to 500°C, but the inlet and exhaust pressures remained the same, what would be the change, if any, in the turbine work output per kilogram of steam?

5-9. What is the total entropy (S) for 5 kg of steam of 50 percent quality at 0.2 MPa pressure?

5-10. What is a reversible process?

5-11. What is the Carnot cycle? Describe its T–s diagram.

5-12. In general terms what does the Carnot cycle indicate about the thermal efficiency of a thermodynamic cycle; that is, if one wanted to increase the thermal efficiency of a cycle, what would have to be done?

5-13. A power plant is being considered for operating on thermal gradients in the ocean. What would be the maximum possible thermal efficiency for the cycle for the following conditions?

(a) If the surface water temperature were 20°C and the water temperature at a usable depth were 10°C.

(b) If the surface water temperature remained at 20°C and the cold temperature were reduced to 5°C.

(c) If the surface temperature were increased to 25°C and the cold temperature were 10°C.

5-14. An orbiting space station has a heat engine with a Carnot cycle thermal efficiency of 40 percent. If the space radiators provide a heat sink temperature of 200°C, what heat source temperature is required?

5-15. If the space radiator heat sink temperature in Problem 5-14 were 250°C, what heat source temperature would be required for a Carnot cycle thermal efficiency of 40 percent?

5-16. An orbiting space station has a heat engine with a Carnot cycle. If the heat source is 1000°C and the heat rejection radiators operate at 300°C, what is the heat engine cycle efficiency?

5-17. If the heat rejection radiators in Problem 5-16 degrade so they operate at 350°C, (a) what would the Carnot cycle efficiency be if the heat source remains at 1000°C?

5-18. A power-generating plant is proposed for operating using the temperature difference found in a certain location in the ocean. The power out is 10,000 kW using a warm surface water inlet temperature of 26°C and a cold deep-water temperature of 15°C. On the basis of a 3Δ°C drop in the temperature of the warm water and a 3Δ°C rise in the temperature of the cold water due to the amount of heat removed and added, calculate the following:

(a) The Carnot cycle efficiency based on the warm and cold water outlet temperatures.

(b) The amount of heat required from the warm water to produce 10,000 kW at the Carnot cycle efficiency for the system.

(c) The mass flow of warm water required to obtain the required heat from a 3Δ°C temperature drop in the warm water.

(d) The mass flow of cold water required to absorb the rejected heat with a 3Δ°C temperature rise in the cold water.

(e) The power required to pump the cold deep water to the surface and through the system heat exchange if the required pumping pressure drop were 7 kPa ($\simeq$ 1 lb/in.^{2}).

CHAPTER 6

CHARACTERISTICS AND PARAMETERS OF GASES

In order to actually apply the principles of thermodynamics to specific processes and cycles, a knowledge of the characteristics of the thermodynamic substance or working fluid must be known. The following chapter discusses these characteristics and various parameters for gases.

6-1 KINETIC THEORY

The *kinetic theory* of gases explains the properties of temperature and pressure, and develops the gas constant as discussed in the following paragraphs.

First there is the model of the structure of the gas mass. The kinetic theory regards the mass M of the gas as being composed of minute, freely moving particles called molecules. These molecules are relatively free to move around, colliding with each other as well as the surface of a confining wall if any exists. The molecules have kinetic energy of motion of the molecule as well as kinetic energy of motion of the internal subparticles comprising the molecular structure.

The temperature T of the gas is a measure of the molecular velocity (internal energy U of the gas). The molecular velocity increases as the temperature increases.

The pressure p of the gas results from the change in momentum of the molecules colliding per unit area per unit time. This concept of pressure based on Newtonian mechanics can be derived from the units of pressure [Pa] as follows.

$$Pa = \mathrm{N/m^2} = [\mathrm{kg}][\mathrm{m/s}]/[\mathrm{m^2}][\mathrm{s}]$$

$$p = \frac{(M)(V)}{(A)(t)} \qquad [\mathrm{Pa}] \tag{6-1}$$

where

$$(M)(\mathbf{V}) = \text{mass times velocity} = \text{momentum}$$

$$A = \text{area}$$

$$t = \text{time}$$

Therefore,

$$\text{pressure} = \text{momentum per unit area per unit time}$$

The pressure for a fixed population density of gas molecules (mass per unit volume) will increase if the velocity $\mathbf{V}$ of the molecules (internal energy U) increases. Consequently, the pressure of a gas with a constant mass per unit volume (specific volume) will increase as the temperature increases.

The gas pressure at a constant temperature (constant internal energy) increases as the molecular population density (mass per unit volume) increases.

The gas constant R relates the volume per unit mass of gas to the pressure and temperature.

The *STATE* of a gas, because it is a single-phase substance, is described by the three independent properties of mass M, pressure p, and temperature T.

6-2 EQUATIONS OF STATE

In thermodynamic processes involving a gas, the relation between the pressure, specific volume, and temperature of the vapor phase (not in contact with the liquid phase) may be expressed by an equation that is called an *equation of state*. This statement is based on the fact that a pure substance has only two independent properties.

The state of a substance in the gaseous phase is defined by the two independent intensive properties temperature T and pressure p, and the independent extensive property mass M. The volume V of the gas for a defined state is related to the temperature T, pressure p, and mass M by the gas constant R. The relationships involving these parameters can be expressed by an equation that is called the equation of state.

Many equations of state have been developed from experimental data, while others have been based on kinetic theory and/or thermodynamic considerations. However, all are empirical in that the constants were determined from experimental observations, and the equations are valid only for interpolation from these data.

6-3 THE PERFECT (IDEAL) GAS LAW

If it existed, a substance in the gaseous phase whose molecules are:

- Linked only by collision forces.
- Not distorted by collision.

would be a perfect or "ideal" gas. The equation of state for a perfect gas (often called the perfect gas law) is as follows:

$$pV = MRT \qquad [\mathrm{J}] \tag{6-2}$$

where

p = pressure (absolute) [Pa]

V = volume [m^3]

M = mass [kg]

R = gas constant [$\mathrm{kJ/kg \cdot \Delta_1 K}$]

T = temperature (absolute) [K]

When written for a unit mass of gas, Equation 6-2 becomes

$$pv = RT \qquad [\mathrm{kJ/kg}] \tag{6-3}$$

where

v = specific volume [m^3/kg]

By the use of the perfect gas law, the gas volume V at a particular state (M, p, and T) can be determined by the following arrangement of the perfect gas law.

$$V = MR\left[\frac{T}{p}\right] \qquad [\mathrm{m}^3] \tag{6-2a}$$

and

$$v = R\left[\frac{T}{p}\right] \qquad [\mathrm{m}^3/\mathrm{kg}] \tag{6-2b}$$

6-4 THE GAS CONSTANT

The gas constant R as indicated in Equation 6-2a is a factor that relates the volume of a specific gas to its properties of state (M, p, and T). The gas constant R has been found to be

- Of different numerical values for different gases.
- A "constant numerical value" for each perfect gas as it is independent of the level of temperature and pressure.
- A function of the molecular weight of the gas.

Investigators also found that if a gas constant $\bar{R}$ were determined per mol(*) of gas, the value of $\bar{R}$ was the same for all perfect gases with a value of 8314.3 [J/kg·mol^{-1}·Δ_1K]. Consequently, the gas constant $\bar{R}$ was called the "universal" gas constant.

The value of the gas constant for most real gases does not vary significantly over a wide range of temperatures and pressures.

The gas constant R is a function of the molecular weight of the perfect gas, and is related to the universal gas constant $\bar{R}$ as follows:

$$R=\frac{\bar{R}}{\mathfrak{M}} \quad [\mathrm{J/kg\cdot\Delta_1 K}] \tag{6-4}$$

where

$\bar{R}$ = universal gas constant

8314.3 J/kg·mol^{-1}·Δ_1K

$\mathfrak{M}$ = molecular weight [mol^{-1}]

Equation 6-4 can be written in the more convenient form of

$$R=\frac{8.3143}{\mathfrak{M}} \quad [\mathrm{kJ/kg\cdot\Delta_1 K}] \tag{6-4a}$$

Values of the molecular weight $\mathfrak{M}$ of some common gases are given in Appendix A, Table A-9, with units of [mol^{-1}].

The relationship given by Equation 6-4 can be derived in the following manner. When Equation 6-2a is written in terms of moles

$$\bar{R}=\frac{pV}{MT/\mathfrak{M}} \quad [\mathrm{J/kg.mol^{-1}\cdot\Delta_1 K}]$$

*From chemistry, a mole is defined as an amount of mass numerically equal to the molecular weight of the substance.

where

$$\mathfrak{M} = \text{molecular weight} \qquad [\text{mol}^{-1}]$$

Then

$$\frac{\bar{R}}{\mathfrak{M}} = \frac{pV}{MT}$$

Equating this with the pV/MT of Equation 6-2a,

$$\frac{pV}{MT} = R = \frac{\bar{R}}{\mathfrak{M}} \qquad [\text{J/kg}\cdot\Delta_1\text{K}] \qquad \text{(Eq. 6-4)}$$

Using Equation 6-4, the gas constant R can be determined for any gas if the molecular weight of that gas is known.

Example Problem 6-1

Determine the gas constant for air.

Equation

$$R = \frac{\bar{R}}{\mathfrak{M}} \qquad [\text{J/kg}\cdot\Delta_1\text{K}] \qquad \text{(Eq. 6-4)}$$

Parameters

$\bar{R} = 8314.3$ J/kg.mol$\cdot\Delta_1$K (universal gas constant)

$\mathfrak{M} = 28.966$ mol^{-1} (Appendix A, Table A-9)

Substitution

$$R = \frac{8314.3}{28.966} \qquad \frac{[\text{J/kg}\cdot\text{mol}\cdot\Delta_1\text{K}]}{[\text{mol}^{-1}]} = [\text{J/kg}\cdot\Delta_1\text{K}]$$

Answer

$$\mathbf{R = 287.04\ J/kg\cdot\Delta_1 K}$$

The perfect gas laws can be used to determine various parameters of the state of a gas as illustrated by the following sample problems.

Example Problem 6-2

Nitrogen at an absolute pressure of 689.5 kPa is used to fill a container having a volume of 0.002 m^3. The filling process is very slow, and the gas in the container attains the ambient (room) temperature of 20°C (293.15 K). Determine the amount of gas in this container.

Equation

$$pV = MRT \qquad [\text{kJ/kg}] \qquad (\text{Eq. 6-3})$$

$$\text{M} = \frac{\text{pV}}{\text{RT}}$$

Parameters

$$p = 689.5 \text{ kPa} = 689\,500 \text{ Pa} = (\text{N/m}^2)$$

$$V = 0.002 \text{ m}^3 \qquad (\text{given})$$

$$T = 293.15 \text{ K} \qquad (\text{given})$$

$$R = \frac{\overline{R}}{\mathfrak{M}} \qquad [\text{J/kg}\cdot\Delta_1\text{K}] \qquad (\text{Eq. 6-4})$$

$$\overline{R} = 8314.3 \text{ J/kg}\cdot\text{mol}\cdot\Delta_1\text{K} \qquad (\text{universal gas constant})$$

$$\mathfrak{M} = 28 \text{ mol}^{-1} \qquad (\text{Appendix A, Table A-9})$$

$$R = \frac{(8314.3)}{(28)} \qquad \frac{[\text{J/kg}\cdot\text{mol}\cdot\Delta_1\text{K}]}{[\text{mol}^{-1}]}$$

$$= 293.15 \text{ J/kg}\cdot\Delta_1\text{K}$$

Substitution

$$M = \frac{(689\,500)(0.002)}{(296.94)(293.15)} \qquad \frac{[\text{N/m}^2][\text{m}^3]}{[\text{J/kg}\cdot\Delta_1\text{K}][\text{K}]} = \frac{[\text{Nm}][\text{kg}]}{[\text{J}]} = [\text{kg}]$$

Answer

$$\boldsymbol{M = 0.01584 \text{ kg}}$$

Example Problem 6-3

A container having a volume of 0.2 m^3 contains nitrogen at a pressure of 100 kPa and a temperature of 15°C. If 0.2 kg of nitrogen is pumped into this container, calculate the final pressure after the gas has returned to its initial temperature. (Assume it to be a perfect gas.)

Equation

$$p_2 = \frac{M_2 RT}{V} \qquad [\text{Pa}] \qquad (\text{Eq. 6-3x})$$

Parameters

$$M_2 = M_1 + \Delta M \qquad [\text{kg}]$$

$$M_1 = \frac{p_1 V}{RT} \qquad [\text{kg}] \qquad (\text{Eq. 6-3x})$$

$$p_1 = 100\ \text{kPa} = 100\ 000\ \text{Pa} = \text{N/m}^2$$

$$V = 0.2\ \text{m}^3 \qquad (\text{given})$$

$$R = \frac{\bar{R}}{\mathfrak{M}} \qquad [\text{J/kg}\cdot\Delta_1\text{K}] \qquad (\text{Eq. 6-4})$$

$$\bar{R} = 8314.3\ \text{J/kg}\cdot\text{mol}\cdot\Delta_1\text{K} \qquad (\text{universal gas constant})$$

$$\mathfrak{M} = 28.013\ \text{mol}^{-1} \qquad (\text{Appendix A, Table A-8})$$

$$R = \frac{8314.3}{28.013} \qquad \frac{[\text{J/kg}\cdot\text{mol}\cdot\Delta_1\text{K}]}{[\text{mol}^{-1}]} = [\text{J/kg}\cdot\Delta_1\text{K}]$$

Therefore,

$$R = 296.94\ \text{J/kg}\cdot\Delta_1\text{K}$$

$$T = 15^\circ\text{C} = 15 + 273.15 = 288.15\ \text{K} \qquad (\text{given})$$

Substitution in Equation for M_1

$$M_1 = \frac{(100\ 000)(0.2)}{(296.94)(288.15)} \qquad \frac{[\text{N/m}^2][\text{m}^3]}{[\text{J/kg}\cdot\Delta_1\text{K}][\text{K}]} = \frac{[\text{Nm}][\text{kg}]}{[\text{J}]} = [\text{kg}]$$

$$M_1 = 0.234\ \text{kg}$$

$$\Delta M = 0.2\ \text{kg} \qquad (\text{given})$$

Substitution in Equation for M_2

$$M_2 = 0.234 + 0.2 \qquad [\text{kg}] + [\text{kg}] = [\text{kg}]$$

$$= 0.434\ \text{kg}$$

Substitution in Equation for p_2

$$p_2 = \frac{(0.434)(296.94)(288.15)}{(0.2)} \qquad \frac{[\text{kg}][\text{J/kg}\cdot\Delta_1\text{K}][\text{K}]}{[\text{m}^3]} = [\text{J/m}^3]$$

$$= [\text{Nm/m}^3] = [\text{N/m}^2] = [\text{Pa}]$$

Answer

$$\mathbf{p_2 = 185{,}700\ Pa = 185.7\ kPa = 0.1857\ MPa}$$

Example Problem 6-4

A mass of 0.01 kg of a certain perfect gas occupies a volume of 0.003 m^3 at a pressure of 600 kPa and a temperature of 73°C. Calculate the molecular weight of the gas. When the gas is allowed to expand until the pressure is 100 kPa and the final volume is 0.02 m^3, calculate the final temperature.

(a) Equation for $\mathfrak{M}$

$$\mathfrak{M}=\frac{\bar{R}}{R} \qquad [\text{mol}^{-1}] \qquad (\text{Eq. 6-4})$$

$$R=\frac{pV}{MT} \qquad [\text{J/kg}\cdot\Delta_1\text{K}] \qquad (\text{Eq. 6-3x})$$

Therefore,

$$\mathfrak{M}=\frac{\bar{R}MT}{pV} \qquad [\text{mol}^{-1}]$$

Parameters

$$\bar{R}=8314.3 \text{ J/kg}\cdot\text{mol}\cdot\text{K} \qquad (\text{universal gas constant})$$

$$M=0.01 \text{ kg} \qquad (\text{given})$$

$$T=73°\text{C}=73+273.15=345.15 \text{ K} \qquad (\text{given})$$

$$p=600 \text{ kPa}=600\,000 \text{ Pa} \qquad (\text{given})$$

$$V=0.003 \text{ m}^3 \qquad (\text{given})$$

Substitution in the Equation for *M*

$$\mathfrak{M}=\frac{(8314.3)(0.01)(346.15)}{(600\,000)(0.003)} \qquad \frac{[\text{J/kg}\cdot\text{mol}\cdot\Delta_1\text{K}][\text{kg}][\text{K}]}{[\text{Pa}][\text{m}^3]}$$

$$=\frac{[\text{J}]}{[\text{Pa m}^3][\text{mol}]}=\frac{[\text{J}]}{[\text{Nm}][\text{mol}]}=[\text{mol}^{-1}]$$

Answer

$$\boldsymbol{\mathfrak{M}=15.99 \text{ mol}^{-1}}$$

(b) Equation for T

$$T=\frac{pV}{MR} \qquad [\text{K}] \qquad (\text{Eq. 6-3x})$$

Parameters

$$p = 100 \text{ kPa} = 100{,}000 \text{ Pa} \qquad \text{(given)}$$

$$V = 0.02 \text{ m}^3 \qquad \text{(given)}$$

$$M = 0.01 \text{ kg} \qquad \text{(given)}$$

$$R = \frac{\bar{R}}{\mathfrak{M}} \qquad [\text{J/kg}\cdot\Delta_1\text{K}] \qquad \text{(Eq. 6-4)}$$

$$\bar{R} = 8314.3 \text{ J/kg}\cdot\text{mol}\cdot\text{K}$$

$$\mathfrak{M} = 15.99 \text{ mol}^{-1} \qquad \text{(from Example Problem 6-4a)}$$

$$R = \frac{(8314.3)}{(15.99)} \qquad \frac{[\text{J/kg}\cdot\text{mol}\cdot\Delta_1\text{K}]}{[\text{mol}^{-1}]} = [\text{J/kg}\cdot\Delta_1\text{K}]$$

$$= 520.0 \text{ J/kg}\cdot\Delta_1\text{K}$$

Substitution in the Equation for T

$$T = \frac{(100\,000)(0.02)}{(0.01)(520.0)} \qquad \frac{[\text{Pa}][\text{m}^3]}{[\text{kg}][\text{J/kg}\cdot\Delta_1\text{K}]} = \frac{[\text{Nm}][\text{K}]}{[\text{J}]} = [\text{K}]$$

Answer

$$\mathbf{T = 384.62 \text{ K}}$$

6-5 REAL GASES AND VAPORS

In a strict sense the gaseous phase of a substance is termed a vapor only when it is in equilibrium with the saturate liquid phase below the critical point. However, often by custom (such as with steam, for example) when heat is added to the saturate vapor, causing it to be no longer in equilibrium with the liquid phase of the substance, the resulting gaseous phase is referred to as a superheated vapor (rather than a gas, which it is).

The state of most common gases can be described by the perfect gas equations of state throughout a wide range of temperature and pressure. Significant deviations from the perfect gas equations are encountered as the gas approaches saturated vapor, on one hand, or the critical point on the other. The deviation encountered is in the volume of the gas for a given state (M, p, T), which may be greater or less than the value calculated by the perfect gas law

$$V = R\left[\frac{MT}{p}\right] \qquad [\text{m}^3] \qquad \text{(Eq. 6-2a)}$$

6-6 THE COMPRESSIBILITY FACTOR

To correct for this deviation, most authors have added a factor **Z** to the perfect gas equation (6-2a).

$$V=(\mathbf{Z})(R)\left[\frac{MT}{p}\right] \qquad [\mathrm{m}^3] \tag{6-5}$$

The factor **Z** is 1.00 for a perfect gas and less than 1.00 for a real gas, except at high pressures; **Z** is often termed the *compressibility factor*. The value of **Z** can be calculated from measurements of mass, pressure, volume, and temperature as indicated by Equation 6-5a.

$$\mathbf{Z}=\frac{pV}{MRT}=\frac{pv}{RT} \qquad [-] \tag{6-5a}$$

The variation of the value of the **Z** factor from 1.0 can be significant in the region near the saturation line, as illustrated in Figure 6-1. Values can be 0.7 or less.

Example Problem 6-5

Determine the compressibility factor for steam at a pressure of 15 MPa at a temperature of 650°C if $R=461.5$ J/kg·K.

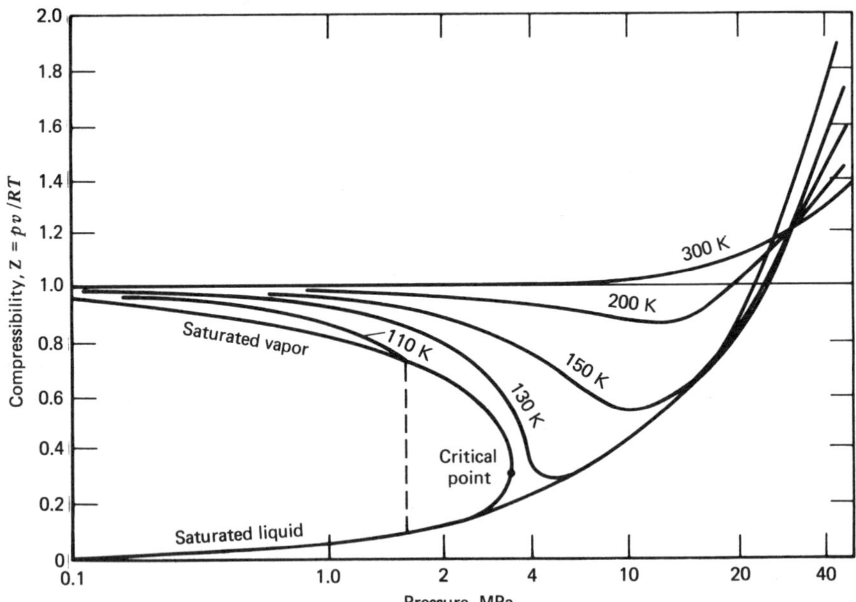

Figure 6-1 Compressibility factors for nitrogen.

Equation

$$Z = \frac{pv}{RT} \quad [—] \quad \text{(Eq. 6-5a)}$$

Parameters

$p = 15 \times 10^6$ [Pa] (given)

$v = 0.0268\ \text{m}^3/\text{kg}$ (steam tables at 15 MPa and 650°C)

$R = 461.5\ \text{J/kg}\cdot\Delta_1\text{K}$ (given)

$T = 650°\text{C} \rightarrow 923\ \text{K}$ (given)

Substitution

$$Z = \frac{(15 \times 10^6)(0.0268)}{(461.5)(923)} = \frac{[\text{N/m}^2][\text{m}^3/\text{kg}]}{[\text{J/kg}\cdot\Delta_1\text{K}][\text{K}]}$$

$$= \frac{[\text{Nm}]}{[\text{J}]} = [—]$$

Answer

$$\mathbf{Z = 0.944}\ [—]$$

A plot of the compressibility factor for nitrogen is shown in Figure 6-1. Lines of constant temperature are plotted on coordinates of compressibility factor and pressure. For temperatures above 300 K the compressibility factor will be close to 1.00 over a wide range of pressures.

From Figure 6-1 it can be seen that the compressibility factor for nitrogen is essentially unity for the temperatures and pressures usually encountered in thermodynamic problems. Consequently, the perfect gas equations can be used for the equations of state of nitrogen throughout the applicable region.

Example Problem 6-6

A rigid tank having a volume of 0.5 m^3 contains 25 kg of nitrogen at a temperature of 250 K. Determine the pressure of the gas when $Z = 0.985$.

Equation

$$pV = ZMRT \quad [\text{J}] \quad \text{(Eq. 6-5x)}$$

Thus

$$p = \frac{ZMRT}{V} \quad [\text{Pa}]$$

Parameters

$\mathbf{Z}=0.985 \quad [\text{—}] \quad \text{(given)}$

$R=0.2968\ \text{kJ/kg}\cdot\Delta_1\text{K} \quad \text{(Appendix A, Table A-9)}$

$T=250\ \text{K} \quad \text{(given)}$

$V=0.5\ \text{m}^3 \quad \text{(given)}$

$M=25\ \text{kg} \quad \text{(given)}$

Substitution

$$p=\frac{(0.985)(25)(0.2968)}{(0.5)}\ \frac{[\text{—}][\text{kg}][\text{kJ/kg}\cdot\Delta_1\text{K}][\text{K}]}{[\text{m}^3]}=\frac{[\text{kg}]}{[\text{m}^3]}$$

$$=\frac{[\text{kNm}]}{[\text{m}^3]}=[\text{kPa}]$$

Answer

$$\boldsymbol{p=3654.35\ \textbf{kPa}=3.65\ \textbf{MPa}}$$

In order to have an equation of state that represents the pV product relationship for a particular gas over a wide range of pressures and temperatures, empirical data have been used to develop equations of state different from the perfect gas laws. One basic or simplified form of such an equation of state is as follows:

$$p=A+BT+\frac{C}{T^2} \qquad [\text{kPa}] \tag{6-6}$$

where the constants A, B, and C are functions of the specific volume v.

Probably the best-known and most useful is the Beattie–Bridgeman equation of state:

$$p=\frac{\bar{R}T(1-\varepsilon)}{v^2}(v+B)-\frac{A}{v^2} \qquad [\text{kPa}] \tag{6-7}$$

where

$p=$ absolute pressure $\quad [\text{kPa}]$

$\bar{R}=$ universal gas constant $\quad (8314.3\ \text{J/kg}\cdot\text{mol}\cdot\Delta_1\text{K})$

$T=$ absolute temperature $\quad [\text{K}]$

$v=$ specific volume $\quad [\text{m}^3/\text{kg}\cdot\text{mol}]$

$A=A_0(1-a/v)$

$B=B_0(1-a/v)$

$\varepsilon=\dfrac{c}{vT^3}$

Table 6-1 Constants of the Beattie–Bridgman Equation of State

Gas	A_0	a	B_0	b	$10^{-4}c$
Helium	2.1886	0.05984	0.01400	0.0	0.0040
Argon	130.7802	0.02328	0.03931	0.0	5.99
Hydrogen	20.0117	−0.00506	0.02096	−0.04359	0.0504
Nitrogen	136.2315	0.02617	0.05046	−0.00691	4.20
Oxygen	151.0857	0.02562	0.04624	0.004208	4.80
Air	131.8441	0.01931	0.04611	−0.001101	4.34
Carbon dioxide	507.2836	0.07132	0.10476	0.07235	66.00

Note: Pressure in kilopascals; specific volume in m^3/k·mol; temperature in Kelvin; $\bar{R}=$ 8.31434 kJ/kg·mol·K.

and the constants A_0, a, B_0, b, and c have been determined experimentally for each substance. The values of these constants for various substances are given in Table 6-1.

6-7 SPECIFIC HEAT PARAMETERS

The specific heat of a solid or liquid is usually defined as the heat (thermal energy) required to change the temperature of a unit mass one degree. The SI (metric) units for specific heat are [kJ/kg·Δ_1K]. The equation for the amount of heat Q to change the temperature T of a mass M is

$$Q=Mc\Delta T \qquad [\text{kJ}] \tag{6-8}$$

where

$Q=$ heat required [kJ]

$M=$ mass [kg]

$c=$ specific heat [kJ/kg·Δ_1K]

$\Delta T=$ temperature rise [ΔK]

For an increase in temperature ($+\Delta T$) of the mass M, heat must be added ($+Q$). For a decrease in temperature ($-\Delta T$) of the mass M, heat must be removed ($-Q$).

When heat is added to a gas, a wide variety of pressure and volume conditions can be obtained, resulting in a range of specific heat values. By custom two "basic" specific heat parameters are used, namely the specific heat at constant volume c_v and the specific heat at constant pressure c_p. Heat added at constant volume increases the internal energy of the gas ($q=\Delta u$). Heat added at constant pressure increases both the internal energy and the flow work, $p\Delta V$ ($q=\Delta u+p\Delta v$). Consequently, the specific heat at constant pressure c_p has a larger value than the specific heat at constant volume c_v.

Now by definition for a reversible nonflow process at constant pressure, the specific heat parameter c_p is defined by the following equation:

$$Q=Mc_p\Delta T \qquad [\mathrm{J}] \tag{6-9}$$

or

$$Q=Mc_p(T_2-T_1) \qquad [\mathrm{J}]$$

and

$$q=c_p\Delta T \qquad [\mathrm{kJ/kg}] \tag{6-9a}$$

and similarly for a reversible nonflow process at constant volume, the specific heat parameter c_v is defined by the equation:

$$Q=Mc_v\Delta T \qquad [\mathrm{J}] \tag{6-10}$$

or

$$Q=Mc_v(T_2-T_1) \qquad [\mathrm{J}]$$

and

$$q=c_v\Delta T \qquad [\mathrm{kJ/kg}] \tag{6-10a}$$

For a perfect gas, the values for c_p and c_v are constant for any one gas at all pressures and temperatures. For real gases, c_p and c_v vary with temperature. An arithmetic average value over the temperature range of interest is often used to compensate for the variations in values with temperature.

6-7.1 THE RELATIONSHIP BETWEEN SPECIFIC HEAT PARAMETERS

There is a unique relationship between c_p and c_v for a perfect gas. using the closed system (nonflow) equation for the first law (Equation 3-8), we find that

$$Q-Wk=(U_2-U_1) \qquad [\mathrm{J}]$$

Substituting Equation 6-10a into Equation 3-8, we obtain the following equation for the change in internal energy:

$$(Q-Wk)=Mc_v(T_2-T_1) \qquad [\mathrm{J}] \tag{6-11}$$

In a constant pressure process, the work done by the expanding fluid is given by the following equation:

$$Wk=p(V_2-V_1) \qquad [\mathrm{J}] \tag{6-12}$$

where

$$Wk = \text{work output} \qquad [\text{J}]$$

$$p = \text{fluid pressure} \qquad [\text{Pa}]$$

$$V = \text{fluid volume} \qquad [\text{m}^3]$$

Now, using the perfect gas equation of state (Equation 6-2), we have

$$pV_1 = MRT_1 \qquad [\text{J}] \qquad \text{and} \qquad pV_2 = MRT_2 \qquad [\text{J}]$$

Combining these relationships with Equation 6-12, we obtain

$$Wk = (pV_2 - pV_1)$$
$$= (MRT_2 - MRT_1)$$

and

$$Wk = MR(T_2 - T_1) \qquad [\text{J}] \qquad (6\text{-}13)$$

Now substituting Equation 6-13 into Equation 6-11 for a constant volume process, we find that

$$Q - MR(T_2 - T_1) = Mc_v(T_2 - T_1)$$

and

$$Q = Mc_v(T_2 - T_1) + MR(T_2 - T_1)$$

or

$$Q = M(c_v + R)(T_2 - T_1) \qquad [\text{J}] \qquad (6\text{-}14)$$

Previously, for a constant pressure process, we obtained

$$Q = Mc_p(T_2 - T_1) \qquad [\text{J}] \qquad (\text{Eq. } 6\text{-}9)$$

Equating these two relationships, we get

$$Mc_p(T_2 - T_1) = M(c_v + R)(T_2 - T_1)$$
$$c_p = c_v + R$$

or

$$(c_p - c_v) = R \qquad [\text{kJ/kg}\cdot\Delta_1\text{K}] \qquad (6\text{-}15)$$

Example Problem 6-7

Determine the gas constant and molecular weight of a perfect gas whose specific heat values are $c_p = 0.846$ kJ/kg·Δ_1K and $c_v = 0.657$ kJ/kg·Δ_1K.

(a) Equation

$$R = (c_p - c_v) \quad [\text{kJ/kg}\cdot\Delta_1\text{K}] \quad (\text{Eq. 6-15})$$

Parameters

$$c_p = 0.846 \text{ kJ/kg}\cdot\Delta_1\text{K} \quad (\text{given})$$

$$c_v = 0.657 \text{ kJ/kg}\cdot\Delta_1\text{K} \quad (\text{given})$$

Substitution

$$R = (0.846 - 0.657)$$

Answer

$$\mathbf{R = 0.189\ KJ/kg\cdot\Delta_1 K}$$

(b) Equation

$$\mathfrak{M} = \frac{\bar{R}}{R} \quad [\text{mol}^{-1}] \quad (\text{Eq. 6-4})$$

Parameters

$$\bar{R} = 8314.3 \text{ J/kg}\cdot\text{mol}\cdot\Delta_1\text{K} \quad (\text{universal gas constant})$$

$$R = 0.189 \text{ kJ/kg}\cdot\Delta_1\text{K} \quad (\text{from Example Problem 6-7a})$$

Substitution

$$\mathfrak{M} = \frac{(8314.3)}{(189.0)} \quad \frac{[\text{J/kg}\cdot\text{mol}\cdot\Delta_1\text{K}]}{[\text{J/kg}\cdot\Delta_1\text{K}]} = [\text{mol}^{-1}]$$

Answer

$$\mathbf{\mathfrak{M} = 44\ mol^{-1}}$$

6-7.2 THE RATIO OF SPECIFIC HEAT PARAMETERS

The ratio of the specific heat at constant pressure to the specific heat at constant volume is γ (the Greek letter gamma) and is always greater than unity. Thus

$$\gamma = \frac{c_p}{c_v} \quad [-] \qquad (6\text{-}16)$$

In general, $\gamma = 1.67$ for monatomic gases such as argon, krypton, xenon, helium, and neon. Similarly, $\gamma = 1.40$ for diatomic gases such as air, nitrogen, oxygen, hydrogen, and carbon monoxide. Triatomic gases, such as carbon dioxide and sulfur dioxide, have a specific heat ratio of about 1.29, while for certain hydrocarbons the value is very low (e.g., isobutane, C_4H_{10}, $\gamma = 1.11$).

Some useful relationships between c_p, c_v, R, and γ can be derived. Starting with Equation 6-15,

$$(c_p - c_v) = R$$

and dividing each side of the equation by c_v, we obtain

$$\left[\frac{(c_p - c_v)}{c_v}\right] = \frac{R}{c_v}$$

Simplifying gives

$$\left[\frac{c_p}{c_v} - 1\right] = \frac{R}{c_v}$$

or

$$(\gamma - 1) = \frac{R}{c_v}$$

Therefore, the specific heat parameters in terms of the gas constant R and the ratio of specific heats γ are as follows:

$$c_v = \frac{R}{(\gamma - 1)} \qquad [\text{kJ/kg}\cdot\Delta_1\text{K}] \qquad (6\text{-}17)$$

Also

$$c_p = \gamma c_v$$

or

$$c_p = \frac{\gamma R}{(\gamma - 1)} \qquad [\text{kJ/kg}\cdot\Delta_1\text{K}] \qquad (6\text{-}18)$$

These relationships are important to the isentropic processes, which are discussed in Section 7-4.

Example Problem 6-8

A perfect gas has a molecular weight of 28 mol^{-1} and a value of $\gamma = 1.40$. Calculate the heat rejected per kilogram of mass if this gas is contained in a rigid vessel initially at 300 kPa and 300°C, and is then cooled until the pressure is decreased to 150 kPa.

Equation

$$\Delta q = c_v(T_2 - T_1) \quad [\text{kJ/kg}] \quad (\text{Eq. 6-10a})$$

Parameters

$$T = 300°\text{C} = 573.15\ \text{K} \quad (\text{given})$$
$$\mathfrak{M} = 28\ \text{mol}^{-1} \quad (\text{given})$$
$$\gamma = 1.40\ [—] \quad (\text{given})$$
$$p_1 = 300\ \text{kPa} \quad (\text{given})$$
$$p_2 = 150 \quad (\text{given})$$

For a rigid vessel $V_1 = V_2$ (constant volume).

Using the perfect gas equation of state (Eq. 6-2) for states 1 and 2 with

$$T_2 = T_1\left(\frac{p_2}{p_1}\right) \quad [\text{K}]$$

we obtain

$$M_1 = M_2 = M$$
$$p_1V_1 = MRT_1$$
$$p_2V_2 = MRT_2$$

Dividing one equation into the other gives

$$T_2 = \frac{(573.15)(150)}{(300)} = 286.58\ \text{K}$$

$$c_v = \frac{R}{(\gamma - 1)} \quad [\text{J/kg}\cdot\Delta_1\text{K}] \quad (\text{Eq. 6-17})$$

$$R = \frac{\overline{R}}{\mathfrak{M}} \quad [\text{J/kg}\cdot\Delta_1\text{K}] \quad (\text{Eq. 6-4})$$

$$\overline{R} = 8314.3\ \text{J/kg}\cdot\text{mol}\cdot\Delta_1\text{K} \quad (\text{universal gas constant})$$

$$R = \frac{(8314.3)}{28} = 296.94\ \text{J/kg}\cdot\Delta_1\text{K}$$

$$c_v = \frac{(296.94)}{(1.4 - 1)} = 742.35\ \text{J/kg}\cdot\Delta_1\text{K}$$

Substitution

$$\Delta q = (742.35)(286.58 - 573.15) \quad [\text{J/kg}\cdot\Delta_1\text{K}][\Delta\text{K}] = [\text{J/kg}]$$
$$= -212735\ \text{J/kg}$$

Answer

$$\mathbf{\Delta q = -212.74\ kJ/kg}$$

The negative sign indicates that heat is leaving the system.

6-8 JOULE'S LAW

Joule's law states that the internal energy of a perfect gas is a function of the absolute temperature only. Thus, according to this law,

$$U=\phi(T)$$

Using the nonflow energy equation and noting that $Wk=0$, we obtain

$$Q=\Delta U=U_2-U_1 \qquad [\mathrm{J}] \qquad (6\text{-}19)$$

or

$$q=\Delta u=u_2-u_1 \qquad [\mathrm{kJ/kg}] \qquad (6\text{-}19a)$$

For a perfect gas at constant volume, we can rewrite Equation 6-10 as follows:

$$\frac{Q}{M}=\frac{Mc_v\Delta T}{M}$$

giving

$$q=c_v\Delta T \qquad [\mathrm{kJ/kg}] \qquad (\text{Eq. } 6\text{-}10a)$$

Substituting from Equation 6-19a, we find that

$$\Delta u=c_v\Delta T \qquad [\mathrm{kJ/kg}] \qquad (6\text{-}20)$$

and integrating gives

$$u=c_vT+K \qquad [\mathrm{kJ/kg}] \qquad (6\text{-}20a)$$

where K is a constant of integration.

Since internal energy can be made zero at any arbitrary reference temperature, let us assume that $u=0$ when $T=0$ K for a perfect gas. Then it follows that $K=0$, and for a perfect gas internal energy is shown to be directly proportional to the absolute temperature.

$$u=c_vT \qquad [\mathrm{kJ/kg}] \qquad (6\text{-}21)$$

However, the absolute value of internal energy is not as important as the change in internal energy for applications to the first law. If the mass, M, is introduced to both sides of Equation 6-20, we obtain

$$\Delta U=Mc_v\Delta T \qquad [\mathrm{J}] \qquad (6\text{-}22)$$

In any process for a perfect gas between state 1 and state 2, we can rewrite Equation 6-10 as follows:

$$\Delta U=(U_2-U_1)=Mc_v(T_2-T_1) \qquad [\mathrm{J}] \qquad (6\text{-}22a)$$

The increase in internal energy for a perfect gas between two states is applicable for any process whether reversible or irreversible.

Example Problem 6-9

Determine the change in internal energy of 2.0 kg of oxygen gas as it changes temperature from 20 to 40°C. (Assume that oxygen is a perfect gas and that the specific heat value is constant over this temperature range.)

Equation

$$\Delta U=Mc_v(T_2-T_1) \qquad [\mathrm{J}] \qquad (\text{Eq. } 6\text{-}22a)$$

Parameters

$$M=2.0\ \mathrm{kg} \qquad (\text{given})$$
$$c_v=0.6618\ \mathrm{kJ/kg\cdot\Delta_1 K} \qquad (\text{Appendix A, Table A-9})$$
$$T_2=40^\circ\mathrm{C}=313.15\ \mathrm{K} \qquad (\text{given})$$
$$T_1=20^\circ\mathrm{C}=293.15\ \mathrm{K} \qquad (\text{given})$$

Substitution

$$\Delta U=(2.0)(0.6618)(313.15-293.15) \qquad [\mathrm{kg}][\mathrm{kJ/kg\cdot\Delta_1 K}][\Delta_1\mathrm{K}]=[\mathrm{kJ}]$$
$$=(2.0)(0.6618)(20)$$

Answer

$$\mathbf{\Delta U=+26.472\ kJ}$$

The sign indicates an increase in internal energy.

6-9. ENTHALPY OF A GAS

From the definition of the property *enthalpy* we have

$$h=u+pv$$

so that

$$\Delta h=\Delta u+\Delta pv$$

Also, from the perfect gas equation of state we have

$$pv=RT \qquad [\mathrm{kJ/kg}] \qquad (\text{Eq. } 6\text{-}3)$$

Thus

$$\Delta pv = \Delta RT = R\,\Delta T$$

and from Joule's law

$$\Delta u = c_v \Delta T \qquad [\text{kJ/kg}] \qquad (\text{Eq. 6-20})$$

Combining these equations, we obtain the following:

$$\Delta h = c_v \Delta T + R\,\Delta T$$

or

$$\Delta h = (c_v + R)\,\Delta T \qquad [\text{kJ/kg}] \qquad (6\text{-}23)$$

But from Equation 6-19 we can obtain

$$c_p = (c_v + R)$$

so that, for a perfect gas,

$$\Delta h = c_p \Delta T \qquad [\text{kJ/kg}] \qquad (6\text{-}24)$$

where

Δh = specific enthalpy [kJ/kg]

c_p = specific heat at constant pressure [$\text{kJ/kg}\cdot\Delta_1\text{K}$]

ΔT = absolute temperature change [ΔK]

By introducing the mass, M, we obtain

$$M\Delta h = Mc_p \Delta T$$

or

$$\Delta H = Mc_p \Delta T \qquad [\text{J}] \qquad (6\text{-}25)$$

Example Problem 6-10

Determine the change in the specific enthalpy of methane gas as it changes the temperature from 400 to 300°C. (Assume that methane behaves as a perfect gas with a constant value for the specific heat over that temperature range.)

Equation

$$\Delta h = c_p \Delta T = c_p(T_2 - T_1) \qquad [\text{kJ/kg}] \qquad (\text{Eq. 6-24})$$

Parameters

$$T_2 = 300°C = 573.15\ K \qquad \text{(given)}$$

$$T_1 = 400°C = 673.15\ K \qquad \text{(given)}$$

$$c_p = 2.2537\ kJ/kg \cdot \Delta_1 K \qquad \text{(Appendix A, Table A-9)}$$

Substitution

$$\Delta h = (2.2537)(573.15 - 673.15) \qquad [kJ/kg \cdot \Delta_1 K][\Delta K] = [kJ/kg]$$

Answer

$$\mathbf{\Delta h = -225.37\ kJ/kg} \qquad \textbf{or a decrease in enthalpy}$$

Example Problem 6-11

A perfect gas has a molecular weight of 16 mol^{-1} and a value of $\gamma = 1.30$. Calculate the heat added per kilogram of gas if it enters a pipe at 30°C and flows steadily to the exit where the gas temperature is 290°C. (Neglect changes in velocity and elevation.)

Equation

$$q - \overset{0}{\cancel{Wk}} = \left[(h_2 - h_1) + \overset{0}{\cancel{\frac{(V_2^2 - V_1^2)}{2g_c}}} + \overset{0}{\cancel{(Z_2 - Z_1)}} \right] \qquad \text{(Eq. 3-25ax)}$$

$$q = (h_2 - h_1) = \Delta h$$

but

$$\Delta h = c_p(T_2 - T_1) \qquad [kJ/kg] \qquad \text{(Eq. 6-24)}$$

Combining the above equations, we obtain

$$q = c_p(T_2 - T_1) \qquad [kJ/kg]$$

Parameters

$$T_1 = 30°C = 303.15\ K \qquad \text{(given)}$$

$$T_2 = 290°C = 563.15\ K \qquad \text{(given)}$$

$$\mathfrak{M} = 16\ mol^{-1} \qquad \text{(given)}$$

$$\gamma = 1.30\ [—] \qquad \text{(given)}$$

$$c_p = \frac{\gamma R}{(\gamma - 1)} \qquad [J/kg \cdot \Delta_1 K] \qquad \text{(Eq. 6-18)}$$

$$R = \frac{\bar{R}}{\mathfrak{M}} \qquad [J/kg \cdot \Delta_1 K] \qquad \text{(Eq. 6-4)}$$

$$\bar{R} = 8314.3\ J/kg \cdot mol \cdot \Delta_1 K \qquad \text{(universal gas constant)}$$

$$R = \frac{(8314.3)}{16} = 519.64\ J/kg \cdot \Delta_1 K$$

$$c_p = \frac{(1.30)(519.64)}{(1.30 - 1)} = 2251.79\ J/kg \cdot \Delta_1 K$$

Substitution

$$q=(2251.79)(563.15-303.15) \qquad [\text{J/kg}\cdot\Delta_1\text{K}][\Delta\text{K}]=[\text{J/kg}]$$
$$= +585461\ \text{J/kg}$$

Answer

$$\mathbf{q = +585.46\ kJ/kg}$$

The positive sign indicates that heat is being added to the system.

6-10 CHANGE OF ENTROPY FOR A GAS

Since the change in entropy depends on the end states and not on the process itself, we can develop a relationship for an ideal gas in a reversible (ideal) process. Using calculus, we can define entropy as

$$T\,ds = du + p\,dv \qquad (6\text{-}26)$$

and dividing each term by T and substituting the following equations:

$$du = c_v\,dT \qquad [\text{kJ/kg}] \qquad (\text{Eq. 6-9})$$

and

$$pv = RT \qquad [\text{kJ/kg}] \qquad (\text{Eq. 6-3})$$

we obtain

$$ds = c_v\frac{dT}{T} + R\frac{dv}{v}$$

Integrating this equation results in the following relationship:

$$\Delta s = c_v \ln\left[\frac{T_2}{T_1}\right] + R\ln\left[\frac{v_2}{v_1}\right] \qquad [\text{kJ/kg}\cdot\Delta_1\text{K}] \qquad (6\text{-}27)$$

We can also obtain a similar equation in terms of T and p by using another definition for entropy

$$T\,ds = dh - v\,dp$$

which results in the following equations:

$$\Delta s = c_p \ln\left[\frac{T_2}{T_1}\right] - R\ln\left[\frac{p_2}{p_1}\right] \qquad [\text{kJ/kg}\cdot\Delta_1\text{K}] \qquad (6\text{-}28)$$

Both Equations 6-27 and 6-28 assume constant specific heat values.

6-11 PROBLEMS

6-1. Describe the nature and/or mechanism of (a) the temperature and (b) the pressure in a gas according to the kinetic theory.

6-2. What does an equation of state do? Give an equation of state.

6-3. Give an equation that states the perfect gas law in terms of specific volume.

6-4. Explain how the "perfect gas law" is an "equation of state."

6-5. Define or describe the gas constant R. What is the universal gas constant $\bar{R}$ and how does it relate to the gas constant R?

6-6. If 1 kg of gas at 101.3 kPa pressure and 80°C occupies a volume of 1 m^3, what is its gas constant?

6-7. If 1 kg of gas at 1.3 MPa pressure at 42°C has a volume of 1 m^3, what is its gas constant R? What gas might it be?

6-8. A cooling system with a volume of 0.2 m^3 is filled with helium at 25°C to a pressure of 2 MPa. What mass of helium is required?

6-9. Air is stored in a tank with a volume of 1 m^3. When the pressure is 1 MPa and the air temperature is 25°C, what mass of air is in the tank?

6-10. What volume will 1 kg of air at 25°C and 100 kPa occupy?

6-11. If the volume of Problem 6-10 were filled with a stoichiometric mixture of hydrogen and oxygen (2 mol of hydrogen to 1 mol of oxygen) at 25°C and 100 kPa, what mass of hydrogen would it contain?

6-12. If the volume of Problem 6-10 were filled with a stoichiometric mixture of hydrogen and air at 25°C and 100 kPa, what mass of hydrogen would it contain? Assume that air is 23.2 percent by weight oxygen (O_2). For a stoichiometric mixture of hydrogen and air, 2 mol of hydrogen (H_2) are required for 1 mol of oxygen (O_2).

6-13. How do real gases vary from perfect gases? In what region are significant deviations to be expected? What magnitudes may these deviations reach?

6-14. Calculate the compressibility factor Z for steam at 0.20 MPa pressure and a temperature of 150°C. Use $R = 0.4615$ kJ/kg·Δ_1K.

6-15. Calculate the compressibility factor Z for steam at 10 MPa pressure and 325°C.

6-16. Compare the compressibility factor Z of steam at a pressure of 0.100 MPa for the following conditions:
(a) Saturate vapor.
(b) Superheated vapor at 200°C.
(c) Superheated vapor at 300°C.
Use $R = 0.4615$ kJ/kg·Δ_1K.

6-17. Compare the compressibility factor Z of F-12 vapor at a pressure of 0.15 MPa for the following conditions:
(a) Saturate vapor.
(b) Superheated vapor at 30°C.
(c) Superheated vapor at 70°C.
Use $R = 0.073$ kJ/kg·Δ_1K.

6-18. Air is heated at a constant volume from 25 to 100°C. How much heat must be added?

6-19. If the air in Problem 6-18 were heated at constant pressure, how much heat must be added? Is this amount of heat different from that required in Problem 6-18? If so, explain why.

6-20. If the mass of air occupying a volume of 1 m^3 at 25°C and 100 kPa pressure received 1000 kJ of heat, what would be the final state of the air for the following processes?
(a) A constant volume.
(b) A constant pressure.

6-21. A perfect gas has the following specific heats:

$$c_v = 0.6618 \text{ kJ/kg}\cdot\Delta_1\text{K}$$
$$c_p = 0.9216 \text{ kJ/kg}\cdot\Delta_1\text{K}$$

What is the gas constant R for this gas and the molecular weight $\mathfrak{M}$?

6-22. In real gases the gas constant is found to be constant over a wide range in temperatures. As air is heated the c_p increases. At a temperature where the c_p has increased 10% from the nominal value in Appendix, A, Table A-9, what is the value of c_v and γ?

6-23. A gas has a molecular weight of 17.03 [mol^{-1}] and a $\gamma = 1.304$ [—]. Calculate the specific heats (c_v and c_p).

6-24. If in an Otto cycle engine the air temperature at the completion of the compression stroke is 416°C, and the ensuing combustion is considered as 1500 kJ/kg of energy added to the air at constant volume, what would be the air temperature when this energy is absorbed by the air and the average value of $c_v = 0.7490$ kJ/kg·Δ_1K over this temperature range?

6-25. If the air in Problem 6-24 contained 0.05 kg of water vapor per kilogram of air and the energy added is 1500 kJ/kg per kilogram of dry air, what would be the resultant equilibrium temperature?

6-6 – $R = \frac{PV}{mT}$

6-9 – $m = \frac{PV}{RT}$

6-16 a) $Z = \frac{Pv_g}{RT}$ b.) $Z = \frac{Pv_g}{RT}$ c.) $Z = \frac{Pv_g}{RT}$

6-21 – $R = (C_p - C_v)$

$\mathfrak{M} = \frac{\bar{R}}{R} = \frac{8.3143}{.2598} = 32.00 \text{ mol}^{-1}$

6-24 – $Q = mC\Delta T \Rightarrow q = C_v \Delta T \Rightarrow \Delta T = \frac{q}{C_v}$

$\Delta T = T_2 - T_1 \Rightarrow T_2 = T_1 + \Delta T$

6-25 – $Q = (n_A C_{v_A} \Delta T) + (n_W C_{v_W} \Delta T) \Rightarrow \Delta T = \frac{Q}{n C_{v_A} + n C_{v_W}}$

$\Delta T = T_2 - T_1 \Rightarrow T_2 = T_1 + \Delta T$

II PROJECT: $Wk_p = v_f^{oc}(P_2 - P_1)$; $\rho = \frac{1}{v_f^{oc}}$, $P_w = \gamma Q H$; $\gamma = \rho g$

CHAPTER 7
PERFECT GAS PROCESSES

The state of a thermodynamic substance is changed from one state to another through a thermodynamic process. As previously discussed, during a process heat may be absorbed or rejected, or work input or output may be involved. Consequently, thermodynamic processes are the elements of thermodynamic cycles for producing power, refrigeration, or other effects. Thermodynamic processes changing the state of thermodynamic substances from an initial state (state 1) to another state (state 2) are controlled by the laws of thermodynamics. Consequently, each process must follow a specific behavior.

An understanding of these processes is a key to the analysis of thermodynamic cycles. Chapter 7 treats gas processes based upon the perfect gas laws; Chapter 8 considers two-phase processes commonly encountered in steam powerplant cycles.

The initial discussion gives equations of state and the work equation for gas processes commonly encountered in thermodynamic gas cycles. These processes include:

- Constant pressure (isobaric).
- Constant volume (isometric).
- Constant temperature (isothermal).
- Adiabatic.
- Polytropic.

Each process is discussed, giving the relationships between pressure–temperature–volume, the change in entropy, and a summary of the work for (1) closed systems and (2) open systems are summarized, followed by the derivations of the equations based upon the perfect gas laws.

It is important for the student to note that for all processes the pressure–temperature–volume relationships and the equation for the change in entropy is the same for closed and open systems; however, the work equations for some processes are different between closed and open systems.

In the following sections at the beginning of each process a summary is given that presents:

- Process definition
- Relationships between properties of state
- Heat requirements
- Work requirements
- Change in entropy

7-1 CONSTANT PRESSURE (ISOBARIC) PROCESS

Probably the most common thermodynamic process is the constant pressure or isobaric process. The combustion chamber in a gas turbine system is one example of a device which approximates the constant pressure process. In the isobaric process the pressure–temperature–volume relationships for a perfect gas are:

Process Definition

$$p_1 = p_2$$

Relationships Between Properties of State

$$p = MR\left[\frac{T}{V}\right] = \text{constant} \qquad [\text{kPa}] \qquad (7\text{-}1)$$

Therefore,

$$T_2 = T_1\left[\frac{V_2}{V_1}\right] \qquad [\text{K}] \qquad (7\text{-}1\text{a})$$

$$V_2 = V_1\left[\frac{T_2}{T_1}\right] \qquad [\text{m}^3] \qquad (7\text{-}1\text{b})$$

Heat Requirements

$$q = c_p(T_2 - T_1) \qquad [\text{kJ/kg}] \qquad (\text{Eq. 6-9a})$$

Work Requirements

$$\text{Closed system } wk_c = p(V_2 - V_1) \qquad [\text{kJ}] \qquad (7\text{-}2)$$

$$\text{Open system } wk_0 = 0 \qquad (7\text{-}3)$$

Change in Entropy

$$\Delta s = c_p \ln\left[\frac{T_2}{T_1}\right] \qquad [\text{kJ/kg}\Delta_1\text{K}] \qquad (7\text{-}4)$$

7-1.1 GAS PARAMETER DERIVATION

T_2/T_2 Relationship

From the perfect gas law

$$pV = MRT$$

since

$$p = \text{constant} = C$$
$$CV = MRT$$

Consequently,

$$\frac{CV_2}{CV_1} = \frac{MRT_2}{MRT_1}$$

Therefore,

$$T_2 = T_1\left[\frac{V_2}{V_1}\right] \qquad [\text{K}] \qquad (\text{Eq. 7-1})$$

Δs Equation

The Δs equation is derived from Equation 6-28.

$$\Delta s = c_p \ln\left[\frac{T_2}{T_1}\right] - R\ln\left[\frac{p_2}{p_1}\right] \qquad [\text{kJ/kg}\cdot\text{K}] \qquad (\text{Eq. 6-28})$$

since $p_2 = p_1$

$$\ln\left[\frac{p_2}{p_1}\right] = 0$$

Therefore,

$$\Delta s = c_p \ln\left[\frac{T_2}{T_1}\right] \qquad [\text{kJ/kg}\cdot\Delta_1\text{K}] \qquad (\text{Eq. 7-4})$$

Example Problem 7-1

Air enters a combustion chamber at 25°C. Consider the combustion as adding 1500 kJ/kg of heat to the air. What is the ratio of the volume of hot air V_2 to the initial volume V_1?

Equation

$$\frac{V_2}{V_1} = \frac{T_2}{T_1} \qquad [—] \qquad \text{(Eq. 7-1x)}$$

Parameters

$$T_1 = 25°\text{C} \rightarrow 298 \text{ K}$$

$$T_2 = T_1 + \Delta T \qquad [\text{K}]$$

$$\Delta T = \frac{\Delta q}{c_p} \qquad [\Delta\text{K}] \qquad \text{(Eq. 6-9a)}$$

$$\Delta q = 1500 \qquad \text{kJ/kg} \qquad \text{(given)}$$

$$c_p = 1.0035 \qquad \text{kJ/kg}\cdot\Delta_1\text{K} \qquad \text{(Appendix A, Table A-9 at 300 K)}$$

$$\Delta T = \frac{1500}{1.0035} \qquad \frac{[\text{kJ/kg}]}{[\text{kJ/kg}\cdot\Delta\text{K}]} = [\Delta\text{K}]$$

$$\Delta T = 1495 \qquad \Delta\text{K}$$

$$T_2 = 298 + 1495 \qquad [\text{K}] + [\Delta\text{K}] = [\text{K}]$$

$$T_2 = 1793 \text{ K}$$

Substitution

$$\frac{V_2}{V_1} = \frac{1793}{298} \qquad \frac{[\text{K}]}{[\text{K}]} = [—]$$

Answer

$$\boldsymbol{\frac{V_2}{V_1} = 6.02\ [—]}$$

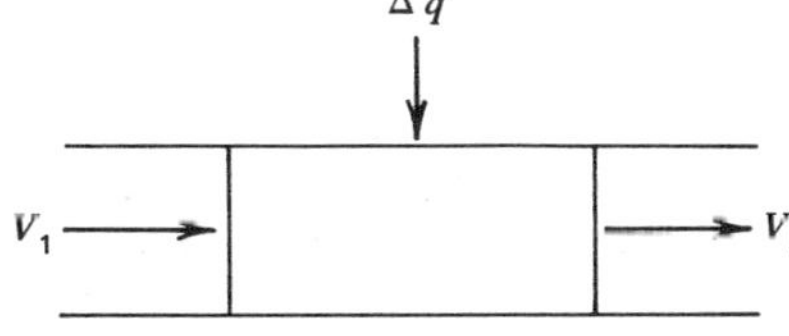

Figure 7-1 Sketch for Example Problem 7-1.

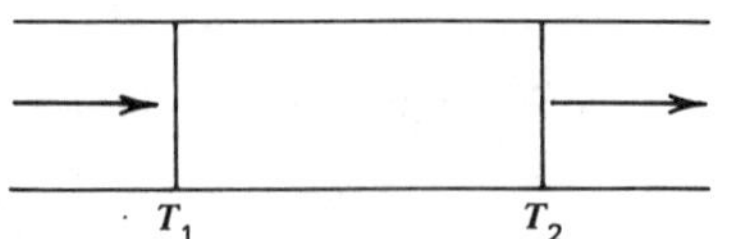

Figure 7-2 Sketch for Example Problem 7-2.

Example Problem 7-2

What is the change in specific entropy (Δs) during the combustion process in Example Problem 7-1? (See Figure 7-2.)

Equation

$$\Delta s = c_p \ln\left[\frac{T_2}{T_1}\right] \qquad [\text{kJ/kg}\cdot\Delta_1\text{K}] \qquad (\text{Eq. 7-4})$$

$$c_p = 1.0035 \qquad \text{kJ/kg}\cdot\Delta_1\text{K} \qquad (\text{Appendix A, Table A-9 at 300 K})$$

$$\frac{T_2}{T_1} = 6.02\ [\text{—}] \qquad (\text{from Example Problem 7-1})$$

$$\ln\left[\frac{T_2}{T_1}\right] = 1.795\ [\text{—}]$$

Substitution

$$\Delta s = (1.0035)(1.795) \qquad [\text{kJ/kg}\cdot\Delta_1\text{K}][\text{—}] = [\text{kJ/kg}\cdot\Delta_1\text{K}]$$

Answer

$$\boldsymbol{\Delta s = 1.80 \qquad \text{kJ/kg}\cdot\Delta_1\text{K}}$$

7-1.2 THE CLOSED ISOBARIC PROCESS

A closed system experiencing a change in volume against atmospheric pressure is involved in a constant pressure process.

The Work Equation

The equation for work (Wk_c) is

$$Wk_c = p(V_2 - V_1) \qquad [\text{J}] \qquad (\text{Eq. 7-2})$$

where

p = absolute pressure [kPa]

V_2 = final volume [m^3]

V_1 = initial volume [m^3]

Equation 7-3 is applicable for any substance, whether gas or liquid; however, most of the equations in this chapter are for gases.

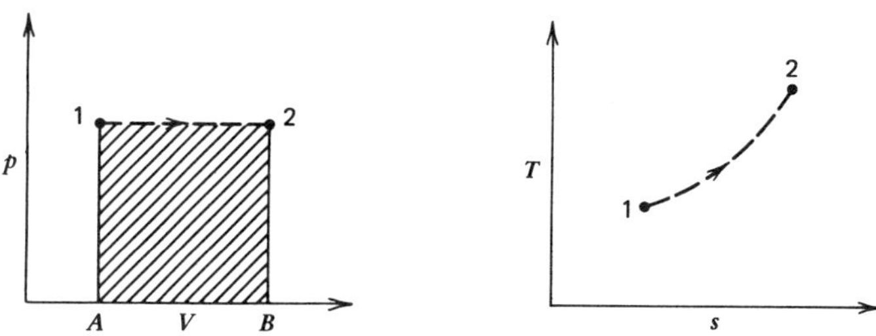

Figure 7-3 Thermodynamic diagrams for a closed constant pressure process.

The equation for work is derived from the definition of work.

$$Wk_c = p\int_{V_1}^{V_2} dV$$

by integrating

$$Wk_c = p(V_2 - V_1) \qquad [\mathrm{J}] \qquad (\text{Eq. 7-2})$$

and

$$wk_c = p(v_2 - v_1) \qquad [\mathrm{kJ/kg}] \qquad (7\text{-}3a)$$

The *p–V* and *T–s* Diagrams

The closed isobaric process is shown graphically by the p–V and T–s diagram in Figure 7-3.

The reversible (ideal) work for the isobaric process in the closed system is represented by the area $A-1-2-B-A$ under the p–V curve.

Example Problem 7-3

A 20-cm-diameter piston-cylinder contains a gas that, during a constant pressure process, moves the piston 7.5 cm. Determine the work done during this process when the gas pressure is 500 kPa. (See Figure 7-4.)

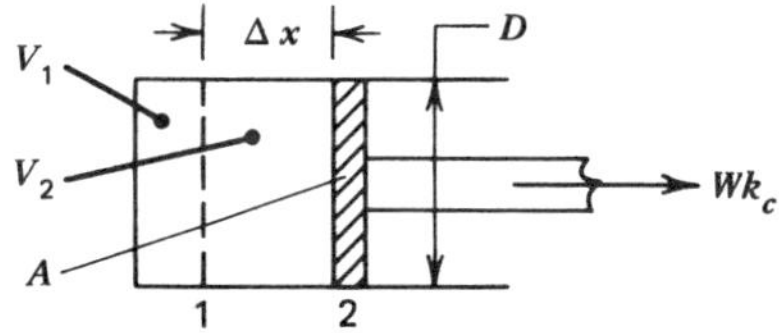

Figure 7-4 Sketch for Example Problem 7-3.

Equation

$$Wk_c = p(V_2 - V_1) \qquad [\mathrm{J}] \qquad (\text{Eq. 7-2})$$

$$(V_2 - V_1) = (A)(\Delta x)$$

Therefore,

$$Wk_c = (p)(A)(\Delta x) \qquad [\text{J}]$$

Parameters

$$p = 500\ \text{kPa} = 500\ \text{kN/m}^2 \qquad \text{(given)}$$

$$A = \frac{\pi D^2}{4} \qquad [\text{m}^2]$$

$$D = 0.20\ \text{m} \qquad \text{(given)}$$

$$A = \pi(0.20)^2/4 \qquad [-][\text{m}^2]/[-] = [\text{m}^2]$$

$$A = 0.0314\ \text{m}^2$$

$$\Delta x = 0.075\ \text{m} \qquad \text{(given)}$$

Substitution

$$Wk_c = (500)(0.0314)(0.075) \qquad [\text{kN/m}^2][\text{m}^2][\text{m}] = [\text{kN}\cdot\text{m}] = [\text{kJ}]$$

Answer

$$\mathbf{Wk_c = 1.178\ kJ}$$

7-1.3 THE OPEN ISOBARIC PROCESS

An open system can have a constant pressure process such as the combustion chamber in an aircraft jet engine.

The Work Equation

The equation for work (Wk_o) is

$$Wk_o = 0 \qquad [\text{J}] \qquad \text{(Eq. 7-3)}$$

This equation for work is derived from the first law of thermodynamics.

$$Wk_o = Q + (H_1 - H_2) \qquad [\text{J}] \qquad \text{(Eq. 3-27)}$$

Since

$$Q = Mc_p(T_2 - T_1) \qquad \text{(Eq. 6-9)}$$

$$H_1 = U_1 + p_1V_1 \qquad \text{(Eq. 3-24a)}$$

$$H_2 = U_2 + p_2V_2 \qquad \text{(Eq. 3-24a)}$$

Substituting in Eq. 3-27

$$Wk_o = Mc_p(T_2 - T_1) + (U_1 - U_2) + p_1V_1 - p_2V_2$$

Since

$$U_1 - U_2 = Mc_v(T_1 - T_2) \qquad \text{(Eq. 6-22a)}$$

Then

$$\begin{aligned} Wk_o &= Mc_p(T_2 - T_1) - Mc_v(T_2 - T_1) + p_1V_1 - p_2V_2 \\ &= M(c_p - c_v)(T_2 - T_1) + p_1V_1 - p_2V_2 \\ &= MR(T_2 - T_1) - (p_2V_2 - p_1V_1) \end{aligned}$$

Since

$$pV = MRT \qquad \text{(Eq. 6-2)}$$

$$Wk_o = MR(T_2 - T_1) - MR(T_2 - T_1)$$

Therefore,

$$Wk_o = 0 \qquad [\text{J}] \qquad \text{(Eq. 7-3)}$$

The p–V and T–s Diagrams

The open isobaric process is shown graphically by the p–V and T–s diagrams in Figure 7-5. The reversible (ideal) work for the open isobaric process is represented by the line A–1, in the p–V diagram. The area represented by this line equals zero.

Example Problem 7-4

How much work is obtained from air passing through a combustor at constant pressure?

Equation

$$Wk_o = 0 \qquad [\text{J}] \qquad \text{(Eq. 7-4)}$$

Answer

$$\mathbf{Wk_o = 0\ J}$$

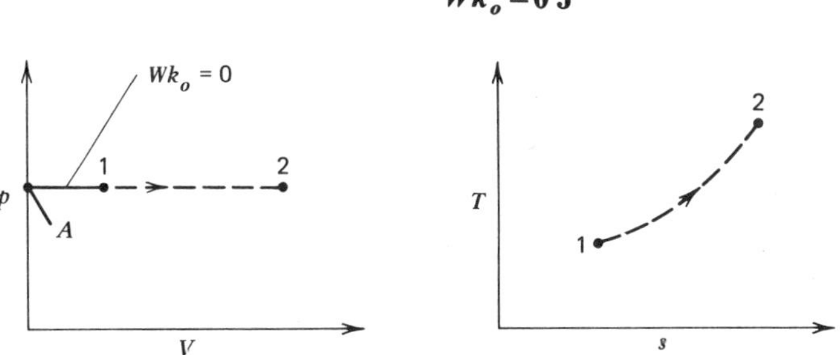

Figure 7-5 Thermodynamic diagrams for an open constant pressure process.

7-2 THE CONSTANT VOLUME (ISOMETRIC) PROCESS

The constant volume or isometric process is frequently encountered in real thermodynamic cycles involving both closed and open systems. In the isometric process the pressure–temperature–volume relationships with a perfect gas are

Process Definition

$$V_1 = V_2$$

Relationships Between Properties of State

$$V = MR\left[\frac{T}{p}\right] = \text{constant} \qquad [\text{m}^3] \tag{7-5}$$

Therefore,

$$T_2 = T_1\left[\frac{p_2}{p_1}\right] \qquad [\text{K}] \tag{7-5a}$$

$$p_2 = p_1\left[\frac{T_2}{T_1}\right] \qquad [\text{kPa}] \tag{7-5b}$$

Heat Requirements

$$q = c_v[T_2 - T_1] \qquad [\text{kJ/kg}] \qquad (\text{Eq. 6-10x})$$

Work Requirements

Closed system: $Wk = 0$ (7-6)

Open system: $Wk_o = -V(p_2 - p_1)$ [kJ] (7-7)

Change in Entropy

$$\Delta s = c_v \ln\left[\frac{T_2}{T_1}\right] \qquad [\text{kJ/kg}\cdot\Delta_1\text{K}] \tag{7-8}$$

7-2.1 GAS PARAMETER DERIVATION

T_2/T_1 Relationship

From the perfect gas law

$$pV=MRT$$

Since $V=\text{constant}=C$

$$pC=MRT$$

consequently

$$\frac{p_2C}{p_1C}=\frac{MRT_2}{MRT_1}$$

Therefore,

$$T_2=T_1\left[\frac{p_2}{p_1}\right] \quad [\text{K}] \qquad (\text{Eq. 7-5a})$$

The Δs Equation

This equation is derived from Equation 6-27.

$$\Delta s=c_v\ln\left[\frac{T_2}{T_1}\right]+R\ln\left[\frac{V_2}{V_1}\right] \qquad [\text{kJ/kg}\cdot\Delta_1\text{K}]$$

In the isometric process $V_2=V_1$ so $\ln\left[\frac{V_2}{V_1}\right]=0$

Therefore,

$$\Delta s=c_v\ln\left[\frac{T_2}{T_1}\right] \qquad [\text{kJ/kg}\cdot\Delta_1\text{K}] \qquad (\text{Eq. 7-8})$$

Example Problem 7-5

A tank is filled with air at 25°C to a pressure for 200 kPa and then sealed. What will be the pressure if the air is heated to 100°C? (See Figure 7-6.)

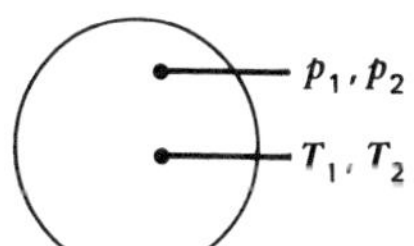

Figure 7-6 Sketch for Example Problem 7-5.

Equation

$$p_2 = p_1\left[\frac{T_2}{T_1}\right] \quad [\text{Pa}] \qquad (\text{Eq. 7-5b})$$

Parameters

$p_1 = 200$ kPa (given)

$T_2 = 100°\text{C} \rightarrow 373$ K (given)

$T_1 = 25°\text{C} \rightarrow 298$ K (given)

Substitution

$$p_2 = 200\left[\frac{373}{298}\right] \quad [\text{kPa}]\frac{[\text{K}]}{[\text{K}]} = [\text{kPa}]$$

Answer

$$\mathbf{p_2 = 250\ kPa}$$

Example Problem 7-6

Air is heated at constant volume from 20°C to 150°C. Determine the change in specific entropy for this isometric process. (Assume that air is a perfect gas.)

Equation

$$\Delta s = c_v \ln\left[\frac{T_2}{T_1}\right] \quad [\text{kJ/kg}\cdot\Delta_1\text{K}] \qquad (\text{Eq. 7-8})$$

Parameters

$T_1 = 20°\text{C} = 293.15$ K (given)

$T_2 = 150°\text{C} = 423.15$ K (given)

$c_v = 0.7165$ kJ/kg·Δ_1K (Appendix A, Table A-9)

Substitution

$$\Delta s = (0.7165)\left[\ln\frac{(423.15)}{(293.15)}\right] \quad [\text{kJ/kg}\cdot\Delta_1\text{K}][—] = [\text{kJ/kg}\cdot\Delta_1\text{K}]$$

$$= (0.7165)(\ln 1.443)$$

$$= (0.7165)(0.3670)$$

Answer

$$\mathbf{\Delta s = 0.263\ kJ/kg\cdot\Delta_1 K}$$

7-2.2 THE CLOSED ISOMETRIC PROCESS

A typical closed isometric process is a substance stored in a tank, such as butane.

The Work Equation

The equation for work, Wk_c, is

$$Wk_c = 0 \qquad [\mathrm{J}] \qquad (\text{Eq. 7-6})$$

The equation for work (Wk_c) is derived from the definition of work.

$$Wk_c = \int_{V_1}^{V_2} p\, dV$$

Since

$$V_1 = V_2 (dV = 0)$$

by integrating

$$Wk_c = 0 \qquad [\mathrm{J}] \qquad (\text{Eq. 7-6})$$

and

$$wk_c = 0 \qquad [\mathrm{kJ/kg}] \qquad (7\text{-}6a)$$

The *p–V* and *T–s* Diagrams

The closed isometric process is shown graphically by the p–V and T–s diagrams in Figure 7-7.

The reversible (ideal) work for the closed isometric process is represented by the line A-1 in the p–V diagram of Figure 7-7. The area represented by this line equals zero.

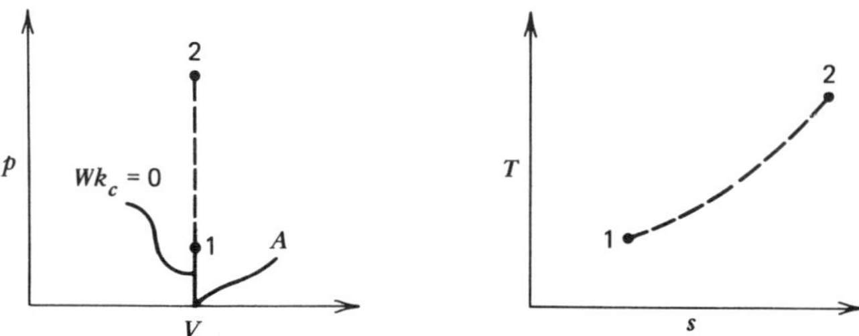

Figure 7-7 Thermodynamic diagrams for a closed constant volume process.

7-2.3 THE OPEN ISOMETRIC PROCESS

The Work Equation

$$Wk_o = -V(p_2 - p_1) \qquad [\mathrm{J}] \qquad (\text{Eq. 7-7})$$

This equation is good for incompressible fluids as well as for gases. This equation for work is derived from the first law of thermodynamics.

$$Wk_o = Q + (H_1 - H_2) \qquad [\mathrm{J}] \qquad (\text{Eq. 3-27})$$

$$Q = Mc_v(T_2 - T_1) \qquad (\text{Eq. 6-10})$$

$$H_1 = U_1 + p_1V_1 \qquad (\text{Eq. 3-24a})$$

$$H_2 = U_2 + p_2V_2 \qquad (\text{Eq. 3-24a})$$

$$Wk_o = Mc_v(T_2 - T_1) + (U_1 - U_2) + p_1V_1 - p_2V_2$$

$$U_1 = Mc_vT_1 \qquad (\text{Eq. 3-22a})$$

$$U_2 = Mc_vT_2 \qquad (\text{Eq. 3-22a})$$

$$Wk_o = Mc_v(T_2 - T_1) + Mc_v(T_1 - T_2) + p_1V_1 - p_2V_2$$

$$Wk_o = p_1V_1 - p_2V_2$$

since

$$V_1 = V_2 = V$$

$$Wk_o = -V(p_2 - p_1) \qquad [\mathrm{J}] \qquad (\text{Eq. 7-7})$$

and

$$wk_o = -v(p_2 - p_1) \qquad [\mathrm{kJ/kg}] \qquad (7\text{-}7\mathrm{a})$$

The *p–V* and *T–s* Diagrams

The open isometric process is shown graphically by the p–V and T–s diagrams in Figure 7-8. The reversible (ideal) work for the open isometric process is represented by the area A–1–2–B–A.

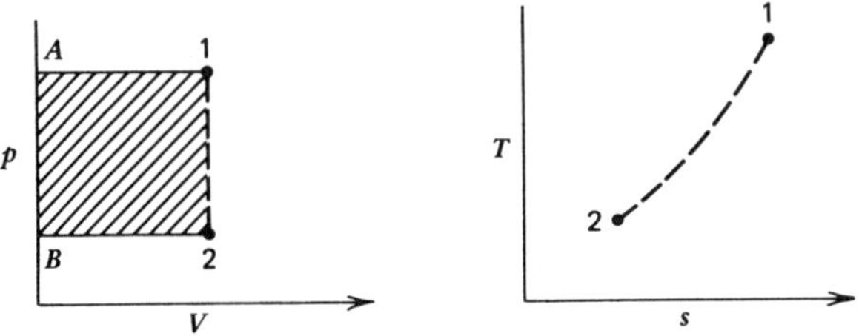

Figure 7-8 Thermodynamic diagrams for an open constant volume process.

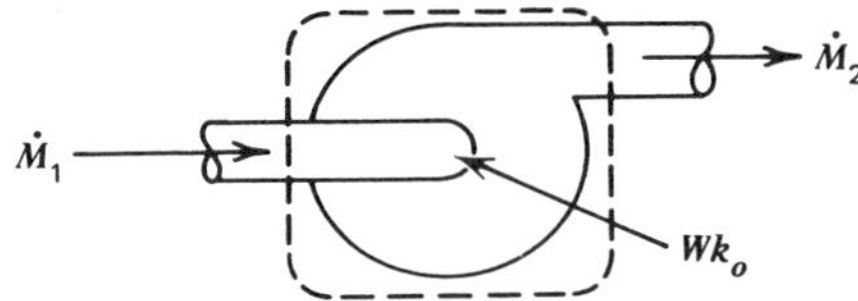

Figure 7-9 Sketch for Example Problem 7-7.

Example Problem 7-7

A pump system under steady flow operation delivers water at 25°C from an inlet pressure of 100 kPa to an exit pressure of 1200 kPa. Determine the work done per unit mass. (Assume an isometric process.) (See Figure 7-9.)

Equation

$$wk_o = -v(p_2 - p_1) \quad [\text{kJ/kg}] \quad (\text{Eq. 7-7a})$$

Parameters

$$p_1 = 100 \text{ kPa} = 100 \text{ kN/m}^2 \quad (\text{given})$$

$$p_2 = 1200 \text{ kPa} = 1200 \text{ kN/m}^2 \quad (\text{given})$$

$$v = 0.001003 \text{ m}^3/\text{kg} \quad (\text{Appendix A; Table A-3.1 at } 25°\text{C})$$

Substitution

$$wk_o = (-0.001003)(1200 - 100) \quad [\text{m}^3/\text{kg}][\text{kN/m}^2]$$
$$= [\text{kNm/kg}] = [\text{kJ/kg}]$$

Answer

$$\mathbf{wk_o = -1.103 \text{ kJ/kg}}$$

The negative sign indicates work input to the system.

7-3 CONSTANT TEMPERATURE (ISOTHERMAL) PROCESS

The constant temperature or isothermal process has many applications in engineering. This type of process will often involve both work and heat transfer as well as other energy changes. When the working fluid or substance can be assumed to be a perfect gas, it can be shown that the enthalpy and internal energy changes are zero.

In the isothermal process the pressure–temperature–volume with a perfect gas are

Process Definition

$$T_1 = T_2$$

Relationship Between Properties of State

$$T = \frac{pV}{MR} = \text{constant}$$

Therefore,

$$V_2 = V_1\left[\frac{p_1}{p_2}\right] \qquad [\text{m}^3] \qquad (7\text{-}9)$$

$$p_2 = p_1\left[\frac{V_1}{V_2}\right] \qquad [\text{kPa}] \qquad (7\text{-}10)$$

Heat Requirements

$$q = wk \qquad [\text{kJ}] \qquad (7\text{-}11)$$

Work Requirements

The work equation is the same for closed and open systems:

$$Wk = C\ln\left[\frac{V_2}{V_1}\right] = C\ln\left[\frac{p_1}{p_2}\right] \qquad [\text{kJ}] \qquad (7\text{-}12)$$

where

$$C = pV = MRT \qquad [\text{kJ}]$$

Change in Entropy

$$\Delta s = +R\ln\left[\frac{p_1}{p_2}\right] = +R\ln\left[\frac{V_2}{V_1}\right] \qquad [\text{kJ/kg}\cdot\Delta_1\text{K}] \qquad (7\text{-}13)$$

7-3.1 GAS PARAMETER DERIVATION

V_2/V_1 Relationship

From the perfect gas law, as T is a constant,

$$p_1V_1 = p_2V_2$$

Therefore,

$$V_2=V_1\left[\frac{p_1}{p_2}\right] \quad [\mathrm{m}^3] \qquad \text{(Eq. 7-9)}$$

p_2/p_1 Relationship

Since $pV_1=p_2V_2$ from Equation 7-9,

$$p_2=p_1\left[\frac{V_1}{V_2}\right] \quad [\mathrm{Pa}] \qquad \text{(Eq. 7-10)}$$

The Δs Equation

The Δs equation is derived from Equation 6-28.

$$\Delta s=c_p\ln\left[\frac{T_2}{T_1}\right]-R\ln\left[\frac{p_2}{p_1}\right] \qquad [\mathrm{kJ/kg\cdot\Delta_1 K}]$$

Since $T_2=T_1,\quad \ln\left[\frac{T_2}{T_1}\right]=0$

Therefore,

$$\Delta s=-R\ln\left[\frac{p_2}{p_1}\right]=R\ln\left[\frac{p_1}{p_2}\right] \qquad [\mathrm{kJ/kg\cdot\Delta_1 K}] \qquad [\text{Eq. 7-13)}$$

7-3.2 THE CLOSED ISOTHERMAL PROCESS

The Work Equation

Using the definition of work

$$Wk_c=\int_{p_1}^{p_2} p\,dV$$

and applying the perfect gas laws, as T is a constant,

$$p_1V_1=p_2V_2=C$$

Therefore,

$$p=\frac{C}{V}$$

Then

$$Wk_c = C\int_{V_1}^{V_2} \frac{dV}{V}$$

Integrating this expression, we obtain

$$Wk_c = C \ln\left[\frac{V_2}{V_1}\right] \qquad [\text{J}] \qquad (\text{Eq. 7-12})$$

The *p–V* and *T–s* Diagrams

The closed isothermal process is shown graphically by the p–V and T–s diagrams in Figure 7-10. The shaded areas $A-1-2-B-A$ represent the work (Wk_c) involved in the process.

Example Problem 7-8

A unit mass of butane is kept at a constant temperature of 90°C while its volume expands to three times its initial volume. Determine the specific work output from this isothermal process. (See Figure 7-11, page 159.)

Equation

$$wk_c = C \ln\left[\frac{V_2}{V_1}\right] \qquad [\text{kJ/kg}] \qquad (\text{Eq. 7-12})$$

Parameters

$C = RT$ (isothermal definition)

$R = 0.2968\ \text{kJ/kg}\cdot\Delta_1\text{K}$ (Appendix A, Table A-9)

$T = 90°\text{C} = 363.15\ \text{K}$ (given)

$\frac{V_2}{V_1} = 3$ [—] (given)

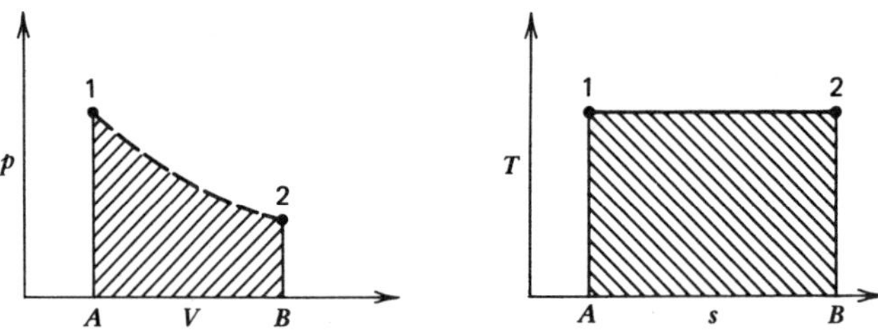

Figure 7-10 Thermodynamic diagrams for a closed isothermal process.

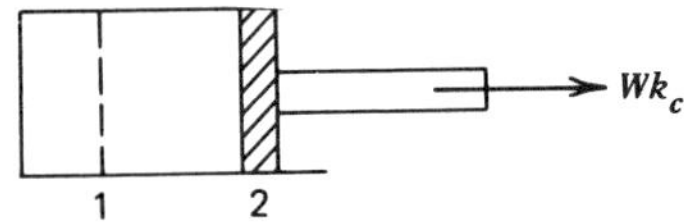

Figure 7-11 Sketch for Example Problem 7-8.

Substitution

$$wk_c = (0.2968)(363.15)(\ln 3) \qquad [\text{kJ/kg}\cdot\Delta_1\text{K}][\text{K}][—] = [\text{kJ/kg}]$$
$$= (0.2968)(363.15)(1.0986)$$

Answer

$$\boldsymbol{Wk_c = +118.4\ \text{kJ/kg}}$$

Work is produced from isothermal expansion.

7-3.3 THE HEAT REQUIRED

In an isothermal process heat Q must be added or removed to maintain the constant temperature. To determine the amount of heat Q involved, with no change in potential, kinetic, or strain energy, the first law of thermodynamics may be applied.

For a closed system,

$$Wk_c = Q_c - \Delta U \qquad [\text{J}] \qquad (\text{Eq. 3-8})$$

For an isothermal process $\Delta U = 0$. Therefore,

$$Q_c = Wk_c$$

since

$$Wk_c = C \ln\left[\frac{V_2}{V_1}\right] \qquad [\text{J}] \qquad (\text{Eq. 7-12})$$

$$Q_c = C \ln\left[\frac{V_2}{V_1}\right] \qquad [\text{J}] \qquad (7\text{-}11)$$

7-3.4 THE OPEN ISOTHERMAL PROCESS

The Work Equation

The work equation can be derived from the first law of thermodynamics.

$$Wk_o = Q_o + (H_1 - H_2) \qquad [\text{J}] \qquad (\text{Eq. 3-27})$$
$$H_1 = U_1 + p_1V_1$$
$$H_2 = U_2 + p_2V_2$$
$$H_1 - H_2 = (U_1 - U_2) + p_1V_1 - p_2V_2$$

For an isothermal system

$$U_1 = U_2$$
$$p_1 V_1 = p_2 V_2$$

Therefore,

$$H_1 - H_2 = 0$$
$$Wk_o = Q_o$$

Because the Q required to maintain the gas temperature constant for a given process pressure ratio is the same for an open or closed process,

$$Q_o = Q_c$$

Since

$$Q_c = C\ln\left[\frac{V_2}{V_1}\right] \quad [\mathrm{J}] \qquad (\text{Eq. 7-11})$$

and

$$Q_o = C\ln\left[\frac{V_2}{V_1}\right] \quad [\mathrm{J}]$$

$$Wk_o = C\ln\left[\frac{V_2}{V_1}\right] \quad [\mathrm{J}] \qquad (\text{Eq. 7-12})$$

This equation can be converted to the pressure ratio, as for a perfect gas

$$\left[\frac{V_2}{V_1}\right] = \left[\frac{p_1}{p_2}\right]$$

so that

$$\ln\left[\frac{V_2}{V_1}\right] = \ln\left[\frac{p_1}{p_2}\right]$$

Therefore,

$$Wk_o = C\ln\left[\frac{V_2}{V_1}\right] = C\ln\left[\frac{p_1}{p_2}\right] \quad [\mathrm{J}] \qquad (\text{Eq. 7-12})$$

The *p–V* and *T–s* Diagrams

The open system isothermal process is shown on the p–V and T–s diagrams in Figure 7-12 for comparison with the closed system in Figure 7-10.

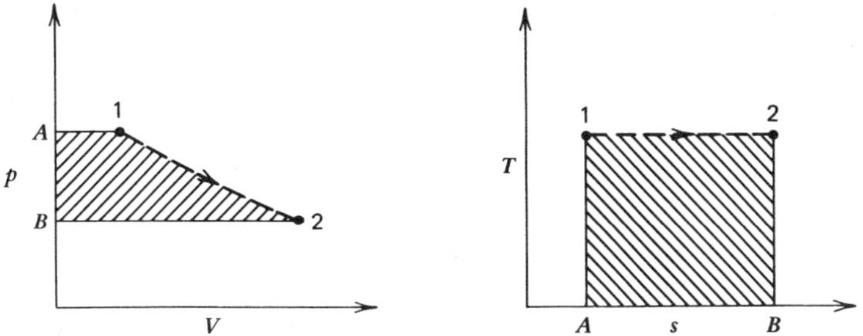

Figure 7-12 Thermodynamic diagrams for an open isothermal process.

The shaded areas $A-1-2-B-A$ represent the work Wk involved in the process. Because $C\ln(V_2/V_1) \neq C\ln(p_1/p_2)$ except for the case of "ideal" gases, the work area is drawn to the pressure ordinate for the open system, in contrast to the volume abscissa for the closed system in the p–V diagram. The T–s diagram is the same for both systems.

Example Problem 7-9

An isothermal compressor is used to increase the pressure of air from 125 to 1500 kPa. If the operating temperature is kept constant at 250°C, determine the work required per unit mass for this process.

Equation

$$wk_o = C\ln\left[\frac{p_1}{p_2}\right] \quad [\text{kJ/kg}] \quad (\text{Eq. 7-12})$$

Parameters

$$p_1 = 125 \text{ kPa} \quad (\text{given})$$

$$p_2 = 1500 \text{ kPa} \quad (\text{given})$$

$$T = 250°\text{C} = 523.15 \text{ K} \quad (\text{given})$$

$$C = RT \quad [\text{kJ/kg}]$$

$$R = 0.287 \text{ kJ/kg}\cdot\Delta_1\text{K} \quad (\text{Appendix A, Table A-9})$$

Substitution

$$wk_o = (0.287)(523.15)\left[\ln\left(\frac{125}{1500}\right)\right] \quad [\text{kJ/kg}\cdot\Delta_1\text{K}][\text{K}][—] = [\text{kJ/kg}]$$

$$= (0.287)(523.15)(\ln 0.0833)$$

$$= (0.287)(523.15)(-2.485)$$

Answer

$$\boldsymbol{wk_o = -373.1 \text{ kJ/kg}}$$

Work is required for isothermal compression.

7-4 THE REVERSIBLE ADIABATIC ISENTROPIC PROCESS

The adiabatic process by definition is one in which the heat transfer Q is zero. A reversible adiabatic process is an isentropic process. Although no actual processes are completely adiabatic, it is a simplifying assumption that can be used to approximate many real processes such as expansions in efficient turbines, compression in efficient turbocompressors, and so forth.

The temperature–pressure relationship and the associated volume–pressure relationship (in accordance with the perfect gas laws) are the key characteristics of the adiabatic process. These relationships are:

Process Definition

$$p_1V_1^{\gamma}=p_2V_2^{\gamma} \qquad \text{(Eq. 7-21)}$$

Relationship Between Properties of State

$$T_2=T_1\left[\frac{p_2}{p_1}\right]^{(\gamma-1)/\gamma}=T_1\left[\frac{V_1}{V_2}\right]^{\gamma-1} \qquad [\text{K}] \qquad (7\text{-}14)$$

$$p_2=p_1\left[\frac{V_1}{V_2}\right]^{\gamma}=p_1\left[\frac{T_2}{T_1}\right]^{\gamma/(\gamma-1)} \qquad [\text{kPa}] \qquad (7\text{-}15)$$

$$V_2=V_1\left[\frac{p_1}{p_2}\right]^{1/\gamma}=V_1\left[\frac{T_1}{T_2}\right]^{1/(\gamma-1)} \qquad [\text{m}^3] \qquad (7\text{-}16)$$

Heat Requirements

$$q=0 \qquad [\text{kJ}] \qquad \text{(Adiabatic Definition)}$$

Work Requirements

Closed system

$$Wk_c=\left[\frac{MR}{1-\gamma}\right](T_2-T_1) \qquad [\text{kJ}] \qquad (7\text{-}17)$$

$$Wk_c=\left[\frac{1}{1-\gamma}\right](p_2V_2-p_1V_1) \qquad [\text{kJ}] \qquad (7\text{-}18)$$

Open system

$$Wk_o=\left[\frac{\gamma MR}{1-\gamma}\right](T_2-T_1) \qquad [\text{kJ}] \qquad (7\text{-}19)$$

$$Wk_o=\left[\frac{\gamma}{1-\gamma}\right](p_2V_2-p_1V_1) \qquad [\text{kJ}] \qquad (7\text{-}20)$$

Change in Entropy

$$\Delta s = 0 \qquad [\text{kJ/kg}\cdot\Delta_1\text{K}] \tag{7-21}$$

7-4.1 GAS PARAMETER DERIVATION

The Basic Adiabatic Process Equation

The basic equation from the first law of thermodynamics for the closed adiabatic process is

$$-\Delta U = Wk_c \qquad [\text{J}] \qquad (\text{Eq. 3-9})$$

$$Wk_c = \int p\, dV \qquad (\text{definition})$$

We substitute a V term for p, using the perfect gas law

$$p = \frac{MRT}{V}$$

so that

$$Wk_c = MR\int T\frac{dV}{V}$$

To evaluate ΔU in terms of temperature, Equation 6-22 can be used

$$\Delta U = Mc_v\Delta T \qquad [\text{J}] \qquad (\text{Eq. 6-22})$$

This can be written

$$\Delta U = Mc_v\int dT$$

Substituting these values in Equation 3-9 results in

$$Mc_v\int dT = -MR\int T\frac{dV}{V}$$

Dividing by (MT), we get

$$c_v\int\frac{dT}{T} = -R\int\frac{dV}{V}$$

and then substitute specific heats for R.

$$R = (c_p - c_v) \qquad (\text{kJ/kg}\cdot\Delta_1\text{K}] \qquad (\text{Eq. 6-19})$$

Thus

$$c_v \int \frac{dT}{T} = -(c_p - c_v) \int \frac{dV}{V}$$

Dividing by c_v we get

$$\int \frac{dT}{T} = -\left(\frac{c_p}{c_v} - 1\right) \int \frac{dV}{V}$$

since

$$\gamma = \frac{c_p}{c_v} \qquad [-] \qquad \text{(Eq. 6-20)}$$

and

$$\int \frac{dT}{T} = (1-\gamma) \int \frac{dV}{V}$$

and integrating both sides of this equation we get

$$\ln T = (1-\gamma) \ln V + \text{constant} \tag{7-22}$$

The Relationship Between Temperature and Pressure

To obtain the relationship between temperature and pressure, the perfect gas law is used, solving for V.

$$V = \frac{MRT}{p}$$

Thus, substituting in Equation 7-22

$$\ln T = (1-\gamma) \ln \frac{MRT}{p} + \text{constant}$$

so that

$$\ln T = (1-\gamma)[\ln T - \ln p] = \text{constant}$$

$$(1-\gamma) \ln p = -\gamma \ln T + \text{constant}$$

Dividing by $-\gamma$ gives

$$-\left[\frac{1-\gamma}{\gamma}\right] \ln p = \ln T + \text{constant}$$

Rewriting in exponential form, we obtain

$$p^{-[(1-\gamma)/\gamma]}=CT=p^{[(\gamma-1)/\gamma]}$$

From this, we find that

$$\frac{CT_2}{CT_1}=\frac{p_2^{(\gamma-1)/\gamma}}{p_1^{(\gamma-1)/\gamma}}$$

so that

$$\frac{T_2}{T_1}=\left[\frac{p_2}{p_1}\right]^{(\gamma-1)/\gamma} \qquad [-] \qquad \text{(Eq. 7-14)}$$

The Relationship Between Volume and Pressure

The obtain the relationship between volume and pressure, the perfect gas law is used solving for T.

$$T=\frac{pV}{MR}$$

Thus, substituting in Equation 7-22

$$\ln V+\ln p=(1-\gamma)\ln V+\text{constant}$$
$$\gamma\ln V=-\ln p+\text{constant}$$
$$\ln V=-\frac{1}{\gamma}\ln p+\text{constant}$$

Rewriting in exponential form, we obtain

$$V=C\frac{1}{(p)^{1/\gamma}}$$

$$V_2=C\frac{1}{(p_2)^{1/\gamma}}$$

$$V_1=C\frac{1}{(p_1)^{1/\gamma}}$$

$$\frac{V_2}{V_1}=\frac{C(p_1)^{1/\gamma}}{C(p_2)^{1/\gamma}}$$

$$\frac{V_2}{V_1}=\left[\frac{p_1}{p_2}\right]^{1/\gamma} \qquad [-] \qquad \text{(Eq. 7-16a)}$$

The p–V Relationship

To obtain the p–V product relationship, the perfect gas law is used solving for T.

$$T=\frac{pV}{MR}$$

Thus, substituting in Equation 7-22 gives

$$\ln\left[\frac{pV}{MR}\right]=(1-\gamma)\ln V+\text{constant}$$

or

$$\ln p+\ln V=\ln V-\gamma\ln V+C$$

Simplifying the equation,

$$\ln p+\gamma\ln V=C$$

and rewriting in exponential form:

$$pV^{\gamma}=C \qquad [\text{kJ}] \qquad \text{Eq. 7-21}$$

The Equation for Δs

The equation showing $\Delta s=0$ (Equation 7-14) is derived from Equation 6-28.

$$\Delta s=c_p\ln\left[\frac{T_2}{T_1}\right]-R\ln\left[\frac{p_2}{p_1}\right] \qquad [\text{kJ/kg}\cdot\Delta_1\text{K}] \qquad (\text{Eq. 6-28})$$

since

$$\left[\frac{T_2}{T_1}\right]=\left[\frac{p_2}{p_1}\right]^{(\gamma-1)/\gamma}$$

$$\Delta s=c_p\ln\left[\frac{p_2}{p_1}\right]^{(\gamma-1)/\gamma}-R\ln\left[\frac{p_2}{p_1}\right]$$

$$=c_p\left(\frac{\gamma-1}{\gamma}\right)\ln\left[\frac{p_2}{p_1}\right]-R\ln\left[\frac{p_2}{p_1}\right]$$

$$\frac{\gamma-1}{\gamma}=\frac{(c_p/c_v)-(c_v-c_v)}{c_p/c_v}=\frac{c_p-c_v}{c_p}$$

$$\Delta s=c_p\left[\frac{c_p-c_v}{c_p}\right]\ln\left[\frac{p_2}{p_1}\right]-R\ln\left[\frac{p_2}{p_1}\right]$$

$$=(c_p-c_v)\ln\left[\frac{p_2}{p_1}\right]-R\ln\left[\frac{p_2}{p_1}\right]$$

and since

$$R=(c_p-c_v) \qquad [\text{kJ/kg}\cdot\Delta_1\text{K}] \qquad \text{(Eq. 6-19)}$$

$$\Delta s=R\ln\left[\frac{p_2}{p_1}\right]-R\ln\left[\frac{p_2}{p_1}\right]$$

thus,

$$\Delta s=0 \qquad [\text{kJ/kg}\cdot\Delta_1\text{K}] \qquad \text{(Eq. 7-21)}$$

7-4.2 THE CLOSED ADIABATIC PROCESS

The Equation for Work

$$Wk_c=\int p\,dV$$

From Equation 7-13

$$p=\frac{C}{V^\gamma}$$

Thus

$$Wk_c=\int_{V_1}^{V_2}\frac{C\,dV}{V^\gamma}$$

and integrating, we get

$$Wk_c=C\left[\frac{1}{1-\gamma}\right](V_2^{1-\gamma}-V_1^{1-\gamma})$$

but since

$$C=pV^\gamma=p_1V_1^\gamma=p_2V_2^\gamma$$

we can substitute $p_iV_i^\gamma$ for C in the above equation for Wk_c and obtain

$$Wk_c=\left[\frac{1}{1-\gamma}\right]([p_2V_2^\gamma]V_2^{1-\gamma}-[p_1V_1^\gamma]V_1^{1-\gamma})$$

$$Wk_c=\left[\frac{1}{1-\gamma}\right](p_2V_2-p_1V_1) \qquad [\text{kJ}] \qquad \text{(Eq. 7-18)}$$

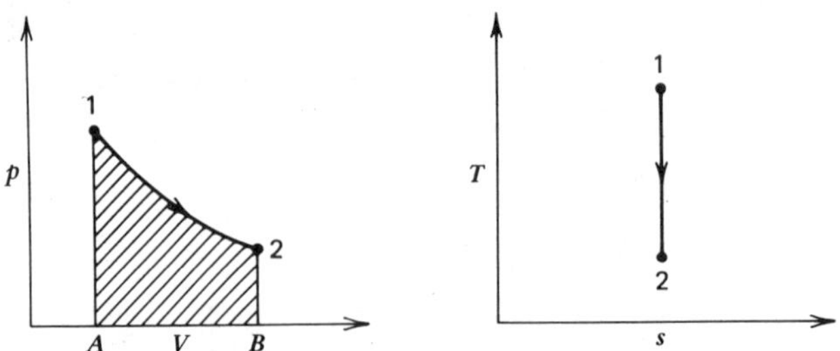

Figure 7-13 Thermodynamic diagrams for a closed system reversible, adiabatic (ideal) process with a perfect gas.

introducing the perfect gas law

$$Wk_c = \left[\frac{MR}{1-\gamma}\right](T_2 - T_1) \qquad [\text{kJ}] \qquad (\text{Eq. 7-17})$$

or

$$Wk_c = Mc_v(T_1 - T_2) \qquad [\text{kJ}] \qquad (\text{Eq. 7-17a})$$

The *p–V* and *T–s* Diagrams

The closed adiabatic process is shown graphically by the p–V and T–s diagrams in Figure 7-13. Since this is a reversible process, it is considered to be isentropic (constant entropy). Therefore, the process is a straight vertical line on the T–s diagram. The specific heat ratio parameter γ is often called the isentropic exponent. The reversible (ideal) work for the adiabatic process in a closed system is represented by the area $A-1-2-B-A$ under the p–V curve in Figure 7-13.

Example Problem 7-10

If 2 m^3 of air at a temperature of 25°C and a pressure of 100 kPa, were adiabatically compressed to a pressure of 400 kPa, what would be the volume of the compressed air? What would be its temperature?

Solution

Use Equations 7-14 and 7-15 and obtain the needed air property from Appendix A, Table A-9. (See Figure 7-14 and Table 7-1, see page 169.)

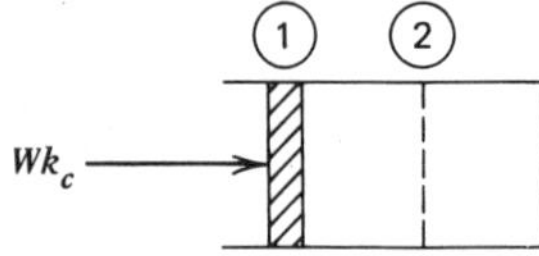

Figure 7-14 Closed system isentropic process sketch for Example Problem 7-10.

(a) Equation

$$V_2 = V_1\left[\frac{p_1}{p_2}\right]^{1/\gamma} \qquad [\text{m}^3] \qquad (\text{Eq. 7-15x})$$

Table 7-1 Table of State

Substance: Air	①	②
Mass M [kg]	—	—
Temperature T [K]	298 (g)	—
Pressure p [kPa]	100 (g)	400 (g)

Note: g (given).

Parameters

$$\gamma = 1.40\ [—] \qquad \text{(Appendix A, Table A-9)}$$

$$V_1 = 2\ \text{m}^3 \qquad \text{(given)}$$

$$p_1 = 100\ \text{kPa} \qquad \text{(given)}$$

$$p_2 = 400\ \text{kPa} \qquad \text{(given)}$$

Substitution

$$V_2 = (2)\left(\frac{100}{400}\right)^{1/1.4} \qquad [\text{m}^3]\frac{[\text{kPa}]}{[\text{kPa}]} = [\text{m}^3]$$

$$= (2)(0.25)^{0.7143}$$

$$= (2)(0.3715) \qquad [\text{m}^3]$$

Answer

$$\mathbf{V_2 = 0.743\ m^3}$$

(b) Equation

$$T_2 = T_1\left[\frac{p_2}{p_1}\right]^{(\gamma-1)/\gamma} \qquad [\text{K}] \qquad \text{(Eq. 7-14x)}$$

Parameters

$$\gamma = 1.40\ [—] \qquad \text{(Appendix A, Table A-9)}$$

$$T_1 = 298\ \text{K} \qquad \text{(given)}$$

$$p_1 = 100\ \text{kPa} \qquad \text{(given)}$$

$$p_2 = 400\ \text{kPa} \qquad \text{(given)}$$

Substitution

$$T_2=(298)\left(\frac{400}{100}\right)^{(1.4-1)/1.4} \qquad [\text{K}]\frac{[\text{kPa}]}{[\text{kPa}]}=[\text{K}]$$

$$=(298)(4)^{0.2857}$$

$$=(298)(1.486) \qquad [\text{K}]$$

Answer

$$\mathbf{T_2=442.8\ K}$$

Example Problem 7-11

One kilogram of air in a closed system expands adiabatically from a temperature of 1000 to 100°C. Calculate the reversible work done by this process. (Assume air to be a perfect gas.)

Equation

$$Wk_c=\left[\frac{MR}{1-\gamma}\right](T_2-T_1) \qquad [\text{J}] \qquad (\text{Eq. 7-17})$$

Parameters

$M=1.0$ kg (given)

$T_1=1000°\text{C}=1273.15$ K (given)

$T_2=100°\text{C}=373.15$ K (given)

$\gamma=1.40$ [—] (Appendix A, Table A-9)

$R=0.287$ kJ/kg·Δ_1K (Appendix A, Table A-9)

Substitution

$$Wk_c=\frac{(1.0)(0.287)}{(1-1.40)}(373.15-1273.15) \qquad \frac{[\text{kg}][\text{kJ/kg}\cdot\Delta_1\text{K}][\Delta_1\text{K}]}{[—]}=[\text{kJ}]$$

$$=\frac{(10.)(0.287)}{(-0.40)}(-900) \qquad [\text{kJ}]$$

Answer

$$\mathbf{Wk_c=+645.8\ kJ}$$

7-4.3 THE OPEN ADIABATIC PROCESS

The Equation for Work

For an open, reversible (ideal) system with a perfect gas, the first law equation can be written as

$$Wk_o = H_1 - H_2 \quad [\mathrm{J}] \quad (\text{Eq. 3-29})$$

$$H_1 = U_1 + p_1V_1 \quad (\text{definition})$$

$$H_2 = U_2 + p_2V_2$$

$$Wk_o = (U_1 - U_2) = (p_1V_1 - p_2V_2)$$

$$U_1 = Mc_vT_1 \quad [\mathrm{J}] \quad (\text{Eq. 6-10a})$$

$$U_2 = Mc_vT_2$$

$$U_1 - U_2 = Mc_v(T_1 - T_2)$$

$$p_1V_1 = MRT_1 \quad [\mathrm{J}] \quad (\text{Eq. 6-2})$$

$$p_2V_2 = MRT_2$$

$$p_1V_1 - p_2V_2 = MR(T_1 - T_2)$$

$$Wk_o = Mc_v(T_1 - T_2) + MR(T_1 - T_2)$$

$$= MR\left(\frac{c_v}{R}\right)(T_1 - T_2) + MR(T_1 - T_2)$$

$$= MR\left(\frac{c_v}{R} + 1\right)(T_1 - T_2)$$

$$\frac{c_v}{R} + 1 = \left[\frac{c_p - R + R}{R}\right] = \frac{c_p}{R} = \left[\frac{c_p}{c_p - c_v}\right]$$

By dividing both the numerator and denominator by c_p, we obtain

$$\frac{c_v}{R} + 1 = \frac{1}{1 - (1/\gamma)} = \frac{\gamma}{\gamma - 1}$$

$$Wk_o = MR\left[\frac{\gamma}{\gamma - 1}\right](T_1 - T_2) = MR\left[\frac{\gamma}{1 - \gamma}\right](T_2 - T_1)$$

Therefore,

$$Wk_o = \left[\frac{\gamma MR}{1 - \gamma}\right](T_2 - T_1) \quad [\mathrm{J}] \quad (\text{Eq. 7-19})$$

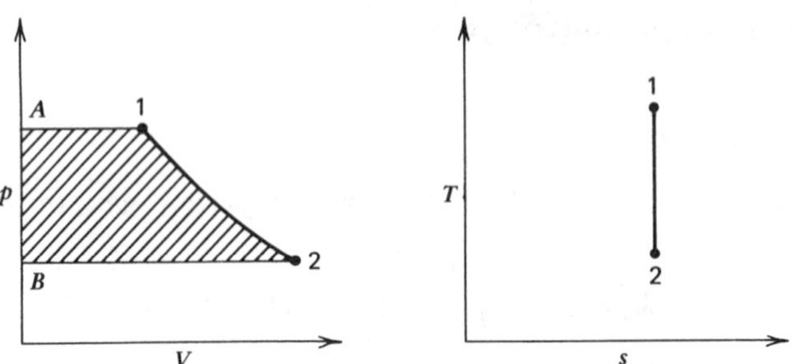

Figure 7-15 Thermodynamic diagrams for an open system, reversible adiabatic (ideal) process with a perfect gas.

Applying the perfect gas law gives

$$T=\frac{pV}{MR}$$

Substituting in Equation 7-16, we obtain

$$Wk_o=\left[\frac{\gamma MR}{1-\gamma}\right]\left[\frac{p_2V_2}{MR}-\frac{p_1V_1}{MR}\right]$$

$$Wk_o=\left[\frac{\gamma}{1-\gamma}\right](p_2V_2-p_1V_1) \qquad [\mathrm{J}] \qquad (\text{Eq. 7-20})$$

The *p–V* and *T–s* Diagram

The open adiabatic process is shown graphically by the p–V and T–s diagrams in Figure 7-15. The reversible (ideal) work for the adiabatic process in an open system is represented by the area A–1–2–B–A in the p–V diagram.

Example Problem 7-12

An adiabatically turbine expands hot air from a temperature of 1227 to 527°C. Calculate the work produced per unit mass.

Solution

Use the Wk_o in Equation 7-20 and obtain needed air properties from Appendix A, Table A-9. (See Figure 7-16 and Table 7-2.)

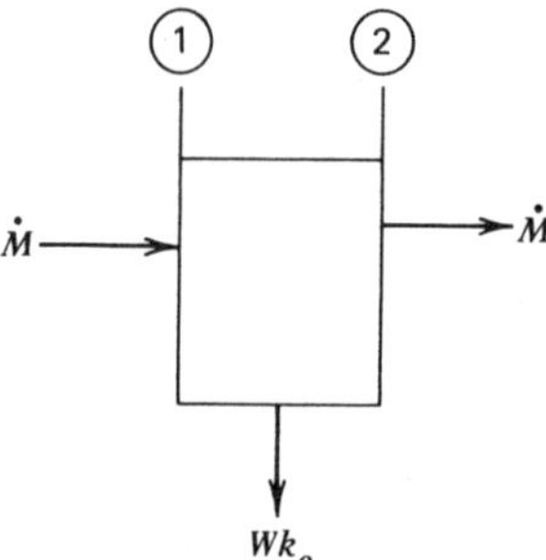

Figure 7-16 Open system isentropic process sketch for Example Problem 7-12.

Table 7-2 Table of State

Substance: Air	①	②
Mass M [kg]	—	—
Temperature T [K]	1500 (g)	800 (g)
Pressure p [kPa]	—	—

Note: g (given).

Equation

$$Wk_o = \left[\frac{\gamma R}{1-\gamma}\right](T_2 - T_1) \qquad [\text{kJ/kg}] \qquad (\text{Eq. 7-19x})$$

Parameters

$$\gamma = 1.40\ [—] \qquad (\text{Appendix A, Table A-9})$$

$$R = 0.2870\ \text{kJ/kg}\cdot\Delta_1\ \text{K} \qquad (\text{Appendix A, Table A-9})$$

$$T_1 \text{ and } T_2 \qquad (\text{Table 7-2, table of state})$$

Substitution

$$Wk_o = \left[\frac{(1.4)(0.2870)}{1-1.40}\right](800-1500) \qquad \frac{[—][\text{kJ/kg}\cdot\Delta_1\text{K}]}{[—]}$$

$$\left[\frac{(1.4)(0.2870)}{-0.40}\right](-700) \qquad [\text{kJ/kg}]$$

Answer

$$\boldsymbol{Wk_o = 703.15\ \text{kJ/kg}}$$

7-5 THE POLYTROPIC PROCESS

When a compression or expansion does not occur in an efficient machine, or when heat is absorbed or rejected during the process, the process is no longer adiabatic and often is not reversible. Such a process is called polytropic and differs from the reversible adiabatic process due to the heat flow into or out of the process or the conversion of mechanical energy into internal energy (mechanical energy losses in a nonreversible adiabatic process) or both.

The equations for the polytropic process work, and p, V, and T relationships are similar to the adiabatic process equations with a polytropic exponent n replacing the adiabatic γ (ratio of specific heats).

Process Definition

$$pV^n = \text{constant} \qquad \text{(process definition)}$$

where

$$n = \text{the polytropic exponent}$$

The polytropic exponent n is determined from the initial and final states of the process.

$$p_1V_1^n = p_2V_2^n \tag{7-23}$$

$$\left[\frac{V_1}{V_2}\right]^n = \frac{p_2}{p_1}$$

$$n\left(\ln\frac{V_1}{V_2}\right) = \ln\frac{p_2}{p_1}$$

$$n = \frac{\ln(p_2/p_1)}{\ln(V_1/V_2)}$$

Relationships Between Properties of State

$$T_2 = T_1\left[\frac{p_2}{p_1}\right]^{(n-1)/n} = T_1\left[\frac{V_1}{V_2}\right]^{n-1} \qquad [\text{K}] \tag{7-23}$$

$$p_2 = p_1\left[\frac{V_1}{V_2}\right]^n = p_1\left[\frac{T_2}{T_1}\right]^{n/(n-1)} \qquad [\text{kPa}] \tag{7-24}$$

$$V_2 = V_1\left[\frac{p_1}{p_2}\right]^{1/n} = V_1\left[\frac{T_1}{T_2}\right]^{1/(n-1)} \qquad [\text{m}^3] \tag{7-25}$$

Heat Requirements

$$q = c_n(T_2 - T_1) \qquad [\text{kJ/kg}\cdot\Delta_1\text{K}] \tag{7-26}$$

where

$$c_n = \frac{c_v(\gamma - n)}{(1-n)} \qquad [\text{kJ/kg}\cdot\Delta_1\text{K}] \tag{7-27}$$

Work Requirements

Closed system $$Wk_c = \left[\frac{MR}{1-n}\right](T_2 - T_1) \qquad [\text{J}] \tag{7-28}$$

$$Wk_c = \left[\frac{1}{1-n}\right](p_2V_2 - p_1V_1) \qquad [\text{J}] \tag{7-29}$$

Open system $$Wk_o = \left[\frac{nMR}{1-n}\right](T_2 - T_1) \qquad [\text{J}] \tag{7-30}$$

$$Wk_o = \left[\frac{n}{1-n}\right](p_2V_2 - p_1V_1) \qquad [\text{J}] \tag{7-31}$$

Change in Entropy

$$\Delta s = c_n \ln\left[\frac{T_2}{T_1}\right] \qquad [\text{kJ/kg}\cdot\Delta_1\text{K}] \tag{7-32}$$

where

c_n = polytropic specific heat parameter

$$c_n = \frac{c_v(\gamma - n)}{(1-n)} \qquad [\text{kJ/kg}\cdot\Delta_1\text{K}]$$

7-5.1 THE RELATIONSHIP BETWEEN TEMPERATURE AND PRESSURE

From the polytropic process definition

$$p_1V_1^n = p_2V_2^n \qquad [\text{kJ}]$$

From the perfect gas laws,

$$V = \frac{MRT}{p}$$

Therefore,

$$p_1\left[\frac{MRT_1}{p_1}\right]^n = p_2\left[\frac{MRT_2}{p_2}\right]^n$$

$$p_1^{1-n}T_1^n = p_2^{1-n}T_2^n$$

$$\frac{T_2}{T_1} = \left[\frac{p_1}{p_2}\right]^{(1-n)/n} = \left[\frac{p_2}{p_1}\right]^{(n-1)/n} \qquad [-] \qquad \text{(Eq. 7-23a)}$$

7-5.2 THE RELATIONSHIP BETWEEN VOLUME AND PRESSURE

Since

$$p_1V_1^n = p_2V_2^n \qquad [\mathrm{kJ}]$$

$$\left[\frac{V_2}{V_1}\right]^n = \frac{p_1}{p_2}$$

Therefore,

$$\frac{V_2}{V_1} = \left[\frac{p_1}{p_2}\right]^{1/n} \qquad [-] \qquad \text{(Eq. 7-25a)}$$

7-5.3 THE POLYTROPIC SPECIFIC HEAT

In the polytropic process the heat involved can be defined as:

$$Q = U_2 - U_1 + Wk_c \qquad [\mathrm{kJ}]$$

$$U_2 - U_1 = Mc_v(T_2 - T_1) \qquad [\mathrm{kJ}]$$

$$Wk_c = \frac{MR}{1-n}(T_2 - T_1) \qquad [\mathrm{kJ}] \qquad \text{(Eq. 7-28)}$$

Substituting in the equation for Q

$$Q = Mc_v(T_2 - T_1) + \frac{MR}{1-n}(T_2 - T_1)$$

$$Q = M\left(c_v + \frac{R}{1-n}\right)(T_2 - T_1)$$

which can be written

$$Q = Mc_n(T_2 - T_1) \qquad [\mathrm{kJ}] \qquad (7\text{-}26)$$

where c_n is called the polytropic specific heat.

Expressing c_n in terms of c_v, γ, and n, since $R = (c_p - c_v)$, we obtain

$$c_n = c_v + \frac{R}{1-n} = \frac{c_v - c_vn + c_p - c_v}{1-n} = \frac{c_p - c_vn}{1-n}$$

$$c_n = \frac{c_v(\gamma - n)}{1-n} \qquad [\mathrm{kJ/kg\cdot\Delta_1K}] \qquad \text{(Eq. 7-27)}$$

From this equation it can be seen that for the adiabatic process ($n = \gamma$) that $c_n = 0$, which means no heat flows in or out of the process. When n is less than γ the sign of c_n is minus ($-$). When n is greater than γ the sign of c_n is plus ($+$).

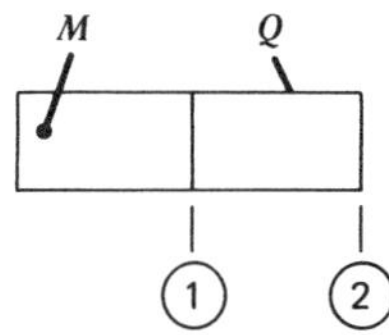

Figure 7-17 Sketch for Example Problem 7-13.

Example Problem 7-13

One kilogram of air undergoes a polytropic expansion with $n=1.25$ from a state of 0.2 MPa and 300 K to a pressure of 0.1 MPa. How much heat is involved? Is it added or removed? (See Figure 7-17.)

Equation

$$Q=Mc_n(T_2-T_1) \quad [\text{J}] \quad (\text{Eq. 7-26x})$$

$$M=1\text{ kg} \quad (\text{given})$$

Parameters

$$c_n=\frac{c_v(\gamma-n)}{1-n} \quad [\text{kJ/kg}\cdot\text{K}] \quad (\text{Eq. 7-27})$$

$$\left.\begin{aligned} c_v&=0.7165\text{ kJ/kg}\cdot\Delta_1\text{K} \\ \gamma&=1.400\ [—] \\ n&=1.25\ [—] \quad (\text{given}) \end{aligned}\right\}\begin{aligned}&\text{Appendix A, Table A-9}\\&\text{at 300 K}\end{aligned}$$

$$c_n=\frac{0.7165(1.400-1.25)}{(1-1.25)}\ \frac{[\text{kJ/kg}\cdot\Delta_1\text{K}]([—][—])}{[—]-[—]}$$

$$=\frac{(0.7165)(0.150)}{(-0.25)}[\text{kJ/kg}\cdot\Delta_1\text{K}]$$

$$=-0.43\text{ kJ/kg}\cdot\Delta_1\text{K}$$

$$T_2=T_1\left[\frac{p_2}{p_1}\right]^{n-1/n} \quad [\text{K}] \quad (\text{Eq. 7-23x})$$

$$T_1=300\text{ K} \quad (\text{given})$$

$$p_1=0.2\text{ MPa} \quad (\text{given})$$

$$p_2=0.1\text{ MPa} \quad (\text{given})$$

$$n=1.25 \quad (\text{given})$$

$$T_2=300\left(\frac{0.1}{0.2}\right)^{(1.25-1)/1.25} \quad [\text{K}]\left[\frac{\text{MPa}}{\text{MPa}}\right]=[\text{K}]$$

$$T_2=300(0.5)^{(0.25/1.25)}=300(0.5)^{0.2}$$

$$=(300)(.871)$$

$$T_2=261\text{ K}$$

Substitution

$$Q=(1)(-0.43)(300-261) \quad [\text{kg}][\text{kJ/kg}\cdot\Delta_1\text{K}][\Delta\text{K}]$$

$$=(0.43)(-39) \quad [\text{kJ}]$$

Answer

$$Q = +16.77 \text{ kJ}$$

The answer has a + sign; therefore, the heat is added.

The Equation for Δs

From the definition for entropy

$$T\,ds = du + p\,dv \qquad \text{(Eq. 6-26)}$$

$$du + p\,dv = c_n\,dT \qquad \text{(definition of } c_n\text{)}$$

then $T\,ds = c_n\,dT$

$$ds = c_n \frac{dT}{T}$$

Integrating this equation results in

$$\Delta s = c_n \ln\left[\frac{T_2}{T_1}\right] \qquad [\text{kJ/kg}\cdot\Delta_1\text{K}] \qquad \text{(Eq. 7-32)}$$

7-5.4 THE CLOSED POLYTROPIC PROCESS

The Equation for Work

$$Wk_c = \int p\,dV \qquad \text{(by definition)} \qquad [\text{kJ}] \qquad \text{(Eq. 6-16)}$$

$$p = \frac{C}{V^n} \qquad \text{(from process definition)} \qquad [\text{kPa}]$$

$$Wk_c = \int_{V_1}^{V_2} \frac{C\,dV}{V^n}$$

Integration yields

$$Wk_c = C\left(\frac{1}{1-n}\right)\left(V_2^{1-n} - V_1^{1-n}\right)$$

Since $C = pV^n = p_1V_1^n = p_2V_2^n$, we find that

$$Wk_c = \left(\frac{1}{1-n}\right)[p_2V_2 - p_1V_1] \qquad [\text{J}] \qquad \text{(Eq. 7-29)}$$

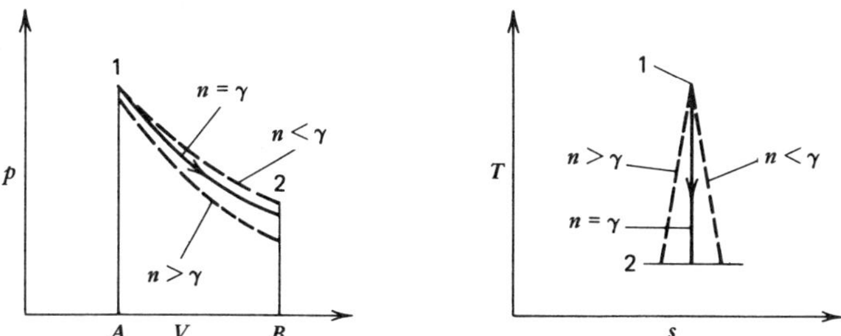

Figure 7-18 Thermodynamic diagrams for a closed polytropic expansion process.

Since $pV = MRT$, we obtain

$$Wk_c = \left[\frac{1}{1-n}\right](MRT_2 - MRT_1)$$

$$= \left[\frac{MR}{1-n}\right](T_2 - T_1) \qquad [\mathrm{J}] \qquad \text{(Eq. 7-28)}$$

The p–V and T–s Diagrams

The closed polytropic process is shown graphically by the p–V and T–s diagrams in Figures 7-18 and 7-19. The differences between these diagrams and those of the closed system adiabatic process lie in the slopes of the lines. The diagrams are different for expansions and compressions.

The reversible (ideal) work for the closed polytropic process is represented by the area A–1–2–B–A in the p–V diagram.

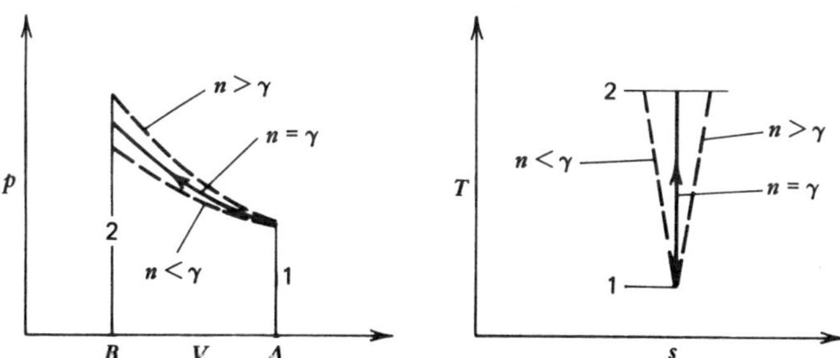

Figure 7-19 Thermodynamic diagrams for a closed polytropic compression process.

7-5.5 THE OPEN POLYTROPIC PROCESS

The Equation for Work

For an open reversible (ideal) system with a perfect gas, the first law equation can be written as

$$Wk_o = H_1 - H_2 + Q \qquad [\mathrm{J}] \qquad \text{(Eq. 3-27)}$$

$$H_1 = U_1 + p_1V_1$$

$$H_2 = U_2 + p_2V_2$$

$$Q = Mc_n(T_2 - T_1)$$

$$Wk_o = U_1 - U_2 + (p_1V_1 - p_2V_2) + Mc_n(T_1 - T_2)$$

$$U_1 = Mc_vT_1$$

$$U_2 = Mc_vT_2$$

$$p_1V_1 = MRT_1$$

$$p_2V_2 = MRT_2$$

$$Wk_o = Mc_v(T_1 - T_2) + MR(T_1 - T_2) - Mc_n(T_1 - T_2)$$

$$= \left[Mc_v + M(c_p - p_v) - Mc_n\right](T_1 - T_2)$$

$$= \left[M(c_p - c_n)\right](T_1 - T_2)$$

$$c_p - c_n = c_p - \left[\frac{c_v(\gamma - n)}{1-n}\right] = \frac{c_p - c_pn - c_p + c_vn}{1-n}$$

$$= \frac{-n(c_p - c_v)}{1-n} = \frac{-nR}{1-n}$$

$$Wk_o = \left[\frac{-nMR}{1-n}\right](T_1 - T_2)$$

$$Wk_o = \left[\frac{nMR}{1-n}\right](T_2 - T_1) \qquad [\mathrm{J}] \qquad \text{(Eq. 7-30)}$$

Substituting into Equation 7-29, we obtain

$$T = \frac{pV}{MR} \qquad \text{(perfect gas law)}$$

$$Wk_o = \left[\frac{nMR}{1-n}\right]\left(\frac{p_2V_2}{MR} - \frac{p_1V_1}{MR}\right)$$

$$Wk_o = \left[\frac{n}{1-n}\right](p_2V_2 - p_1V_1) \qquad [\mathrm{J}] \qquad \text{(Eq. 7-31)}$$

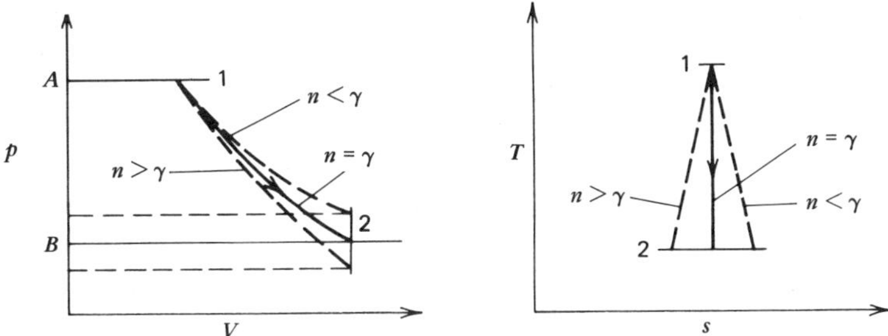

Figure 7-20 Thermodynamic diagrams for an open polytropic expansion process.

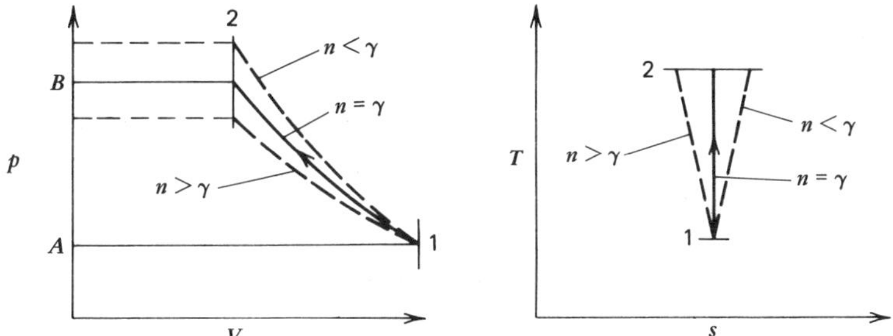

Figure 7-21 Thermodynamic diagrams for an open polytropic compression process.

The *p–V* and *T–s* Diagrams

These diagrams are similar to the closed-cycle polytropic diagrams; however, the area representing work is based on the pressure rather than on the volume, as shown in Figures 7-20 and 7-21.

The reversible (ideal) work for the open polytropic process is represented by the area $A-1-2-B-A$ in the $p-V$ diagram.

7-6 GAS TABLES

The formulas presented for the gas processes discussed in Sections 7-1 through 7-5 are valid when the specific heats (c_v and c_p) and the ratio of the specific heats (γ) is constant. The specific heats of real gases often vary in value, especially at higher temperatures. In temperature regions where the gas specific heat values vary, the end state of a process as obtained from integration (considering the varying specific heat values can be significantly different from the value obtained by the formulas assuming constant specific heat values). In the analysis of power-producing cycles wherein air is compressed, heated, and then expanded through a turbine, relatively large errors in power output may be encountered if the varying specific heat values are not considered.

The thermodynamic properties of a real gas can be presented in tables in the same manner and format as for superheated steam (Appendix A, Table A-3.3). Such a presentation results in a relatively voluminous table to cover all combinations of temperatures and pressures. Its use often necessitates making double interpolations, especially for isentropic processes. Because many real gases behave as perfect gases ($p_v = RT$) over a wide range of temperatures and pressures, the perfect gas relationship can be used to reduce the size and complexity of a table to provide the needed thermodynamic property values of the gas.

Keenan and Kaye[8] developed a simplified table of thermodynamic properties of gases for use with the isentropic process. A typical gas table is shown in Table 7-3 and is valid for air at pressures up to approximately 4 MPa. The table has columns of temperature T, specific internal energy u, specific enthalpy h, with which the reader is now familiar, and two new parameters, relative pressure p_r and relative specific volume v_r.

For a perfect gas it has been established that the specific internal energy, u, and the specific enthalpy, h, are a function of temperature only and are not affected by the pressure level. The columns of specific internal energy, u, and specific enthalpy, h, list the values of these parameters associated with the various temperature levels, T.

Entropy s, which is the key factor in isentropic processes, proves to be a function of pressure as well as temperature for gases. Consequently, a unique value of entropy s for a gas temperature level does not exist. Keenan and Kaye showed that because of the perfect gas relationship between p, v, and T, a relative pressure, p_r, corresponding to a selected pair of temperatures is uniquely associated with a value of entropy regardless of the gas pressure level. By the artifice of selecting one temperature of the temperature pair as absolute zero (0 K), the temperature pair could be identified by the gas temperature level, T. The values of relative pressure p_r (ratio of the state pressure to a fixed datum pressure) were then listed for the various temperature levels. In a similar manner a relative specific volume (v_r) could be used, and its values were listed.

By the definition of relative pressure p_r and relative specific volume v_r, we obtain

$$p_{r_2} = p_{r_1}\left(\frac{p_2}{p_1}\right) \qquad [-] \tag{7-33}$$

and

$$v_{r_2} = v_{r_1}\left(\frac{v_2}{v_1}\right) \qquad [-] \tag{7-34}$$

and

$$v_r = \left(\frac{RT}{p_r}\right) \qquad [-] \tag{7-35}$$

Table 7-3 Thermodynamic Properties of Air at Low Pressure (From Table A-11, Appendix A)

T, [K]	u, [kJ/kg]	h, [kJ/kg]	p_r [—]	v_r [—]
300	214.09	300.19	1.3860	144.32
310	221.27	310.24	1.5546	132.96
320	228.45	320.29	1.7375	122.81
330	235.65	330.34	1.9352	113.70
340	242.86	340.43	2.149	105.51
350	250.05	350.48	2.379	98.11
360	257.23	360.58	2.626	91.40
370	264.47	370.67	2.892	85.31
380	271.72	380.77	3.176	79.77
390	278.96	390.88	3.481	74.71
400	286.19	400.98	3.806	70.07
410	293.45	411.12	4.153	65.83
420	300.73	421.26	4.522	61.93
430	308.03	431.43	4.915	58.34
440	315.34	441.61	5.332	55.02
450	322.66	451.83	5.775	51.96
460	329.99	462.01	6.245	49.11
470	337.34	472.25	6.742	46.48
480	344.74	482.48	7.268	44.04
490	352.11	492.74	7.824	41.76
500	359.53	503.02	8.411	39.64
510	366.97	513.32	9.031	37.65
520	374.39	523.63	9.684	35.80
530	381.88	533.98	10.372	34.07
540	389.40	544.35	11.097	32.45
550	396.89	554.75	11.858	30.92
560	404.44	565.17	12.659	29.50
570	411.98	575.57	13.500	28.15
580	419.56	586.04	14.382	26.89
590	427.17	596.53	15.309	25.70
600	434.80	607.02	16.278	24.58
610	442.43	617.53	17.297	23.51
620	450.13	628.07	18.360	22.52
630	457.83	638.65	19.475	21.57
640	465.55	649.21	20.64	20.674
650	473.32	659.84	21.86	19.828
660	481.06	670.47	23.13	19.026
670	488.88	681.15	24.46	18.266
680	496.65	691.82	25.85	17.543
690	504.51	702.52	27.29	16.857
700	512.37	713.27	28.80	16.205

The reader who is interested in the full derivation of these parameters and the application of the tables to various mixtures of combustion products is referred to Reference 1, which has a detailed explanation of the derivation of the tables and their use.

To illustrate the use of these tables for air, sample problems based on a gas turbine (using the relative pressure parameter p_r) and for a gasoline

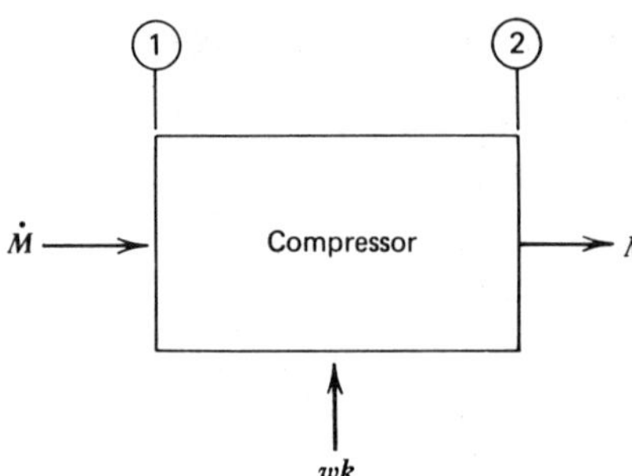

Figure 7-22 Open system isentropic compression process sketch for Example Problem 7-14.

engine (using the relative specific volume parameter v_r) are given. A comparison of the results using the tables based on the real specific heats and the perfect gas law equations using constant specific heats is made. It will be seen that there is little variation between the two methods for the compression process, but a significant difference in the turbine expansion process is encountered.

Example Problem 7-14 (Gas Turbine Engine)

In a gas turbine cycle, ambient air at 27°C is compressed to a pressure ratio of 15:1. The combustion raises the air temperature at constant pressure to 1200 K. The gas expands through the turbine with a 15:1 expansion pressure ratio. Both the compression and expansion processes are isentropic. Calculate the following items using the perfect gas equations with constant specific heat and the gas tables (Table A-11), and compare the results of these two methods of calculation.

(a) Compressor outlet air temperature.
(b) Compressor work input.
(c) Turbine exhaust air temperature.
(d) Turbine work output.
(e) Net work output (turbine work output less the compressor work input).

Solution (a)

See Figure 7-22 and Table 7-4.

Table 7-4 Table of State

Substance: Air	①	②	②′
		Perfect Gas	Perfect Gas
Mass M [kg]	1 (a)	1 (a)	1 (a)
Temperature T [K]	300 (g)	650.3	641.2
Pressure P [kPa]	p_1	$15p_1$	$15p_1$
$\text{Work}_{1-2}\ wk$ [kJ/kg]		−351.9	−349.7

Note: a, assumed; g, given.

Equation

$$T_2 = T_1\left[\frac{p_2}{p_1}\right]^{\gamma-1/\gamma} \qquad [\mathrm{K}] \qquad (\text{Eq. 7-14x})$$

Parameters

$$T_1 = 300\ \mathrm{K} \qquad (\text{given})$$

$$\frac{p_2}{p_1} = 15\ [\text{—}] \qquad (\text{given})$$

$$\gamma = 1.4\ [\text{—}] \qquad (\text{Appendix A, Table A-9})$$

Substitution

$$T_2 = 300\ [15]^{(1.4-1)/1.4} \qquad [\mathrm{K}][\text{—}] = [\mathrm{K}]$$

$$= 300\ [15]^{0.4/1.4} = 300\ [15]^{0.2857}$$

$$= 300\ (2.168)$$

Answer

$$\mathbf{T_2 = 650.3\ K}$$

Solution (b)
Equation

$$wk_C = \left[\frac{\gamma R}{1-\gamma}\right][T_2 - T_1] \qquad [\mathrm{kJ/kg}] \qquad (\text{Eq. 7-20x})$$

Parameters

$$\gamma = 1.4\ [\text{—}] \qquad (\text{Appendix A, Table A-9})$$

$$R = 0.287\ \mathrm{kJ/kg \cdot K} \qquad (\text{Appendix A, Table A-9})$$

$$T_2 \text{ and } T_1 \qquad (\text{Table 7-4, Table of State})$$

Substitution

$$wk_c = \frac{(1.4)(0.287)}{1-1.4}(650.3-300) \qquad \frac{[\text{—}][\mathrm{kJ/kg \cdot K}][\mathrm{K}-\mathrm{K}]}{[\text{—}]} = [\mathrm{kJ/kg}]$$

$$= \frac{(1.4)(0.287)}{-0.4}(+350.3)$$

Answer

$$\mathbf{Wk_C = -351.9\ kJ/kg}$$

Solution (a′)

Equation

$$p_{r_2} = p_{r_1}\left[\frac{p_2}{p_1}\right] \qquad [—] \qquad \text{(Eq. 7-33)}$$

Parameters

$$p_{r_1} = 1.3860\,[—] \qquad \text{(Appendix A, Table A-11, at } T = 300\text{ K)}$$

$$\frac{p_2}{p_1} = 15 \qquad \text{(given)}$$

Substitution

$$p_{r_2} = (1.3860)(15) = [—][—] = [—]$$

Answer

$$\mathbf{p_{r_2} = 20.79\,[—]}$$

Equation

$$T_2 = T_{640} + \Delta T$$

Parameters

Table 7-5 Interpolation Data

T [K]	p_r [—]
640	20.64
T_2	20.79
650	21.86

Substitution

$$T_2 = 640 + (650 - 640)\frac{(20.79 - 20.64)}{(21.86 - 20.64)} \qquad [\text{K}]$$

$$= 640 + (10)\frac{0.15}{1.22} = 640 + 1.23$$

Answer

$$\mathbf{T_2 = 641.2\ K}$$

Solution (b′)

Equation

$$wk_C = h_1 - h_2 \qquad [\text{kJ/kg}] \qquad \text{(Eq. 3-29a)}$$

Parameters

$$h_1 = 300.19 \text{ kJ/kg} \qquad \text{(Appendix A, Table A-11, at 300 K)}$$

Table 7-6 Interpolation Data

h[kJ/kg]	T[K]
649.21	640
h_2	641.2
659.84	650

$$h_2 = 649.21 + (654.84 - 649.21)\frac{(641.2 - 640)}{(650 - 640)} \qquad [\text{kJ/kg}]$$

$$= 649.21 + (5.63)\left(\frac{1.2}{10}\right)$$

$$= 649.21 + 0.68$$

$$= 649.89 \text{ kJ/kg}$$

Substitution

$$wk_C = 300.19 - 649.89 \qquad [\text{kJ/kg} - \text{kJ/kg}] = [\text{kJ/kg}]$$

Answer

$$\boldsymbol{wk_C = -349.70 \text{ kJ/kg}}$$

Solution (c)

Use perfect gas equations (the same method as for the compressor). (See Figure 7-23 and Table 7-7, page 188.)

Equation

$$T_2 = T_1\left[\frac{p_2}{p_1}\right]^{(\gamma-1)/\gamma} \qquad [\text{K}] \qquad \text{(Eq. 7-14x)}$$

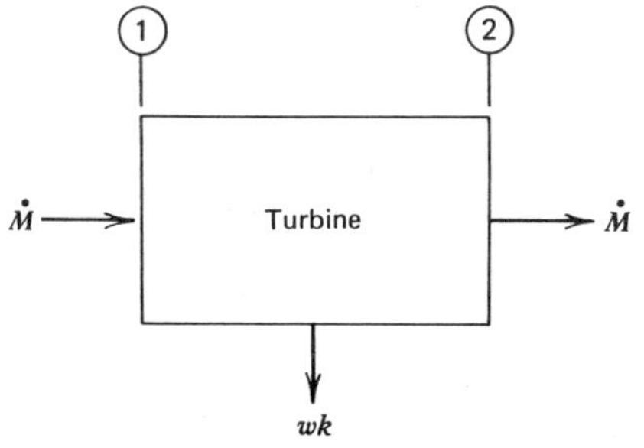

Figure 7-23 Open system isentropic process sketch for Example Problem 7-14c.

Table 7-7 Table of State

Substance: Air	①	②	②′
		Perfect Gas	Perfect Gas
Mass M [kg]	1	1 (a)	1 (a)
Temperature T [K]	1200 (g)	553.5	595.8
Pressure p [kPa]	p_1	$\frac{1}{15} p_1$	$\frac{1}{15} p_1$
Work$_{1-2}$ wk [kJ/kg]		649.4	675.23

Note: g, given; a, assumed.

Parameters

$$T_1 = 1200 \text{ K} \qquad \text{(given)}$$

$$\frac{p_2}{p_1} = \frac{1}{15}[-] \qquad \text{(given)}$$

$$\gamma = 1.4\,[-] \qquad \text{(Appendix A, Table A-9)}$$

Substitution

$$T_2 = 1200\left(\frac{1}{15}\right)\left(\frac{1.4-1}{1.4}\right) \qquad [\text{K}][-] = [\text{K}]$$

$$= 1200\left(\frac{1}{15}\right)^{0.2857} = 1200\frac{1}{2.168}$$

Answer

$$\mathbf{T_2 = 553.5 \text{ K}}$$

Solution (d)

Equation

$$wk_T = \left[\frac{\gamma R}{1-\gamma}\right](T_2 - T_1) \qquad [\text{kJ/kg}] \qquad \text{(Eq. 7-20x)}$$

Parameters

$$\gamma = 1.4\,[-] \qquad \text{(Appendix A, Table A-9)}$$

$$R = 0.287 \text{ kJ/kg}\cdot\Delta_1\text{K} \qquad \text{(Appendix A, Table A-9)}$$

$$T_2 \text{ and } T_1 \qquad \text{(Table 7-7, table of state)}$$

Substitution

$$wk_T = \frac{(1.4)(0.287)}{(1-1.4)}(553.5-1200)\frac{[-][\text{kJ/kg}\cdot\Delta_1\text{K}]}{-} = [\text{kJ/kg}]$$

$$= \frac{(1.4)(0.287)}{(-0.4)}(-6465)$$

Answer

$$\mathbf{wk_T = 649.4\ kJ/kg}$$

Solution (c′)
Use the gas table (the same as for the compressor).
Equation

$$p_{r_2} = p_{r_1}\left(\frac{p_2}{p_1}\right) \qquad [-] \qquad \text{(Eq. 7-33)}$$

Parameters

$$p_{r_1} = 238.0\ [-] \qquad \text{(Appendix A, Table A-11, at 1200 K)}$$

$$\frac{p_2}{p_1} = \frac{1}{15}[-] \qquad \text{(given)}$$

Substitution

$$p_{r_2} = (238.0)\left(\tfrac{1}{15}\right) \qquad [-][-]=[-]$$

Answer

$$\mathbf{p_{r_2} = 15.867\ [-]}$$

Equation

$$T_2 = T_{590} + \Delta T$$

Parameters

Table 7-8 Interpolation Data

T [K]	p_r[—]
590	15.309
T_2	15.867
600	16.278

Substitution

$$T_2 = 590 + (600 - 590)\frac{(15.867 - 15.309)}{(16.278 - 15.309)} \qquad [\text{K}]$$

$$= 590 + (10)\frac{(0.558)}{(0.969)} = 590 + 5.76$$

Answer

$$\mathbf{T_2 = 595.8\ K}$$

Solution (d′)
Equation

$$wk_T = h_1 - h_2 \qquad [\text{kJ/kg}] \qquad (\text{Eq. 3-29a})$$

Parameters

$$h_1 = 1277.79\ \text{kJ/kg} \qquad (\text{Appendix A, Table A-11, at 1200 K})$$

Table 7-9 Interpolation Data

h [kJ/kg]	T [K]
596.53	590
h_2	595.8
607.02	600

Substitution

$$h_2 = 596.53 + (607.02 - 596.53)\frac{595.8 - 590}{600 - 590} \qquad [\text{kJ/kg}]$$

$$= 596.53 + 10.49\left(\frac{5.8}{10}\right) = 596.53 + 6.03$$

$$h_2 = 602.56\ \text{kJ/kg}$$

$$wk_T = 1277.79 - 602.56 \qquad [\text{kJ/kg} - \text{kJ/kg}] = [\text{kJ/kg}]$$

Answer

$$\mathbf{wk_T = 675.23\ kJ/kg}$$

Solution (e)
Equation

$$wk_{\text{net}} = wk_T + wk_C \qquad [\text{kJ/kg}]$$

Parameters

From the perfect gas laws,

$$wk_T = 649.4 \text{ kJ/kg} \quad [(\text{from Solution (d)})]$$

$$wk_C = -351.9 \text{ kJ/kg} \ [(\text{from Solution (b)})]$$

Substitution

$$wk_{net} = 649.4 + (-351.9) \qquad [\text{kJ/kg} + \text{kJ/kg}] = [\text{kJ/kg}]$$

Answer

$$\mathbf{wk_n = 297.5 \text{ kJ/kg}}$$

Solution (e′)

Equation

$$wk_{net} = wk_T + wk_C \qquad [\text{kJ/kg}]$$

Parameters

From the gas tables,

$$wk_T = 675.23 \text{ kJ/kg} \qquad [(\text{from Solution (d′)})]$$

$$wk_C = -347.9 \text{ kJ/kg} \qquad [(\text{from Solution (b′)})]$$

Substitution

$$wk_{net} = 675.23 + (-347.9) \qquad [\text{kJ/kg} + \text{kJ/kg}] = [\text{kJ/kg}]$$

Answer

$$\mathbf{wk_{net} = 327.3 \text{ kJ/kg}}$$

The next work output of the gas turbine (wk_{net}) calculated by the perfect gas laws is 9 percent lower than the net work calculated by the gas tables. The difference is primarily in the turbine work output (the high temperature process).

Example Problem 7-15 (Gasoline Engine)

In a gasoline engine cycle, ambient air at 27°C is compressed to a volume ratio of 1:8. The combustion raises the air temperature at constant volume to 1500 K. The heated air then expands to a volume ratio of 8:1. Both the compression and expansion processes are isentropic.

Calculate the gas temperatures and work as in Example Problem 7-24. (See Figure 7-24 and Table 7-10, page 192.)

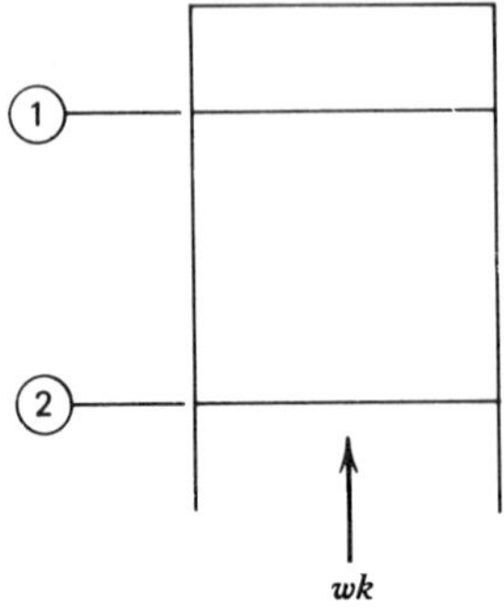

Figure 7-24 Closed system isentropic process (compression stroke), sketch for Example Problem 7-15.

Solution (a)

Use the perfect gas equation to determine T_2 by Equation 7-14.

Equation

$$T_2 = T_1\left(\frac{v_1}{v_2}\right)^{\gamma-1} \quad [\mathrm{K}] \qquad (\text{Eq. 7-14x})$$

Parameters

$$T_1 = 300\ \mathrm{K} \qquad (\text{given})$$

$$\left(\frac{v_1}{v_2}\right) = 8[—] \qquad (\text{given})$$

$$\gamma = 1.4\,[—] \qquad (\text{Appendix A, Table A-9})$$

Substitution

$$T_2 = 300\,(8)^{1.4-1} \qquad [\mathrm{K}][—] = [\mathrm{K}]$$

$$= 300\,(8)^{0.4} = (300)(2.297)$$

Answer

$$\mathbf{T_2 = 689.2\ K}$$

Table 7-10 Table of State

Substance: Air	①	②	②′
		Perfect Gas	Perfect Gas
Mass M [kg]			
Temperature T [K]	300 (g)	689.2	617
Pressure p [kPa]			
Specific volume v [m^3/kg]	v_1	$v_{1/8}$ (g)	$v_{1/8}$ (g)
Work$_{C(1-2)}$ Wk [kJ/kg]		−279.8	−277.2

Note: g (given).

Solution (b)

Use the perfect gas equation to determine *wk* by Equation 7-18.

Equation

$$wk_C=\left[\frac{R}{1-\gamma}\right](T_2-T_1) \qquad [\mathrm{kJ/kg}] \qquad (\text{Eq. 7-18a})$$

Parameters

$R=0.287$ kJ/kg·Δ_1K (Appendix A, Table A-9)

$\gamma=1.4$ [—] (Appendix A, Table A-9)

T_2 and T_1 (Table 7-10, Table of State)

Substitution

$$wk_C=\left(\frac{0.287}{1-1.4}\right)(689.2-300)\frac{[\mathrm{kJ/kg\cdot\Delta_1K}][\mathrm{K-K}]}{[—]}=[\mathrm{kJ/kg}]$$

$$=\left(\frac{0.287}{-0.4}\right)(389.2)$$

Answer

$$\mathbf{wk_C=-279.3\ kJ/kg}$$

Solution (a′)

Use the gas tables to determine T_2 and v_{r_1}. Calculate v_{r_2} from v_{r_1} and v_2/v_1 (Equation 7-33), and determine T_2 for v_{r_2} from the gas tables. To determine wk_C use Equation 3-8x and obtain u_1 and u_2 from gas tables.

Equation

$$v_{r_2}=v_{r_1}\frac{v_2}{v_1} \qquad [—] \qquad (\text{Eq. 7-34})$$

Parameters

$v_{r_1}=144.32$ [—] (Appendix A, Table A-11 at 300 K)

$\dfrac{v_2}{v_1}=\dfrac{1}{8}$[—] (given)

Substitution

$$v_{r_2}=(144.32)\left(\frac{1}{8}\right) \qquad [—][—]=[—]$$

Answer

$$\mathbf{v_{r_2}=18.040[—]}$$

Equation

$$T_2 = T_{670} + \Delta T$$

Parameters

Table 7-11 Interpolation Data

T [K]	v_r [—]
670	18.266
T_2	18.040
680	17.543

Substitution

$$T_2 = 670 + (680 - 670)\frac{(18.266 - 18.040)}{(18.266 - 17.543)} \qquad [\text{K}]$$

$$= 670 + (10)\frac{(0.226)}{(0.723)} = 670 + 3.1$$

Answer

$$\mathbf{T_2 = 673.1}$$

Solution (b′)
Equation

$$wk_C = u_1 - u_2 \qquad [\text{kJ/kg}] \qquad (\text{Eq. 3-9b})$$

Parameters
(See Table 7-12.)

$$u_1 = 214.09 \qquad [\text{kJ/kg}] \qquad (\text{Appendix A, Table A-11, at 300 K})$$

Table 7-12 Interpolation Data

u [kJ/kg]	T [K]
488.88	670
u_2	673.1
496.65	680

$$u_2 = 488.88 + (496.65 - 488.88)\left(\frac{673.1 - 670}{680 - 670}\right)$$

$$= 488.88 + (7.77)(0.31)$$

$$= 488.88 + 2.41$$

$$= 491.29 \text{ kJ/kg}$$

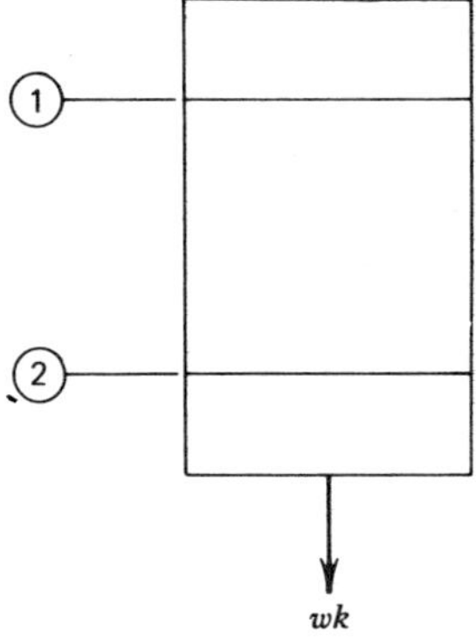

Figure 7-25 Closed system isentropic process (expansion stroke) sketch for Example Problem 7-15.

Substitution

$$wk_C = 214.09 - 491.29 \qquad [\text{kJ/kg} - \text{kJ/kg}] = [\text{kJ/kg}]$$

Answer

$$\mathbf{wk_C = 277.2\ kJ/kg}$$

Solution (c)

Use the same procedures as for the compression stroke.
(See Figure 7-25 and Table 7-13.)

Equation

$$T_2 = T_1\left(\frac{v_1}{v_2}\right)^{\gamma-1} \qquad [\text{K}] \qquad (\text{Eq. 6-14x})$$

Parameters

$$T_1 = 1500\ \text{K} \qquad (\text{given})$$

$$\frac{v_1}{v_2} = \frac{1}{8}[-] \qquad (\text{given})$$

$$\gamma = 1.4[-] \qquad (\text{Appendix A, Table A-9})$$

Table 7-13 Table of State

Substance: Air	①	②	②′
		Perfect Gas	Perfect Gas
Mass M [kg]			
Temperature T [K]	100 (g)	652.9	752
Pressure p [kPa]			
Specific volume v [m^3/kg]	v_1	$8v_1$ (g)	$8v_1$ (g)
Work$_{T(1-2)}$ wk [kJ/kg]		607.79	651.81

Note: g (given).

Substitution

$$T_2 = 1500\left(\frac{1}{8}\right)^{1.4-1} \qquad [\mathrm{K}][—]=[\mathrm{K}]$$

$$= 1500\left(\frac{1}{8}\right)^{0.4} = 1500\frac{1}{2.2974}$$

Answer

$$\mathbf{T_2 = 652.9\ K}$$

Solution (c′)

Equation

$$T_2 = T_{750} + \Delta T \qquad [\mathrm{K}]$$

Parameters

Table 7-14 Interpolation Data

T [K]	v_{r_2} [—]
750	13.391
T_2	13.294
760	12.905

Substitution

$$T_2 = 750 + (760-750)\frac{13.391-13.294}{13.391-12.905} \qquad [\mathrm{K}]$$

$$= 750 + (10)\frac{(0.097)}{(0.486)} = 750 + 2.0$$

Answer

$$\mathbf{T_2 = 752\ K}$$

Solution (d)

Equation

$$wk_T = \left(\frac{R}{1-\gamma}\right)(T_2 - T_1) \qquad [\mathrm{kJ/kg}] \qquad \text{(Eq. 7-18a)}$$

Parameters

$R = 0.287$ kJ/kg·Δ_1K (Appendix A, Table A-9)

$\gamma = 1.4$[—] (Appendix A, Table A-9)

T_2 and T_1 (Table 7-13, Table of State)

Substitution

$$wk_T = \frac{0.287}{(1-1.4)}[652.9-1500] \qquad \frac{[\text{kJ/kg}\cdot\Delta_1\text{K}]}{[—]}[\text{K}] = [\text{kJ/kg}]$$

$$= \frac{0.287}{-0.4}(-847.1)$$

Answer

$$\boldsymbol{wk_T = 607.79\ \text{kJ/kg}}$$

Equation

$$v_{r_2} = v_{r_1}\left(\frac{v_2}{v_1}\right) \qquad [—] \qquad (\text{Eq. 7-34})$$

Parameters

$v_{r_1} = 1.6617$ [—] (Appendix A, Table A-11, at 1500 K)

$\frac{v_2}{v_1} = 8$ [—] (given)

Substitution

$$v_{r_2} = (1.6617)(8) \qquad [—][—] = [—]$$

Answer

$$\boldsymbol{v_{r_2} = 13.294\ [—]}$$

Solution (d′)

Equation

$$wk_T = u_1 - u_2 \qquad [\text{kJ/kg}] \qquad (\text{Eq. 3-9b})$$

Parameters

$u_1 = 1205.47$ kJ/kg (Appendix A, Table A-11, at 1500 K)

Table 7-15 Interpolation Data

u [kJ/kg]	T [K]
552.05	750
u_2	752
560.08	760

$$u_2 = 552.05 + (560.08 - 552.05)\frac{(752-750)}{(760-750)} \qquad [\text{kJ/kg}]$$

$$= 552.05 + (8.03)\frac{(2)}{(10)} = 552.05 + 1.61$$

$$= 553.66\ \text{kJ/kg}$$

Substitution

$$wk_T = 1205.47 - 553.66 \qquad [\text{kJ/kg} - \text{kJ/kg}] = [\text{kJ/kg}]$$

Answer

$$\boldsymbol{wk_T = 651.81\ \text{kJ/kg}}$$

Solution (e)

Equation

$$wk_{\text{net}} = wk_T + wk_C \qquad [\text{kJ/kg}]$$

Parameters

From the perfect gas laws,

$$wk_T = 607.79\ \text{kJ/kg} \qquad [\text{from solution (d)}]$$

$$wk_C = -279.3\ \text{kJ/kg} \qquad [\text{from solution (b)}]$$

Substitution

$$wk_{\text{net}} = 607.79 + (-279.3) \qquad [\text{kJ/kg} + \text{kJ/kg}] = [\text{kJ/kg}]$$

Answer

$$\boldsymbol{wk_{\text{net}} = 328.5\ \text{kJ/kg}}$$

Solution (e′)

Equation

$$wk_{\text{net}} = wk_T + wk_C \qquad [\text{kJ/kg}]$$

Parameters

From the gas tables

$$wk_T = 651.81\ \text{kJ/kg} \qquad (\text{from solution d}')$$

$$wk_C = -277.20\ \text{kJ/kg} \qquad (\text{from solution b}')$$

Substitution

$$wk_{\text{net}} = 651.81 + (-277.20) \qquad [\text{kJ/kg} + \text{kJ/kg}] = [\text{kJ/kg}]$$

Answer

$$\boldsymbol{wk_{\text{net}} = 374.6\ \text{kJ/kg}}$$

The net work output of the gasoline engine wk_{net} calculated by the perfect gas laws is 12 percent lower than the wk_{net} calculated by the gas tables. The difference is in the expansion stroke output (high temperature process).

7-7 PROBLEMS

7-1. List five (5) common gas processes.

7-2. In a constant pressure open process at 200 kPa the volume of air is reduced from 2 m^3 at 300°C to 1 m^3. How much work, if any, is involved? How much heat, if any, is involved?

7-3. If the constant pressure process in Problem 7-2 were a closed process, instead of open, how much work and/or heat, if any, would be involved?

7-4. Explain why the answers to Problems 7-2 and 7-3 are different.

7-5. Treating an aircraft engine afterburner as a duct with a constant flow path cross-sectional area and operating at a constant pressure; if the exhaust gases are raised from 500 to 1000°C, when the afterburner is activated, what will be the increase in the exhaust gas exit velocity?

7-6. If 1 kg of air at 250 kPa pressure and 100°C were to be "expanded" at constant pressure in a closed process to twice the original volume, what would the following be?
(a) The final state of the air.
(b) The amount of heat (if any) involved in the process.
(c) The amount of work (if any) involved in the process.

7-7. If the conditions of Problem 7-6 were applied to an open process, what are the answers to (a), (b), and (c)? If any of these answers are same for Problems 7-6 and 7-7 explain why this is so. If the answers are different, explain why this is so.

7-8. Treating liquid water as an incompressible fluid, determine the specific work (wk/kg) to pump water (saturated liquid) at 20 kPa pressure into a boiler at 3.0 MPa pressure.

7-9. A tank contains nitrogen gas at 20 MPa at 25°C. If the tank is exposed to sunshine and becomes heated to 40°C, what will be the pressure in the tank?

7-10. In a gasoline engine at the end of the compression stroke the air-fuel mixture is ignited and heat is released while the piston is close to "top dead center." Considering this process to be equivalent to compressed air with heat added at constant volume, what would the following be if 1500 kJ/kg of heat is put into the compressed air having an initial temperature of 410°C and a pressure of 1800 kPa?
(a) The temperature.
(b) The pressure.

7-11. An open process isothermal compressor receives air at a temperature of 25°C and a pressure of 100 kPa, discharging the air at 2000 kPa. Describe the state of the air after this compression. Determine the specific work (kJ/kg) associated with this compression.

7-12. In Problem 7-11 is any heat involved? If so, how much?

7-13. If the isothermal compression described in Problem 7-11 were a closed process, what differences, if any, would there be in the following?
(a) The final state of the air.
(b) The work associated with the compression.
(c) The amount of heat involved.

7-14. In an air refrigeration system air at 100°C is expanded adiabatically by a reversible open process to one-quarter the initial pressure. What will the exhaust temperature be? How much specific work (kJ/kg) will be produced?

7-15. Helium at 25°C is expanded to 20 percent the initial pressure by a reversible adiabatic expansion. What will its temperature be? How much specific work (kJ/kg) will be produced?

7-16. The volumetric compression ratio in a gasoline engine is 8:1. If the initial (intake) pressure is 100 kPa, what is the pressure at the end of the compression stroke if an isentropic compression of air is assumed? What is the temperature for an initial air temperature of 25°C?

7-17. If the compression ratio of the gasoline engine in Problem 7-16 were increased to 8.5:1, what would be the pressure at the end of the compression stroke for the same initial intake pressure? What would be the temperature for the same initial air temperature?

7-18. A Diesel engine operating with a 23:1 volumetric compression ratio has the same initial (intake) pressure and temperature as the gasoline engine in Problem 7-16. Assuming an isentropic compression of air, what would be the pressure and temperature at the end of the compression stroke?

7-19. Air in a steady flow open system is compressed isentropically. If the inlet temperature is 20°C and the exit 500°C, determine (a) the compression ratio and (b) the specific work input required.

7-20. If the air in Problem 7-19 were compressed to the same compression ratio, but by a polytropic process with $n = 1.33$, what would be (a) the air temperature after compression and (b) the specific work input required?

7-21. If the air in Problem 7-19 were compressed to the same compression ratio, but by a polytropic process with $n = 1.25$, what would be (a) the compressed air temperature and (b) the specific work input required? How does this compare with the isentropic compression of Problem 7-19?

7-22. Air at 20°C is compressed to a pressure ratio of 20:1, then heated an additional 500°C in a constant pressure process; and then expanded through a 20:1 pressure ratio. Compare the net work obtained (work output from the expansion less the work input to the compression from (a) reversible adiabatic compression and expansion and (b) a polytropic compression and expansion with a value of $n = 1.33$.

7-23. If the compression in Problem 7-22 were isothermal for both (a) and (b) and the expansions were (a) adiabatic and (b) polytropic with $n = 1.33$, determine the net work output for the two cases.

7-24. A gas turbine receives air at 1007°C and expands it isentropically to 507°C. Calculate the specific work (wk/kg) produced by using (a) the equations in Section 7-4; and (b) Keenan and Kays gas tables in Appendix A, Table A-11.

7-25. The gas turbine in Problem 7-24 receives air at 1167°C and expands it isentropically to 567°C. Calculate the specific work (wk/kg) produced by using the following:
(a) The equations in Section 7-4.
(b) The Keenan and Kays gas tables (Appendix A, Table A-11).

7-26. Calculate the pressure expansion ratio of the expansion in Problem 7-24 calculated using the following:
(a) The equations in Section 7-4.
(b) The Keenan and Kays gas tables (Appendix A, Table A-11).

7-27 Calculate the pressure expansion ratio of the expansion in Problem 7-25, calculated using the following:
(a) The equations in Section 7-4.
(b) The Keenan and Kays gas tables (Appendix A, Table A-11).

7-28. Make a table of the various equations for the five (5) gas processes discussed in this chapter. The columns of the table should include the name of the processes, the equation for the process definition, and, between states 1 and 2, the pressure ratio, the volume ratio, the temperature ratio, the work (both open and closed process), and the heat q.

#7-5 - $\frac{P_1 V_1}{T_1} = \frac{P_2 V_2}{T_2}$ SINCE $V = k$, velocity (v) can be used in place of Volume.

$\therefore \frac{v_2}{v_1} = \frac{T_2}{T_1}$

#7-8 - $wk_{sp} = wk_o = -v(P_2 - P_1)$ v on pg 388 $\left[\frac{wk}{kg}\right]$

#7-16 - (a) $V_2 = V_1 \left[\frac{P_1}{P_2}\right]^{1/\gamma} \Rightarrow P_2 = \left[\frac{V_1}{V_2}\right]^{\gamma} (P_1)$

(b) $T_2 = T_1 \left[\frac{P_2}{P_1}\right]^{\gamma-1/\gamma}$

#7-18 - (a) $P_2 = \left[\frac{V_1}{V_2}\right]^{\gamma} (P_1)$ (b) $T_2 = T_1 \left[\frac{P_2}{P_1}\right]^{\gamma-1/\gamma}$

#7-24 - (a) $wk_o = \frac{\gamma R}{1-\gamma}(T_2 - T_1)$ (b) $wk_o = h_1 - h_2$

#7-25 $wk_o = \frac{\gamma R}{1-\gamma}(T_2 - T_1)$ (b) $wk_o = h_1 - h_2$

CHAPTER 8
TWO-PHASE PROCESSES

Two-phase processes are encountered in many thermodynamic cycles such as steam powerplants, mechanical refrigeration, and so forth. Two-phase processes are more complex than single-phase perfect gas processes. When phase changes occur, the single-phase vapor usually has a significant deviation from the perfect gas law.

For describing the change in state in a two-phase process, the use of a temperature–specific entropy (T–s) diagram of the substance involved has been found to be convenient. Such a diagram consists of overlay grids of lines of constant pressure p, specific volume v, and specific enthalpy h on a rectangular coordinate plot of the saturation line with temperature T as the ordinate and the specific entropy s as the abscissa.

The process of interest can then be drawn as a line on the T–s diagram starting at the initial state point, following the appropriate process line (for instance, constant temperature, constant entropy, etc.) to the process endpoint, where the values of the various description of the state properties can be determined.

8-1 THE TWO-PHASE SATURATION LINE TEMPERATURE–ENTROPY PLOT

The temperature–specific entropy diagram is usually drawn on rectangular coordinates with temperature T as the ordinate and specific entropy s as the abscissa. The saturation line (values of specific entropy for saturated liquid and saturated vapor at various temperatures) is plotted on the T–s coordinates. At the critical temperature the specific entropy for saturated liquid and saturated vapor has the same value. This results in a dome-shaped curve as shown in Figure 8-1.

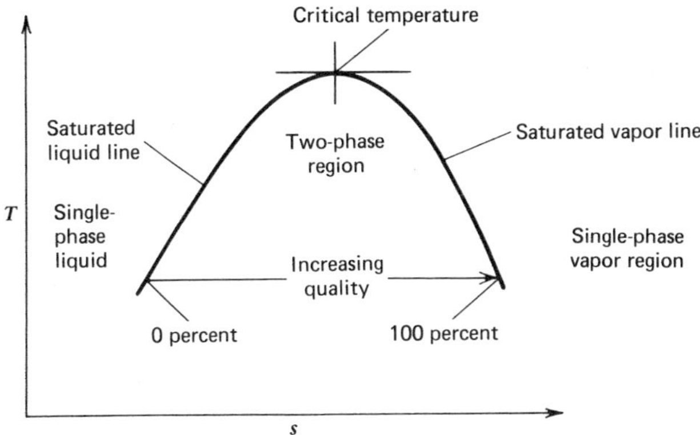

Figure 8-1 Substance saturation line on T–s diagram.

8-2 THE CONSTANT PRESSURE OPEN PROCESS

A common method of putting heat into or removing heat from a two-phase thermodynamic system is boiling (heat input) and condensation (heat removal). Such boiling and condensing processes are usually accomplished at essentially a constant pressure and are characterized by the constant pressure process. In boiling, the heat added converts the liquid to a saturated vapor (which occurs at a constant temperature). If the vapor is subsequently superheated by heat addition, its temperature is increased. A constant pressure process from a saturated liquid (1) to a saturated vapor (2) and a superheated vapor (3) on the T–s diagram is shown in Figure 8-2.

The work involved in the open constant pressure process is zero.

$$Wk_o = 0 \qquad [\mathrm{J}]$$

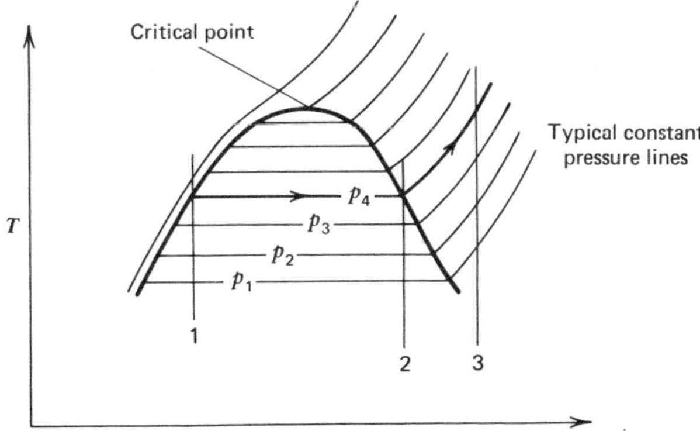

Figure 8-2 A constant pressure process on the temperature-specific entropy diagram.

This is derived from the first law of thermodynamics. For the phase-change region from saturated liquid (1) to saturated vapor (2),

$$wk_{o_{1-2}} = \Delta q + h_1 - h_2 \qquad [\text{kJ/kg}] \qquad (\text{Eq. 3-27a})$$

since

$$\Delta q = h_2 - h_1 = -(h_1 - h_2) \qquad [\text{kJ/kg}] \qquad (\text{Eq. 3-28a})$$

$$wk_{o_{1-2}} = 0 \qquad [\text{kJ/kg}]$$

For the vapor constant pressure process between the saturated vapor (2) and superheated vapor (3),

$$wk_{o_{2-3}} = 0 \qquad [\text{kJ/kg}]$$

as discussed in Section 7-1.3 for the open isobaric gas process.

Condensation is the reverse of boiling with heat rejection instead of heat addition.

8-3 THE ISENTROPIC OPEN PROCESS

A common method of obtaining work from a thermodynamic system is by expansion, such as accomplished by a turbine or an expansion engine. The expansion process through efficient turbines or expansion engines approximates a reversible adiabatic expansion (which is isentropic as discussed in Section 7-4.3). Such an isentropic expansion from superheated vapor into the quality region is illustrated in Figure 8-3.

The work produced by an open isentropic expansion is

$$wk_{o_{1-2}} = h_1 - h_2 \qquad [\text{kJ/kg}] \qquad (\text{Eq. 3-29a})$$

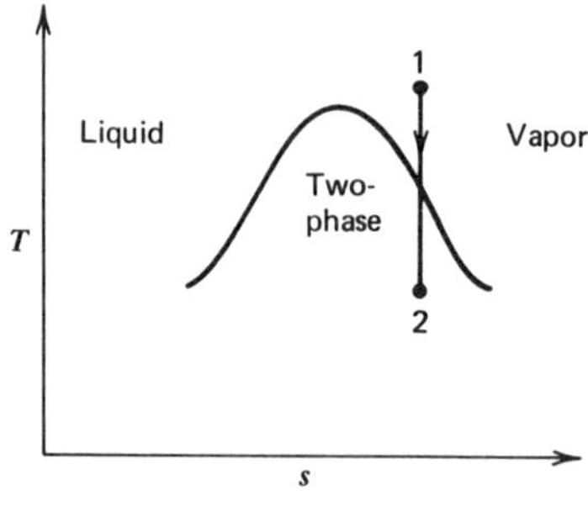

Figure 8-3 An isentropic expansion process on the temperature-specific entropy diagram.

8-4 THE MOLLIER DIAGRAM

To simplify the determination of the work output of an isentropic expansion [$wk=h_1-h_2$], Richard Mollier (1863–1935) plotted the thermodynamic data of a substance on an orthogonal grid of enthalpy h versus entropy s. On this grid were drawn:

- Saturated vapor line.
- Constant quality lines.
- Constant superheat lines.*
- Constant temperature lines.
- Constant pressure lines.

An example of the Mollier diagram for steam is shown in Figure 8-4. An actual (expanded) Mollier diagram for steam is located in Appendix B. This diagram can readily be used to relate the following data for a steam turbine with an isentropic (constant entropy) expansion process:

- Steam inlet state.
- Steam exhaust state.
- Work output.

Various processes trace the lines on the Mollier diagram shown in Figure 8-5, page 207.

For example, if the inlet steam is at a pressure of 3 MPa and 200°C of superheat, the intersection of the constant pressure line for 3 MPa and the constant 200°C superheat define values of 7.02 kJ/kg·Δ_1K for the specific entropy s_1 and 3315 kJ/kg for the specific enthalpy h_1.

An isentropic expansion in the turbine follows the vertical constant specific entropy line ($s=7.02$ kJ/kg·Δ_1K) until the exhaust pressure is reached. If the exhaust pressure is 20 kPa, the intersection of the specific entropy line ($s=7.02$ kJ/kg·Δ_1K) and the constant pressure line ($p=20$ kPa) is at the 2300 kJ/kg specific enthalpy line h_2.

Since the work output is

$$wk=h_1-h_2 \quad [\text{kJ/kg}] \qquad (\text{Eq. 3-29a})$$

$$h_1=3315 \quad \text{kJ/kg}$$

$$h_2=2300 \quad \text{kJ/kg}$$

we find that

$$wk=(3315-2300) \quad [\text{kJ/kg}]-[\text{kJ/kg}]=[\text{kJ/kg}]$$

$$=1015 \quad \text{kJ/kg}$$

*In this figure superheat is defined as the number of degrees of temperature above saturation.

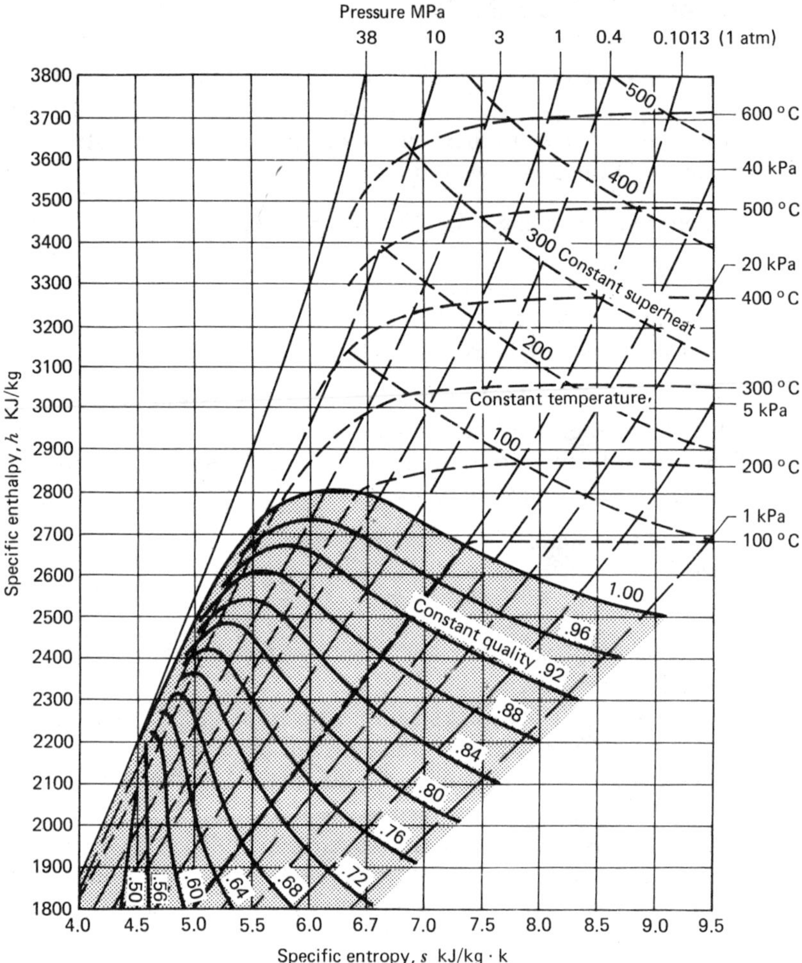

Figure 8-4 Mollier diagram for steam.

The quality x_2 of the exhaust steam from the Mollier diagram is approximately 0.87 [—] or 87 percent.

8-5 THE CONSTANT ENTHALPY (ISENTHALPIC) OPEN PROCESS

The expansion valve in a mechanical refrigeration cycle is a common example of a constant enthalpy process. The thermodynamic substance (refrigerant) is passed through a flow restriction from a high pressure to a low pressure with no heat or work added or removed. The isenthalpic process (throttling process) is not usually a part of power-producing cycles,

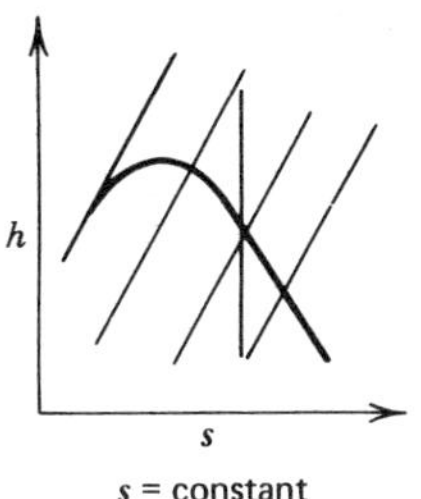

s = constant

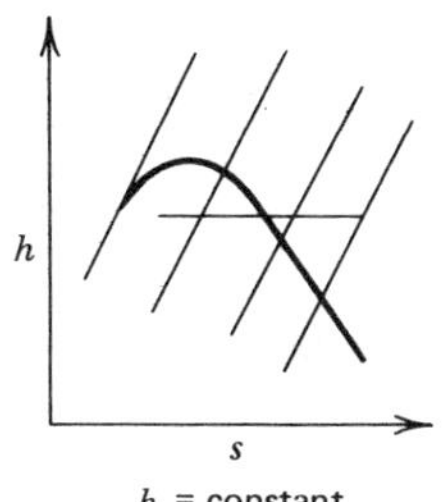

h = constant

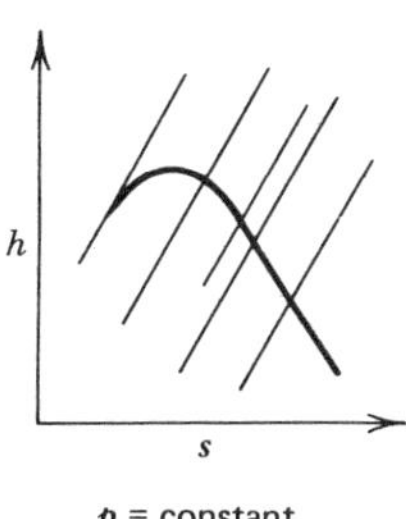

p = constant

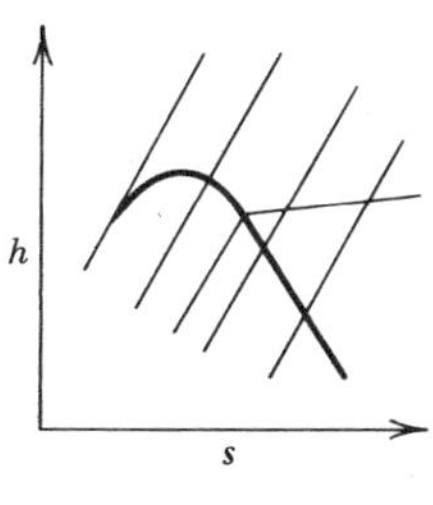

T = constant

Figure 8-5 Various processes on the Mollier diagram.

except as a means of power output control (the throttle valve). The isenthalpic process is illustrated in Figure 8-6.

Since the process is adiabatic at a constant enthalpy value ($h_2 = h_1$), the $wk_{12} = 0$.

$$
\begin{aligned}
wk_{12} = \quad & q_{12} - \Delta h_{12} \\
& q_{12} = 0 \qquad \text{(adiabatic process)} \\
& \Delta h_{12} = 0 \qquad \text{(process definition)}
\end{aligned}
$$

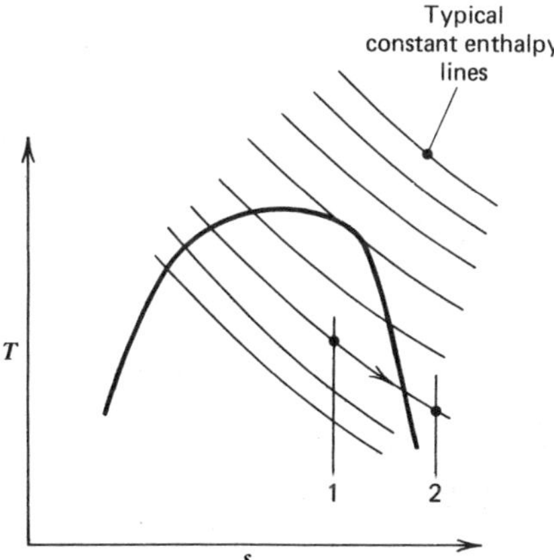

Figure 8-6 A constant enthalpy process on the temperature-specific entropy diagram.

Therefore,

$$wk_{12} = 0 - 0 = 0 \qquad [\text{kJ/kg}]$$

One application of the constant enthalpy process is the determination of the quality of a substance by using a constant enthalpy expansion from the quality state (1) to a superheated vapor state (2), which can be defined by measurements of the two independent properties of pressure p_2 and temperature T_2. By so defining state 2, the value of enthalpy h_2 can be ascertained from thermodynamic tables. Using this value of enthalpy $(h_1 = h_2)$ and a pressure measurement p_1 of state 1, the quality of the substance at state 1 can readily be calculated as indicated by Example problem 8-1.

The instrument for making such a determination of the quality of a substance is called a throttling calorimeter. A diagram of a typical throttling calorimeter is given in Figure 8-7, page 209.

The steam enters the calorimeter through a sampling tube and expands through the orifice at the end of this tube into the chamber of the device. A thermometer and manometer connected to that chamber monitors the final temperature and pressure, respectively. The initial pressure is monitored by a pressure gage connected to the steam pipe just upstream of the sampling tube. The calorimeter is designed so that the flow areas where temperature and pressure are measured approximate each other, thus eliminating the kinetic energy (velocity) terms in the general energy equation. A rule of thumb on the use of these devices is that the degree of superheat at the final state should be a minimum of 6°C.

Example Problem 8-1

A throttling calorimeter installed in a steam pipe has a temperature reading of 110°C and a pressure of 0.1 MPa. If the inlet pressure is 1.0 MPa, determine the moisture in the steam flowing through the pipe. (See Figure 8-8, page 209.)

Equation

$$x_1 = \frac{h_1 - h_{f_1}}{h_{v_1} - h_{f_1}} \qquad [—] \qquad \text{(Eq. 2-26x)}$$

Parameters

$$V_1 = V_2 \qquad \text{(given)}$$

$$h_1 = h_2 \qquad [\text{kJ/kg}] = \text{process definition}$$

where h_2 is obtained from the superheated steam tables, Appendix A, Table A-3.3.

$$T_2 = 100°\text{C} \qquad \text{(given)}$$

$$p_2 = 0.1 \text{ MPa} \qquad \text{(given)}$$

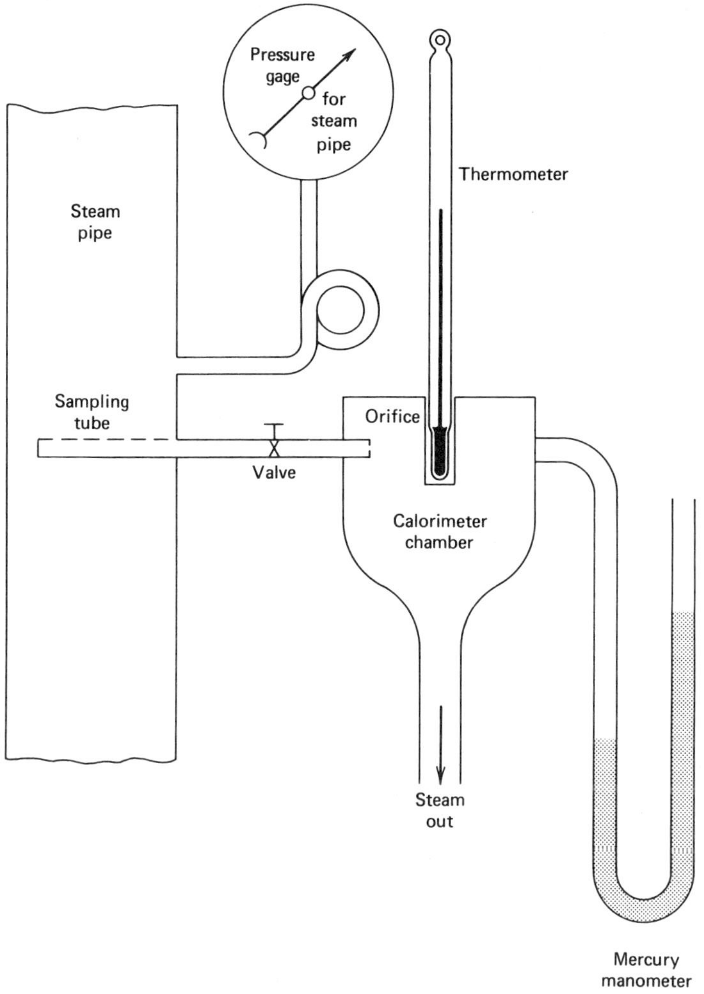

Figure 8-7 Schematic diagram of a throttling calorimeter.

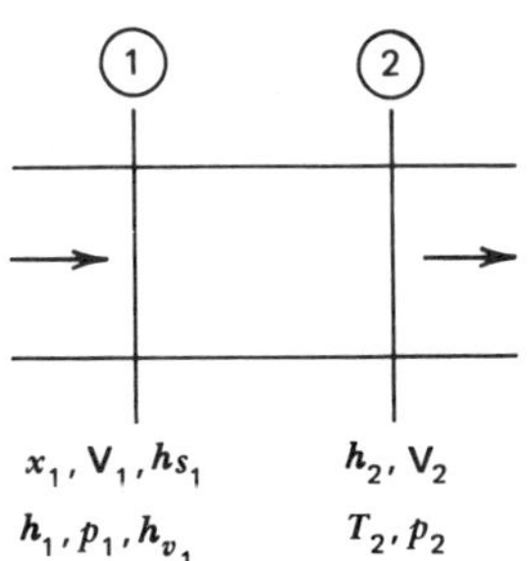

Figure 8-8 Sketch for Example Problem 8-1.

Since the steam table gives values for 100 and 150°C, the value for h_2 (at 110°C) is obtained by interpolation.

$$h_2 = h_{100} + \left[\frac{110-100}{150-100}\right][h_{150} - h_{100}] \qquad [\text{kJ/kg}]$$

$$h_{100} = 2676.2 \text{ kJ/kg}$$

$$h_{150} = 2776.4 \text{ kJ/kg}$$

$$h_2 = 2676.2 + \frac{10}{50}(2776.4 - 2676.2) \qquad [\text{kJ/kg}] + \left[\frac{°\text{C}}{°\text{C}}\right][\text{kJ/kg}]$$

$$= 2672.2 + (0.2)(100.2)$$

$$= 2696.2 \text{ kJ/kg}$$

Therefore,

$$h_1 = 2696.2 \text{ kJ/kg}$$

The values of h_{f_1} and h_{g_1} are obtained.

From the saturated steam tables, Appendix A, Table A-3.2, we find that

$$p_1 = 1.0 \text{ MPa} \qquad (\text{given})$$

$$h_{f_1} = 762.8 \text{ kJ/kg}$$

$$h_{v_1} = 2778.2 \text{ kJ/kg}$$

Substitution

$$x_1 = \frac{2696.2 - 762.8}{2778.1 - 762.8} \qquad \left[\frac{\text{kJ/kg} - \text{kJ/kg}}{\text{kJ/kg} - \text{kJ/kg}}\right] = [—]$$

$$= \frac{1933.4}{2015.3}$$

Answer

$$\boldsymbol{x_1 = 0.96} \; [—]$$

The Mollier diagram (Figure 8-4) can also be used for constant enthalpy expansions. The intersection of the constant pressure and the constant superheat lines for the expanded state determines the value of the enthalpy h_2. Moving to the left along the constant enthalpy line (at the value determined) to its intersection with the constant pressure line of the initial state pressure determines the percent moisture (or quality) of the initial state.

It can be seen from the Mollier diagram that the throttling calorimeter can only be used for evaluating quality close to 1.00, even when isenthalpic expansions are made to vacuum conditions.

8-6 PROBLEMS

8-1. Draw and describe the typical T–s diagram of a thermodynamic substance indicating the saturation line, the critical temperature, the two-phase region, the single phase vapor region, and the single-phase liquid region.

8-2. Draw and describe the Mollier diagram.

8-3. A steam turbine receives steam at 3 MPa and 500°C and expands the steam isentropically to an exhaust pressure of 20 kPa. Calculate the specific work produced (wk/kg) and the exhaust steam quality by (a) equations, and (b) the Mollier diagram. The results should be the same. Explain any difference in your answers.

8-4. If in Problem 8-3 the inlet steam temperature were increased to 600°C, what would be the exhaust steam quality and the specific work produced by the expansion?

8-5. A throttling calorimeter installed on a steam pipe indicates a pressure of 200 kPa and a temperature of 135°C in the calorimeter. What is the quality of the steam flowing in the steam pipe if the steam pipe pressure is 0.85 MPa? Solve by (a) calculations and (b) using the Mollier chart.

8-6. If the steam pipe in Problem 8-5 were 3 MPa, what would be the quality of the steam in the pipe?

8-7. In a refrigeration system saturated liquid Freon-12 at 70°C passes through an expansion valve to a pressure of 219 kPa. What is the quality of the Freon-12 leaving the expansion valve?

8-8. If the saturated liquid Freon-12 in Problem 8-7 were passed through a heat exchange (with no pressure drop) and cooled to 25°C before passing through the expansion valve to 219 kPa pressure, what is the quality of the Freon-12 leaving the expansion valve?

8-9. Assuming that the Freon-12 leaves the refrigerator evaporator as saturated vapor at -10°C, what increase in the refrigerating effect per 1 kg of Freon-12 is realized between the conditions of Problems 8-7 and 8-8?

8-3 - $wk_0 = h_1 - h_2$ $\quad (x) = \frac{s - s_f}{s_{fg}}$ $\quad h_2 = [x(h_{fg})] + h_f$

8-5 - " "

10-3 - $P_2 = P_1 \left[\frac{T_2}{T_1}\right]^{\gamma/\gamma-1}$; $\dot{m} = \frac{C_r}{Q_r}\left(\frac{5\ kW}{h_2 - h_3}\right)$

$Power_{in} = \frac{\gamma R \dot{m}}{1-\gamma}(T_2 - T_1)$ $\quad$ C.O.P. $= \frac{Q_{REJ}}{Power}$ OR $\frac{h_2 - h_3}{h_1 - h_2}$

10-10 - $\dot{m} = \frac{\dot{Q}}{h_1 - h_4}$ $\quad Power_{req.} = \frac{\gamma \dot{m} R}{1-\gamma}(T_2 - T_1)$ $\quad \dot{V} = \dot{m} v$ $\quad v$ @ -10°C

10-17 - $T_2 = T_1 \left[\frac{P_2}{P_1}\right]^{\gamma-1/\gamma}$ $\quad q_{in} = C_p \Delta T$

10-20 - $Q_0 = M_0 C_{p_0} \Delta T_0$ $\quad M_{H_2O} = \frac{Q_0}{\Delta h}$; $\Delta h = h_2 - h_1$ $\quad h_2$ INT

CHAPTER 9

POWER-PRODUCING HEAT ENGINE CYCLES

Power (or work) is obtained from heat by a thermodynamic cycle. The Carnot cycle is theoretically the most efficient thermodynamic cycle operating between two given temperature levels. The Carnot cycle has not yet been successfully implemented; however, the regenerative Stirling and Ericsson cycles approach the efficiency of the Carnot cycle. This chapter describes various power-producing cycles, their component processes, and the parameters affecting the cycle efficiency. The power-producing heat engine cycles discussed in the following sections are

- The "ideal" Carnot cycle.
- The steam powerplant (Rankine cycle).
- The gasoline engine (Otto cycle).
- The diesel engine (Diesel cycle).
- The gas turbine engine (Brayton cycle).
- Other cycles: Stirling and Ericsson.

9-1 THE IDEAL CARNOT CYCLE

The Carnot cycle, discussed in Section 5-4, is the most efficient theoretical cycle operating between two given temperature levels. Consequently, it is often used as a standard of comparison against which the performance of heat engine cycles is measured.

The Carnot cycle is a reversible four-process cycle as indicated by the thermodynamic diagrams in Figure 9-1. The work output of the Carnot cycle is the area 1–2–3–4–1 in this figure.

$$wk = (\Delta T)(\Delta s) \qquad [\text{kJ/kg}]$$

The thermal efficiency of the Carnot cycle is

$$\eta c_{\text{th}} = 1 - \frac{T_L}{T_H} \qquad [-] \qquad \text{(Eq. 5-2)}$$

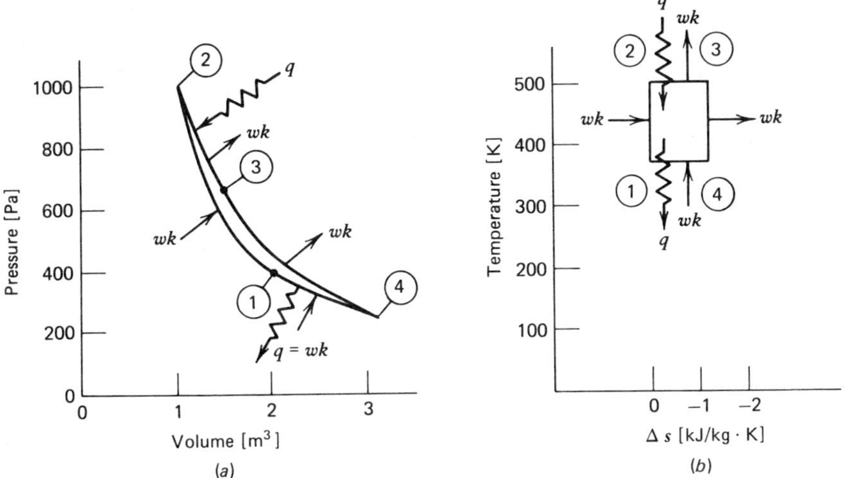

Figure 9-1 Thermodynamic diagrams for the Carnot cycle.

The Carnot cycle comprises two isothermal processes [(1–2) and (3–4)] and two isentropic processes [(2–3) and (4–1)] as illustrated in Figure 9-1. Because the work absorbed and produced in the two isentropic processes is a function of the temperature difference ΔT (see Equations 7-19 and 7-20), the absorbed work and produced work from the isentropic processes are equal, with the result that the net work from the isentropic processes is zero ($Wk_{net} = 0$).

The isothermal process (1–2) absorbs heat equal to the work produced, which is a function of the temperature $T_{1\text{-}2}$ and the pressure ratio p_1/p_2 (see Equation 7-22). Similarly, the heat rejected and the work absorbed by the isothermal process (3–4) is a function of the temperature $T_{3\text{-}4}$ and the pressure ratio p_3/p_4. From the Δs Equation (7-11), if $\Delta s_{1\text{-}2}$ equals $-\Delta s_{3\text{-}4}$, then $p_1/p_2 = p_4/p_3$. It follows that the net work produced from the two isothermal processes is

$$Wk_{net} = MR[(T_{1\text{-}2}) - (T_{3\text{-}4})] \ln \frac{p_1}{p_2} \qquad [\text{kJ}]$$

By rearranging this equation

$$Wk_{net} = [(T_{1\text{-}2}) - (T_{3\text{-}4})] MR \ln \frac{p_1}{p_2}$$

the equation becomes

$$Wk_{net} = (\Delta T)(\Delta s) \qquad [\text{kJ}]$$

The Carnot cycle has not been successfully implemented due to the heat transfer and work input–output requirements of the two isothermal processes. However, the Carnot cycle thermal efficiency has been approached by the Stirling and the Ericsson cycles, which are discussed in Sections 9-4.1 and 9-4.2.

9-2 THE STEAM POWERPLANT (RANKINE CYCLE)

The Rankine cycle is an ideal thermodynamic cycle that approximates the simple steam powerplant cycle. It is of interest not only because of its simplicity, practicability, and historical importance, but also because other more complex cycles may be regarded as modifications of the Rankine cycle. (See Figure 9-2.)

9-2.1 THE SIMPLE STEAM POWERPLANT CYCLE

A block diagram of the typical simple steam powerplant is shown in Figure 9-3.

The flow of the substance (water) around this diagram involves the following steps.

1–2. Pumping the liquid condensate into the boiler.
2–3. Boiling the water to generate steam under pressure.
3–4. Expanding the steam through a turbine.
4–1. Condensing the exhaust steam from the turbine.

Figure 9-2 El Segundo steam powerplant generating station. (Photo courtesy of Southern California Edison Company.)

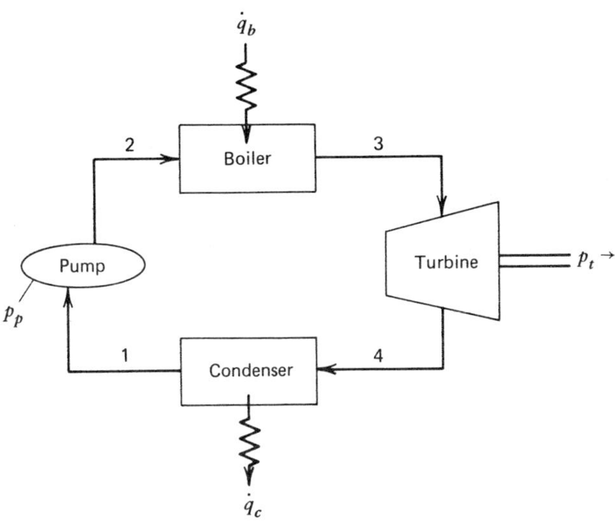

Figure 9-3 Block diagram of a simple steam powerplant.

These steps can be approximated by the open-system thermodynamic processes outlined in Table 9-1 (below). The thermodynamic diagrams connecting these processes to form the Rankine cycle are illustrated in Figure 9-4, page 216.

The pumping process 1–2, assuming a reversible process with an incompressible fluid, is isentropic.

$$\Delta s = \int \frac{du}{T} + \int \frac{p\,dV}{T} \qquad [\text{kJ/kg} \cdot \Delta_1 \text{K}] \qquad (\text{Eq. 5-1})$$

For an incompressible process $dV = 0$. Therefore,

$$\int \frac{p\,dV}{T} = 0$$

Since there is no change in internal energy,

$$du = 0$$

Table 9-1

Cycle Step	Thermodynamic Process Description
1–2	Isentropic compression in pump
2–2′–3	Constant pressure (isobaric) heat addition
3–4	Isentropic expansion in turbine (or other prime mover)
4–1	Constant pressure (isobaric) heat rejection

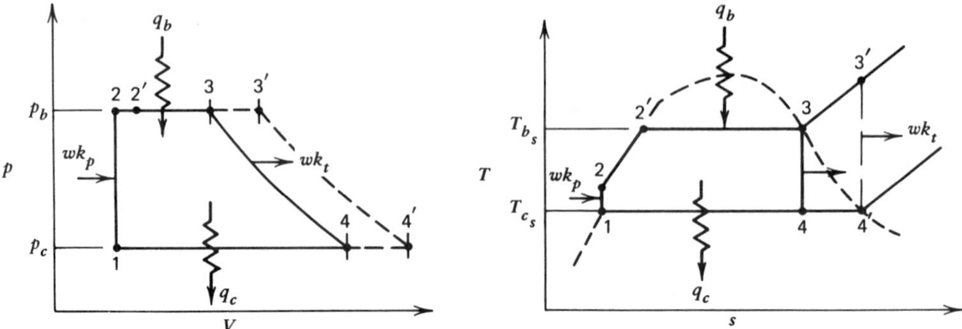

Figure 9-4 Thermodynamic diagrams for the Rankine cycle.

Therefore,

$$\int \frac{du}{T} = 0$$

$$\Delta s = 0 + 0 = 0$$

Hence the process is isentropic.

The work wk_p added to the system is

$$wk_p = -v(p_2 - p_1) \qquad [\text{kJ/kg}] \qquad (\text{Eq. 7-8a})$$

The value of h_2 is the enthalpy of point 1 (h_1) plus the work of the pump.

$$h_2 = h_1 + wk_p \qquad [\text{kJ/kg}] \qquad (\text{Eq. 3-29a})$$

Again the value of h_1 can be obtained from the steam tables for saturated liquid at the condenser pressure p_c.

Next heat is added in the isobaric process 2–3. In the first step (2–2′) the subcooled liquid temperature is increased to saturation temperature T_{bs} at the boiler pressure. Next (2′–3) the latent heat of vaporization is added converting the saturated liquid to the saturate vapor. Following this the saturate vapor may be superheated by adding more heat (3–3′).

The amount of heat added is the amount of the enthalpy increase from point 2 to 3 (or 3′).

$$q_b = (h_3 - h_2) \qquad [\text{kJ/kg}] \qquad (\text{Eq. 3-28a})$$

The value of h_3 can be obtained from the steam tables for the state at 3 (pressure, temperature, and quality or superheat).

Work is produced by the isentropic expansion process (3–4 or 3′–4′). The work produced by this process is

$$wk_t = (h_3 - h_4) \qquad [\text{kJ/kg}] \qquad (\text{Eq. 3-29a})$$

where h_3 has been determined from the steam tables for the state of 3 and h_4 is determined, as discussed in Section 8-3, for a two-phase isentropic expansion to the condenser pressure p_c.

Heat is removed by the isobaric process (4–1). During this process all vapor is converted to saturated liquid at the condenser pressure.

$$q_c=(h_1-h_4) \qquad [\text{kJ/kg}] \qquad (\text{Eq. 3-28a})$$

The value of h_1 was obtained from the steam tables for saturated liquid at the condenser pressure. The value of h_4 was determined from the process (3–4).

The net work output is

$$wk_{\text{net}}=(wk_t-wk_p) \qquad [\text{kJ/kg}]$$

The net power output is

$$p=\dot{M}wk_{\text{net}} \qquad [\text{kW}]$$

where

$$\dot{M}=\text{substance mass flow rate} \qquad [\text{kg/s}]$$

The thermal efficiency of the Rankine cycle can be determined as follows:

$$\eta_{\text{th}}=\frac{wk_{\text{net}}}{q_{\text{input}}}=\left[\frac{wk_t-wk_p}{q_b}\right] \qquad [-] \qquad (9\text{-}1)$$

where

$$wk_t=\text{work output from turbine} \qquad [\text{kJ/kg}]$$
$$wk_p=\text{work input to pump} \qquad [\text{kJ/kg}]$$
$$q_{\text{input}}=\text{heat added in the boiler } (q_b) \qquad [\text{kJ/kg}]$$

Example Problem 9-1

A Rankine powerplant operates on steam at a boiler pressure of 2 MPa and a condenser pressure of 10 kPa. Assuming that the steam enters the turbine as a saturated vapor, calculate the thermal efficiency of this cycle. (See Figure 9-5 and Table 9-2, page 218).

Equation

$$\eta_{\text{th}}=\left[\frac{wk_t-wk_p}{q_b}\right] \qquad [-] \qquad (\text{Eq. 9-1})$$

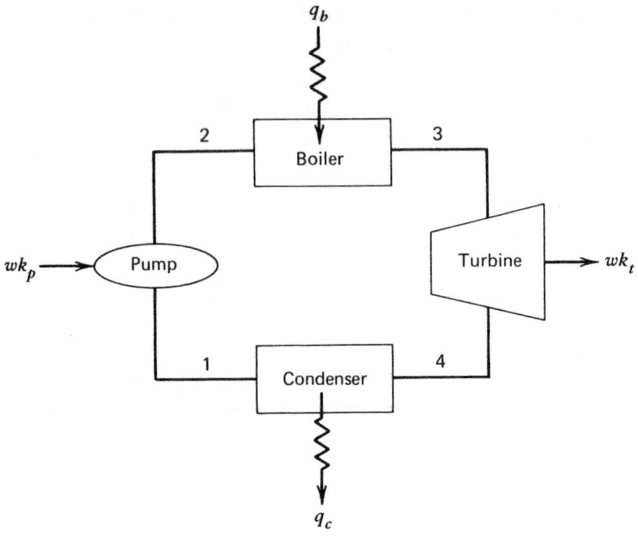

Figure 9-5 Sketch for Example Problem 9-1.

(a) To Obtain the Turbine work wk_t

Equation

$$wk_t = (h_3 - h_4) \qquad \text{[kJ/kg]}$$

Parameters

$$p_3 = 2\ \text{MPa} \qquad \text{(given)}$$

$$\text{saturated vapor} \qquad \text{(given)}$$

$$h_3 = 2799.5\ \text{kJ/kg} \qquad \text{(Appendix A, Table A-3.2)}$$

The quantity h_4 can be obtained from the Mollier diagram for an expansion to 10 kPa or calculated by first determining the exhaust quality x_4 using Equation 2-26 with specific entropies. The equation

$$x_4 = \frac{s_4 - s_{f_4}}{s_{fg_4}} \qquad [-]$$

Table 9-2

Cycle Step	Thermodynamic Process Description
1–2	Isentropic compression
2–3	Isobaric heat input
3–4	Isentropic expansion
4–1	Isobaric heat rejection

is followed by calculating h_4 using Equation 2-25 and specific enthalpies,

$$h_4 = xh_{g_4} + (1-x)h_{f_4} \qquad \text{[kJ/kg]}$$

Calculating x_4, we obtain

$$x_4 = \frac{s_4 - s_{f_4}}{s_{fg_4}} \qquad [—]$$

From the steam tables,

$$p_3 = 2 \text{ MPa} \qquad \text{(given)}$$
$$\text{saturated vapor} \qquad \text{(given)}$$
$$s_3 = 6.3409 \text{ kJ/kg}\cdot\Delta_1\text{K}$$

Because the expansion is isentropic

$$s_4 = s_3$$

Therefore,

$$s_4 = 6.3409 \text{ kJ/kg}\cdot\Delta_1\text{K}$$

Also from the steam tables at

$$p_4 = 10 \text{ kPa} \qquad \text{(given)}$$
$$s_{f_4} = 0.6493 \text{ kJ/kg}\cdot\Delta_1\text{K}$$
$$s_{fg_4} = 7.5009 \text{ kJ/kg}\cdot\Delta_1\text{K}$$

we find that

$$x_4 = \frac{6.3409 - 0.6493}{7.5009} \qquad \frac{[\text{kJ/kg}\cdot\Delta_1\text{K}] - [\text{kJ/kg}\cdot\Delta_1\text{K}]}{[\text{kJ/kg}\cdot\Delta_1\text{K}]} = [—]$$

$$= \frac{5.6916}{7.5009}$$

$$x_4 = 0.7588 \; [—]$$

Then calculating

$$h_4 = h_{f_4} + x_4 h_{fg_4}$$

from the steam tables at

$$p_4 = 10 \text{ kPa} \qquad \text{(given)}$$
$$h_{f_4} = 191.83 \text{ kJ/kg}$$
$$h_{fg_4} = 2392.8 \text{ KJ/kg}$$

we find that

$$x_4 = 0.7588 \text{ [—]} \qquad \text{(calculated above)}$$

$$h_4 = 191.83 + (0.7588)(2392.8) \qquad \text{[kJ/kg]} + \text{[—][kJ/kg]}$$
$$= 191.83 + 1815.66$$
$$= 2007.49 \text{ kJ/kg}$$

Substitution

$$wk_t = 2799.5 - 2007.49 \qquad \text{[kJ/kg]} - \text{[kJ/kg]}$$

Answer

$$\mathbf{wk_t = 792.0 \text{ kJ/kg}}$$

(b) To Obtain the Pump Work wk_p
Equation

$$wk_p = v_{f_1}(p_2 - p_1)$$

Parameters

$$p_1 = 10 \text{ kPa}$$
$$v_{f_1} = 0.00101 \text{ m}^3/\text{kg}$$
$$p_2 = 2 \text{ MPa} \rightarrow 2000 \text{ kPa} \qquad \text{(given)}$$

Substitution

$$wk_p = (0.00101)(2000 - 10) \qquad [\text{m}^3/\text{kg}][\text{kPa}] = \left[\frac{\text{m}^3}{\text{kg}}\right]\left[\frac{\text{kN}}{\text{m}^2}\right] = [\text{kJ/kg}]$$
$$= (0.00101)(1990)$$

Answer

$$\mathbf{wk_p = 2.01 \text{ kJ/kg}}$$

(c) To Obtain the Heat Input to the Boiler q_b
Equation

$$q_b = (h_3 - h_2) \qquad \text{[kJ/kg]}$$

Parameters

$$h_3 = 2799.5 \text{ kJ/kg} \qquad \text{(Appendix A, Table A-3.2)}$$
$$h_2 = h_1 + wk_p \qquad \text{[kJ/kg]}$$
$$p_1 = 10 \text{ kPa} \qquad \text{(given)}$$

for saturated liquid

$$h_1 = 191.83 \text{ kJ/kg} \qquad \text{(Appendix A, Table A-3.1)}$$
$$wk_p = 2.01 \text{ kJ/kg} \qquad \text{from (b) above}$$
$$h_2 = 191.83 + 2.01 \qquad [\text{kJ/kg} + \text{kJ/kg}] = [\text{kJ/kg}]$$
$$h_2 = 193.84 \text{ kJ/kg}$$

Substitution

$$q_b = 2799.5 - 193.84 \qquad [\text{kJ/kg}] - [\text{kJ/kg}] = [\text{kJ/kg}]$$

Answer

$$\mathbf{q_b = 2605.7 \text{ kJ/kg}}$$

Final Substitution

$$\eta_{th} = \frac{792.0 - 2.01}{2605.7} \qquad \frac{[\text{kJ/kg}] - [\text{kJ/kg}]}{[\text{kJ/kg}]} = [—]$$
$$= \frac{790.0}{2605.7}$$

Final Answer

$$\mathbf{\eta_{th} = 0.303 \text{ [—] or } 30.3 \text{ percent}}$$

The thermal efficiency of the Rankine cycle can be increased by decreasing the condenser pressure, increasing the boiler pressure, and by superheating the vapor. The quality of the steam leaving the turbine is increased when superheated steam is supplied at the inlet nozzle, and this is important to the life and performance of steam turbines. For example, in the operation of power station turbines the exhaust steam quality is usually maintained above 88 percent.[9] To accomplish this, a reheat cycle, which is discussed in the following section, is often used.

9-2.2 THE REHEAT RANKINE CYCLE

The objectives of the reheat cycle include:

- Effecting the turbine expansion with high quality or no condensation in the turbine exhaust.
- Increasing the thermal efficiency.

The reheat cycle modifies the simple Rankine cycle by expanding the steam in stages with reheating between the expansion states.

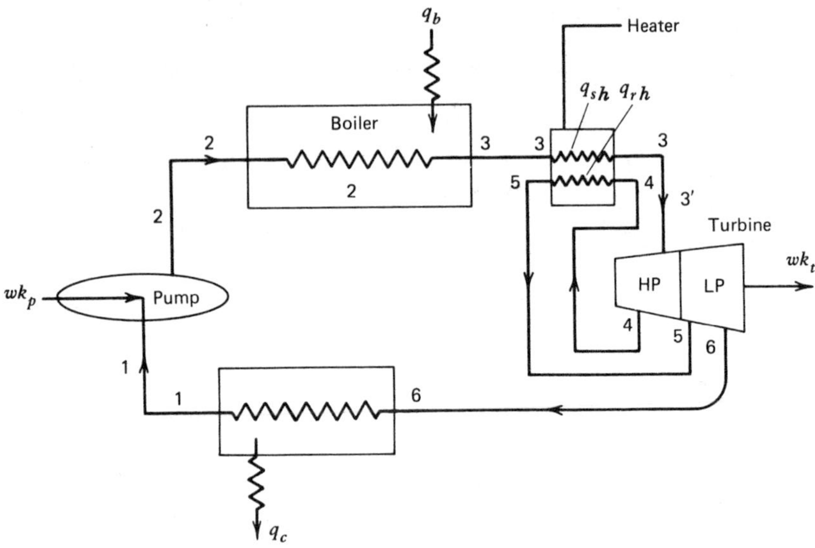

Figure 9-6 Block diagram of a steam powerplant with reheat.

A block diagram of an illustrative two-stage expansion with reheating between the expansion stages is shown in Figure 9-6.

The flow of the substance (H_2O) around this diagram involves the following steps:

1–2. Pumping liquid condensate into the boiler.
2–3. Boiling the water to generate saturate vapor.
3–3′. Superheating the saturate vapor.
3′–4. Expanding the superheated vapor to near-saturate vapor state (high pressure turbine exhaust).
4–5. Reheating high pressure turbine exhaust.
5–6. Expanding the reheated vapor to near-saturate vapor state (low pressure turbine exhaust).
6–1. Condensing the low pressure turbine exhaust steam to saturated liquid.

These steps can be approximated by the following open system thermodynamic process (Table 9-3):

Table 9-3

Cycle Step	Thermodynamic Process Description
1–2	Isentropic compression (liquid) in pump
2–3′	Constant pressure (isobaric) heat addition
3′–4	Isentropic expansion (vapor) in HP turbine
4–5	Constant pressure (isobaric) heat addition
5–6	Isentropic expansion (vapor) in LP turbine
6–1	Constant pressure (isobaric) heat rejection

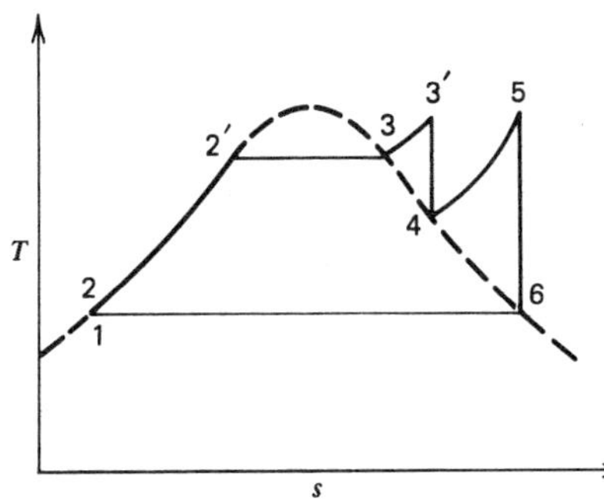

Figure 9-7 *T–s* thermodynamic diagram of a two-stage reheat cycle.

The thermodynamic temperature-specific entropy diagram connecting these processes to form the two-stage Rankine reheat cycle is shown in Figure 9-7.

If the superheating (3–3′) and the reheating (4–5) are accomplished in the same heater, the temperature of 3′ and 5 will probably be approximately the same as shown in Figure 9-7. This results in more work being obtained from the low pressure turbine than from the high pressure turbine. However, the superheat and reheat temperatures, turbine expansion ratios, and so forth, can be tailored to meet the requirements of the cycle design being considered.

The pumping process 1–2 and the heat addition processes 2–3 and 3–3′ are similar to the corresponding processes in the simple Rankine cycle.

The high pressure turbine expansion (3′–4) to near-saturate vapor exhaust state produces work with the same isentropic equations and method of evaluating enthalpy as used for the simple Rankine cycle turbine expansion application.

The high pressure turbine exhaust reheat (4–5) is an isobaric heating process (at a lower pressure), being similar to the superheating process (3–3′).

The low pressure turbine expansion (5–6) to near-saturate vapor exhaust state produces work with the same equations and methods of evaluating enthalpy as used for the high pressure turbine expansion.

The heat removal process (6–1) by the isobaric condensation is similar to the condensation process in the simple Rankine cycle.

The net work output (wk_{net}) of the reheat cycle is

$$wk_{\text{net}} = wk_{\text{HPt}} + wk_{\text{LPt}} - wk_p \qquad [\text{kJ/kg}] \tag{9-2}$$

where

wk_{HPt} = high pressure turbine output work [kJ/kg]

wk_{LPt} = low pressure turbine output work [kJ/kg]

wk_p = pump input work [kJ/kg]

The power output then is

$$p_{net} = \dot{M} wk_{net} \qquad [\text{kW}] \tag{9-3}$$

where

$$\dot{M} = \text{the substance mass flow rate} \qquad [\text{kg/s}]$$

The thermal efficiency of the reheat cycle can be determined by the standard definition:

$$\eta_{th} = \left[\frac{wk_{net}}{q_{input}}\right] = \left[\frac{wk_{HPt} + wk_{LPt} - wk_p}{q_b + q_{sh} + q_{rh}}\right] \quad [-] \tag{9-4}$$

where

wk_{HPt} = high pressure turbine output work [kJ/kg]

wk_{LPt} = low pressure turbine output work [kJ/kg]

wk_p = pump input work [kJ/kg]

q_b = heat input to boiler (to saturate vapor) [kJ/kg]

q_{sh} = heat input to superheating boiler vapor [kJ/kg]

q_{rh} = heat input to reheating high pressure turbine exhaust vapor [kJ/kg]

Example Problem 9-2

A reheat cycle powerplant operates on steam. If superheated steam enters the high pressure turbine at 4 MPa, 450°C, and after expansion to 0.35 MPa is reheated to 450°C and expands in the low pressure turbine to 10 kPa, calculate the thermal efficiency of this cycle. (See Figure 9-6.)

Equation

$$\eta_{th} = \frac{wk_t - |wk_p|}{q_b + q_{sh} + q_{rh}} \qquad [-] \qquad \text{(Eq. 9-4)}$$

(a) To Obtain the Gross Turbine Work *wk*$_t$

Equation

$$wk_t = wk_{HPt} + wk_{LPt} \qquad [\text{kJ/kg}]$$

Parameters

$$wk_{HPt} = (h_{3'} - h_4) \qquad [\text{kJ/kg}]$$

$h_{3'} = 3330.3$ kJ/kg (Appendix A, Table A-3.3 at 4.0 MPa and 450°C)

To determine h_4: $s_4 = s_{3'}$ (Isentropic process)

$s_{3'} = 6.9363$ kJ/kg·Δ_1K (Appendix A, Table A-3.3 at 450°C, 4.0 MPa)

Therefore,

$$s_4 = 6.9363 \text{ kJ/kg}\cdot\Delta_1\text{K}$$

At 0.35 MPa for saturate vapor $s_{g_4} = 6.9405$ kJ/kg·Δ_1K. Since $s_{g_4} > s_4$, state 4 is the quality region

$$x_4 = \frac{s_4 - s_{f_4}}{s_{fg_4}} \quad [\text{—}]$$

$s_4 = 6.9363$ kJ/kg·Δ_1K

$s_{f_4} = 1.7275$ KJ/kg·Δ_1K (Appendix A, Table A-3.3 at 0.35 MPa)

$s_{fg_4} = 5.2130$ kJ/kg·Δ_1K

$$x_4 = \frac{6.9363 - 1.7275}{5.2130} \quad \frac{[\text{kJ/kg}\cdot\Delta_1\text{K} - \text{kJ/kg}\cdot\Delta_1\text{K}]}{[\text{kJ/kg}\cdot\Delta_1\text{K}]} = [\text{—}]$$

$$= \frac{5.2088}{5.2130}$$

$$= 0.999 \ [\text{—}]$$

Round off to

$$x_4 = 1.00 \ [\text{—}]$$

so that

$h_4 - 2732.4$ kJ/kg (Appendix A, Table A-3.3 at 0.35 MPa)

$wk_{\text{HPt}} = 3330.3 - 2732.4$ [kJ/kg − kJ/kg] = [kJ/kg]

$wk_{\text{HPt}} = 597.9$ kJ/kg

$wk_{\text{LPt}} = h_5 - h_6$ [kJ/kg]

$$h_5 = h_{400} + \Delta h$$

Interpolate from Appendix A, Tables A-3.3.
Determine h at 0.30 MPa, 450°C

$$h_{500} - 3487.1 \text{ kJ/kg}$$

$$h_{400} = 3276.6 \text{ kJ/kg}$$

Therefore

$$\Delta h = 210.5$$

$$\tfrac{50}{100}\Delta h = \tfrac{50}{100}(210.5) = 105.3 \text{ kJ/kg}$$

$$\begin{aligned} h_{450} &= h_{400} + \tfrac{50}{100}\Delta h \\ &= (3276.6 + 105.3) \\ &= 3381.9 \text{ kJ/kg} \end{aligned}$$

Determine h at 0.40 MPa, 450°C.

$$h_{500} = 3484.9 \text{ kJ/kg}$$

$$h_{400} = 3273.4 \text{ kJ/kg}$$

Therefore,

$$\Delta h = 211.5 \text{ kJ/kg}$$

$$\tfrac{50}{100}\Delta h = \tfrac{50}{100}(211.5) = 105.8 \text{ kJ/kg}$$

$$\begin{aligned} h_{450} &= h_{400} + \tfrac{50}{100}\Delta h \\ &= (3273.4 + 105.8) \text{ kJ/kg} \\ &= 3379.2 \text{ kJ/kg} \end{aligned}$$

Determine h_5 at 0.35 MPa, 450°C.

$$\begin{aligned} h_5 &= \frac{h_{0.30\ \text{MPa}} + h_{0.40\ \text{MPa}}}{2} \\ &= \frac{3381.9 + 3379.2}{2} = \frac{6761.1}{2} \\ &= 3380.5 \text{ kJ/kg} \end{aligned}$$

Determine h_6 at $p = 10$ kPa (given).

$$s_6 = s_5 \qquad \text{(isentropic process)}$$

To evaluate s_5, follow the similar interpolation procedures used to obtain h_5.
Determine s at 0.30 MPa, 450°C

$$s_{500} = 8.3251$$

$$s_{400} = 8.0330$$

Therefore,

$$\Delta s = 0.2921 \text{ kJ/kg} \cdot \Delta_1\text{K}$$

$$\tfrac{50}{100}\Delta s = \tfrac{50}{100}(0.2921) = 0.1461$$

$$\begin{aligned} s_{450} &= s_{400} + \tfrac{50}{100}\Delta s \\ &= (8.0330 + 0.1461) \\ &= 8.1791 \text{ kJ/kg} \cdot \Delta_1\text{K} \end{aligned}$$

Determine s at 0.40 MPa, 450°C.

$$s_{500}=8.1913$$

$$s_{400}=7.8985$$

Therefore,

$$\Delta s=0.2928\ \text{kJ/kg}\cdot\Delta_1 K$$

$$\tfrac{50}{100}\Delta s=\tfrac{50}{100}(0.2928)=0.1464\ \text{kJ/kg}\cdot\Delta_1\text{K}$$

$$\begin{aligned} s_{450}&=s_{400}+\tfrac{50}{100}\Delta s \\ &=(7.8985+0.1464) \\ &=8.0449\ \text{kJ/kg}\cdot\Delta_1\text{K} \end{aligned}$$

Determine s_5 at 0.35 MPa, 450°C.

$$\begin{aligned} s_5&=\frac{s_{0.30\ \text{MPa}}+s_{0.40\ \text{MPa}}}{2} \\ &=\frac{8.1791+8.0449}{2} \qquad [\text{kJ/kg}\cdot\Delta_1\text{K}]+[\text{kJ/kg}\cdot\Delta_1\text{K}] \\ &=[\text{kJ/kg}\cdot\Delta_1\text{K}] \\ &=\frac{16.2240}{2} \\ &=8.1120\ \text{kJ/kg}\cdot\Delta_1\text{K} \\ s_{g_6}&=8.1502\ \text{kJ/kg}\cdot\Delta_1\text{K} \qquad \text{(Appendix A, Table A-3.2, at 10 kPa)} \end{aligned}$$

Since $s_{g_6}>s_5$, state 6 is in the quality region.

$$x_6=\frac{s_6-s_{f_6}}{s_{fg_6}}\ [\text{—}]$$

$$\begin{aligned} s_6&=s_5=8.1120\ \text{kJ/kg}\cdot\Delta_1\text{K} \\ s_{f_6}&=0.6493\ \text{kJ/kg}\cdot\Delta_1\text{K} \qquad \text{(Appendix A, Table A-3.2, at 10 kPa)} \\ s_{fg_6}&=7.5009\ \text{kJ/kg}\cdot\Delta_1\text{K} \end{aligned}$$

$$\begin{aligned} x_6&=\frac{8.1120-0.6493}{7.5009} \qquad \frac{[\text{kJ/kg}\cdot\Delta_1\text{K}-\text{kJ/kg}\cdot\Delta_1\text{K}]}{[\text{kJ/kg}\cdot\Delta_1\text{K}]}=[\text{—}] \\ &=\frac{7.4627}{7.5009} \\ &=0.995\ [\text{—}] \end{aligned}$$

Round off to $x_6=1.00$ [—]

(Appendix A, Table A-3.2, at 10 kPa)

Therefore,

$$x_6 = 2584.7 \text{ kJ/kg}$$

$$wk_{\text{LPt}} = 3380.5 - 2584.7 \qquad [\text{kJ/kg} - \text{kJ/kg}] = [\text{kJ/kg}]$$
$$= 795.8 \text{ kJ/kg}$$

Substitution

$$wk_t = 597.9 + 795.8 \qquad [\text{kJ/kg} + \text{kJ/kg}] = [\text{kJ/kg}]$$

Answer

$$\boldsymbol{wk_t = 1393.7 \textbf{ kJ/kg}}$$

(b) To Obtain the Pump Work wk_p

Equation

$$wk_p = v_{f_1}(p_1 - p_2) \qquad [\text{kJ/kg}]$$

Parameters

$$v_{f_1} = 0.001079 \text{ m}^3/\text{kg}$$
$$p_1 = 350 \text{ kPa} \qquad \text{(given)}$$
$$p_2 = 4000 \text{ kPa} \qquad \text{(given)}$$

Substitution

$$wk_p = (0.001079)(350 - 4000) \qquad [\text{m}^3/\text{kg}][\text{kPa} - \text{kPa}] = [\text{kJ/kg}]$$
$$= (0.001079)(-3650) \qquad [\text{kJ/kg}]$$
$$= -3.49 \text{ kJ/kg}$$

Answer

$$\boldsymbol{|wk_p| = 3.94 \textbf{ kJ/kg}}$$

(c) To Obtain the Heat Input to the Boiler q_b

Equation

$$q_b = (h_3 - h_2) \qquad [\text{kJ/kg}]$$

Parameters

$$\left.\begin{aligned} h_3 &= 2801.4 \text{ kJ/kg} \\ h_2 &= h_1 + |wk_p| \qquad [\text{kJ/kg}] \end{aligned}\right\} \text{(Appendix A, Table A-3.2, at 4.0 MPa)}$$

$$h_1 = 584.3 \text{ kJ/kg} \qquad \text{(Appendix A, Table A-3.2, at 0.35 MPa)}$$

$$h_2 = 584.3 + 3.9 \qquad [\text{kJ/kg}] + [\text{kJ/kg}] = [\text{kJ/kg}]$$
$$= 588.2 \text{ kJ/kg}$$

Substitution

$$q_b = 2801.4 - 588.2 \qquad [\text{kJ/kg}] - [\text{kJ/kg}] = [\text{kJ/kg}]$$

Answer

$$\mathbf{q_b = 2213.2\ kJ/kg}$$

(d) To Obtain the Heat Input to the Superheater q_{sh}

Equation

$$q_{sh} = (h_{3'} - h_3) \qquad [\text{kJ/kg}]$$

Parameter

$$h_{3'} = 3330.3\ \text{kJ/kg} \qquad \text{(Appendix A, Table A-3.3, at 4.0 MPa and 450°C)}$$

Substitution

$$q_{sh} = 3330.3 - 2801.4 \qquad [\text{kJ/kg} - \text{kJ/kg}] = [\text{kJ/kg}]$$

Answer

$$\mathbf{q_{sh} = 528.9\ kJ/kg}$$

(e) To Obtain the Heat Input to the Reheater q_{rh}

Equation

$$q_{rh} = (h_5 - h_4) \qquad [\text{kJ/kg}]$$

Parameters

$$h_5 = 3380.5\ \text{kJ/kg} \qquad \text{[See calculation (a)]}$$

$$h_4 = 2732.4\ \text{kJ/kg} \qquad \text{[See calculation (a)]}$$

Substitution

$$q_{rh} = 3380.5 - 2732.4 \qquad [\text{kJ/kg} - \text{kJ/kg}] = [\text{kJ/kg}]$$

Answer

$$\mathbf{q_{rh} = 648.1\ kJ/kg}$$

Final Substitution

$$\eta_{\text{th}} = \frac{1393.7 - 3.9}{2213.2 + 528.9 + 648.1} \qquad \frac{[\text{kJ/kg}]}{[\text{kJ/kg}]} = [-]$$

Final Answer

$$\eta_{th} = 0.410\ [—]\ \text{or}\ 41.0\ \text{percent}$$

9-3 INTERNAL COMBUSTION ENGINES

In this book the authors have categorized power-producing thermodynamic cycles as *internal combustion engines* when the thermodynamic cycles are open and the heat input is accomplished by heat release through combustion within the thermodynamic substance. The heat rejection, required by the thermodynamic cycle, is simply accomplished by the discharge of the thermodynamic substance from the system at the heat rejection temperature level.

Because of this mechanism of heat input and rejection, internal combustion engines can produce more power in a given size than engine cycles that require the input heat and rejected heat to be transferred through a surface, such as that of a boiler, superheater, condenser, and so forth. Also, internal combustion engines can be more responsive to changes in operational power output requirements because there is no inertia in the cycle heat input or rejection.

The three major internal combustion cycles are

- The Otto cycle (the gasoline engine)—spark ignition.
- The Diesel cycle (the diesel engine)—compression ignition.
- The Brayton cycle (the gas turbine).

These three cycles are discussed in the following sections. The common analysis of these cycles is to treat them as a series of air processes with the combustion treated as a simple heat addition. This method of analysis is called the air standard power cycle and uses the following assumptions:

1. The working fluid is an ideal gas (air).
2. The specific heat values are constant.
3. Energy addition or removal only is required to achieve the desired state changes (closed system).
4. Each process is reversible. Friction, pressure losses, turbulence, and so forth, is neglected.

Although the results of such an ideal cycle are artificial, they are useful when making parametric studies or comparisons between different cycles. The analysis of real cycles is very difficult, due to the variable amount of mass related to fuel addition, changes in the working fluid, and variations in specific heat values; it requires the use of thermodynamic data for air–fuel mixtures and combustion products for accurate design calculations.

9-4 THE GASOLINE ENGINE (OTTO CYCLE)

The Otto cycle, named for Nicholas A. Otto (1832–1891) who patented this cycle in 1876, is a spark ignition engine. The cycle can be described as follows:

1. A mass of combustible air–fuel mixture is compressed.
2. Combustion of the air–fuel mixture is initiated by a spark.
3. The combustion-heated gas is expanded.
4. The expanded combustion products are exhausted from the chamber (usually a cylinder) and are replaced by a new mass of combustible air–fuel mixture.

These steps can be approximated by the air standard processes listed in Table 9-4.

The thermodynamic diagrams of these processes making this cycle are shown in Figure 9-8.

The following discussion takes the reader through the complete standard air cycle by discussing each of the four processes in sequence.

Process 1–2 is an adiabatic compression of the air–fuel mixture, requiring a work input into the system. The amount of compression is usually expressed as a "compression ratio" r_c, which is based on volumetric measurements

$$r_c = \frac{V_1}{V_2} \qquad [-] \tag{9-5}$$

where

V_1 = initial gas volume $[\mathrm{m}^3]$

V_2 = compressed gas volume $[\mathrm{m}^3]$

In this process (1–2) the main equations are

$$q_{1\text{-}2} = 0$$

$$T_2 = T_1 (r_c)^{\gamma - 1} \qquad [\mathrm{K}] \tag{9-6}$$

$$p_2 = p_1 (r_c)^{\gamma} \qquad [\mathrm{MPa}] \tag{9-7}$$

$$wk_{1\text{-}2} = \left[\frac{R}{1-\gamma}\right](r_c^{\gamma-1} - 1)T_1 \qquad [\mathrm{kJ/kg}] \tag{9-8}$$

Table 9-4

Cycle Step	Thermodynamic process Description
1–2	Adiabatic compression
2–3	Constant volume (isometric) heat input
3–4	Adiabatic expansion
4–1	Constant volume (isometric) heat rejection

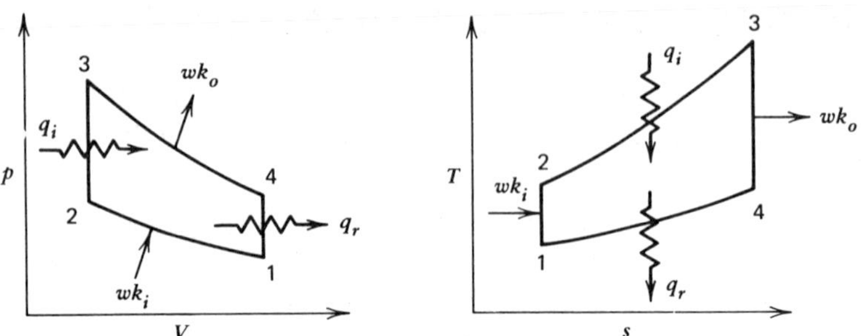

Figure 9-8 Thermodynamic diagrams for the Otto cycle.

For air, Appendix A, Table A-9, lists

$$R = 0.28704 \text{ kJ/kg}\cdot\Delta_1\text{K}$$

$$\gamma = 1.400 \; [-]$$

so that

$$wk_{1\text{-}2} = -0.7176\,(r_c^{0.40} - 1)T_1 \qquad [\text{kJ/kg}] \tag{9-8a}$$

These equations are derived from the adiabatic process equations given in Section 7-4.

Process 2–3 is a constant volume heat addition (spark-ignited combustion) process, which increases the gas temperature with no work involved. The amount of heat added by the combustion will vary according to the heating value of the fuel and the air–fuel mixture ratio. A nominal value often used for the gasoline air standard cycle is

$$q_i = 1500 \text{ kJ/kg of air}$$

In this process (2–3) the main equations are

$$(T_3 - T_2) = \frac{q_i}{c_v} \qquad [\Delta\text{K}] \tag{9-9}$$

where

q_i = heat added per unit mass of air [kJ/kg]

c_v = specific heat at constant volume [kJ/kg·Δ_1K]

T_2 = air temperature before heat addition [K]

T_3 = air temperature after heat addition [K]

Therefore,

$$p_3 = p_2\left[\frac{T_3}{T_2}\right] \qquad [\text{kPa}] \qquad (\text{Eq. 7-5x})$$

$$wk_{2\text{-}3} = 0 \qquad [\text{kJ/kg}] \qquad (\text{Eq. 7-7})$$

Process 3–4 is an adiabatic expansion process that expands the heated gas, producing a work output from the system. In this process the main equations are

$$T_4 = T_3\left(\frac{1}{r_c}\right)^{\gamma-1} \qquad [\text{K}] \qquad (9\text{-}10)$$

or

$$T_4 = T_3(r_c)^{1-\gamma} \qquad [\text{K}]$$

$$p_4 = p_3\left(\frac{1}{r_c}\right)^{\gamma} \qquad [\text{Pa}] \qquad (9\text{-}11)$$

or

$$p_4 = p_3(r_c)^{-\gamma} \qquad [\text{Pa}]$$

The work output wk_o from the expansion is

$$wk_o = \left(\frac{R}{1-\gamma}\right)(r_c^{1-\gamma} - 1)T_3 \qquad [\text{kJ/kg}] \qquad (9\text{-}12)$$

So, for air using $R = 0.28704$ kJ/kg$\cdot\Delta_1$K and $\gamma = 1.400$ Equation 9-12 becomes

$$wk_o = 0.7176(1 - r_c^{-0.4})T_3 \qquad [\text{kJ/kg}] \qquad (9\text{-}12\text{a})$$

This equation is derived in a similar manner to Equation 9-6.

Process 4–1 is a constant volume heat rejection process, which reduces the gas temperature to T_1 with no work involved. The main equations of this process are

$$q_r = c_v(T_1 - T_4) \qquad [\text{kJ/kg}] \qquad (\text{Eq. 6-10a})$$

$$p_1 = p_4\left[\frac{T_1}{T_4}\right] \qquad [\text{Pa}] \qquad (\text{Eq. 7-5x})$$

$$wk_{4\text{-}1} = 0 \qquad [\text{kJ/kg}] \qquad (\text{Eq. 7-7})$$

Two major overall performance characteristics of interest are

- Thermal efficiency (η_{th}).
- Power output (p_o).

These characteristics are discussed in the following sections.

9-4.1 THERMAL EFFICIENCY

The thermal efficiency η_{th} of the Otto cycle (assuming a constant c_v for air throughout the temperature range) is

$$\eta_{th} = 1 - \frac{1}{(r_c)^{\gamma - 1}} \qquad [-] \tag{9-13}$$

It is interesting to note that the thermal efficiency is not affected by the heat input q_i.

This equation indicates an important characteristic of thermal efficiency for an air standard Otto cycle: It increases with compression ratio. A graph of this relationship is shown in Figure 9-9.

In practice the compression ratio is limited by the air–fuel combustion characteristics of detonation and preignition. Both of these abnormal combustion modes degrade the engine performance and can cause physical damage to the engine. Recent automobile engines (1979 models) generally have compression ratios around 8 : 1. Specific examples of 1979 production models are listed in Table 9-5.

A typical automobile gasoline engine is shown in Figure 9-10.

When the air–fuel mixture detonates, the flame front moves through the combustion chamber at a supersonic velocity rather than progress smoothly with a subsonic velocity from the spark ignition source. Extremely high, instantaneous local pressures can damage metal surfaces exposed to such shock fronts. These pressures produce audible "knocks."

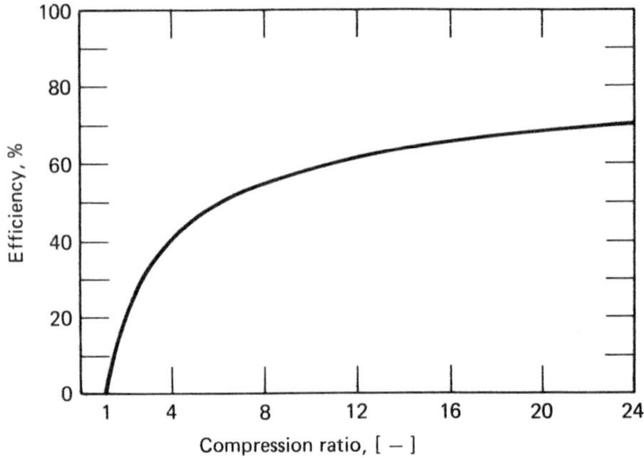

Figure 9-9 Characteristic of thermal efficiency with a compression ratio.

Table 9-5 Typical Automobile Engine Compression Ratios—1979

Manufacturer	Displacement	Number of Cylinders	Compression Ratio
Dodge, Plymouth	225 cu in.	6	8.4:1
Dodge, Plymouth	318 cu in.	8	8.5:1
Dodge, Plymouth	360 cu in.[a]	8	8.0:1
Chevrolet, Standard		8	8.3:1
Ford, Standard		8	8.1:1

[a]High performance model (manufacturer's data).

As the compression ratio increases, increasing the temperature of the compressed air–fuel mixture, hot spots in the engine may act as ignition sources initiating the combustion before the compression process is completed. This is called preignition, which substantially increases the input work of compression and results in a lower net work output from the cycle.

The equation for the thermal efficiency (Equation 9-13) is derived in the following manner:

$$\eta_{\text{th}} = \left[\frac{q_i - |q_r|}{q_i}\right] = 1 - \frac{|q_r|}{q_i} = 1 - \left[\frac{c_v(T_4 - T_1)}{c_v(T_3 - T_2)}\right]$$

Figure 9-10 Typical automobile gasoline engine. (Photo courtesy of Ford Motor Company).

Thus

$$\eta_{th} = 1 - \left[\frac{T_1(T_4/T_1 - 1)}{T_2(T_3/T_2 - 1)} \right]$$

but from isentropic relationships

$$\frac{T_2}{T_1} = \left(\frac{V_1}{V_2} \right)^{\gamma - 1} = \left(\frac{V_4}{V_3} \right)^{\gamma - 1} = \frac{T_3}{T_4}$$

Therefore

$$\frac{T_3}{T_2} = \frac{T_4}{T_1}$$

and

$$\eta_{th} = 1 - \left(\frac{T_1}{T_2} \right)$$
$$= 1 - (r_c)^{1-\gamma}$$

or

$$\eta_{th} = 1 - \frac{1}{(r_c)^{\gamma - 1}} \qquad [-] \qquad \text{(Eq. 9-13)}$$

where

r_c = compression ratio
γ = isentropic exponent

9-4.2 POWER OUTPUT

The power output of an Otto cycle is the product of the net work output (Wk_{net}) per cycle and the number of cycles per second (speed of the engine).

The net work output per cycle can be obtained in several ways including:

- Evaluating the q_{net} from the heat input and heat rejection processes.
- Evaluating the net work from the work of the compression and expansion processes.
- Determining the mean effective pressure (MEP).

Evaluating the net work output from the heat input is calculated by the following method.

$$Wk_{net} = M_c(q_i - q_r) \qquad [kJ] \qquad (Eq. 3\text{-}11x)$$

where

M_c = mass of air in the cycle process [kg]

q_i = heat input [kJ/kg]

q_r = heat rejected [kJ/kg]

Here q_i is a function of the heating values of the fuel. Nominal values of the air standard cycle heat input for several fuels are given in Table 9-6.

The heat input values are in Table 9-6 are modifications of the nominal heat release values of the actual air–fuel mixture ratios so as to give comparable results between the simple air standard analysis (using an average value of 1.0 kJ/kg·Δ_1K for the c_p of air) and the actual combustion products. Although the actual heat release of these air–fuel mixtures is significantly larger than the values shown, the actual specific heats of air and the combustion products at elevated temperatures is also significantly higher. This results in the actual temperatures after combustion being comparable to the temperatures indicated by the simplified air standard analysis using the heat input values listed in Table 9-6.

The heat rejected (q_r) is calculated by the equation

$$q_r = M_c c_v(T_1 - T_4) \qquad [kJ] \qquad (Eq. 6\text{-}10)$$

In order to evaluate q_r, T_4 must be determined. This is done by sequentially calculating T_2 from T_1 and r_c, T_3 from T_2 and q_i; and T_4 from T_3 and r_c as outlined in the earlier discussion of the cycle processes.

Table 9-6 Nominal Values of the Air Standard Cycle Heat Input for Several Fuels

Fuel	Heat Input (q_i) [kJ/kg air]
Alcohol (wood)	690
Butane	1605
Gasoline	1500
Natural gas	1720
Propane	1645

Evaluating the net work from the work of the compression and expansion processes, we use the following closed system equations:

$$Wk_{\text{net}} = Wk_{3-4} - |Wk_{1-2}|$$

$$Wk_{3-4} = M_c\left[\frac{R}{1-\gamma}\right](r_c^{1-\gamma} - 1)T_3 \qquad [\text{kJ}] \qquad (\text{Eq. 9-12x})$$

$$Wk_{1-2} = M_c\left[\frac{R}{1-\gamma}\right](r_c^{\gamma-1} - 1)T_1 \qquad [\text{kJ}] \qquad (\text{Eq. 9-8x})$$

$$Wk_{\text{net}} = \left[\frac{M_c R}{1-\gamma}\right]\left[(r_c^{1-\gamma})T_3 - (r_c^{\gamma-1} - 1)T_1\right] \qquad [\text{kJ}] \qquad (9\text{-}14)$$

so that for the air standard cycle

$$Wk_{\text{net}} = 0.7176 M_c\left[(r_c^{-0.4} - 1)T_3 - (r_c^{0.4} - 1)T_1\right] \qquad [\text{kJ}] \quad (9\text{-}14\text{a})$$

The mass per cycle (M_c) used in both the heat and work equations is determined from the perfect gas laws for the air standard cycle:

$$M_c = \frac{p_1 V_1}{R T_1} \qquad [\text{kg}]$$

The subscripts refer to point 1 on the Otto cycle thermodynamic diagrams in Figure 9-8.

For a particular engine geometry (V_1 = constant) M_c will increase as the inlet (initial) temperature T_1 decreases or the inlet (initial) pressure increases.

To control the work output per cycle, the inlet can be throttled to reduce the mass per cycle M_c by reducing the inlet pressure p_1.

The mean effective pressure (MEP) parameter, often used by authors such as Taylor and Taylor,[(10)] is primarily an experimental parameter that can be related to air standard cycles. The MEP is defined as the characteristic pressure which, when acting on a piston throughout the piston stroke, would produce the amount of work actually developed.

$$\text{MEP} = \frac{Wk_c}{Al} \qquad [\text{kPa}] \qquad (9\text{-}15)$$

where

Wk_c = net work output per cycle [kJ]

A = piston area [m^2]

l = piston stroke length [m]

Then

$$Wk_c = (\text{MEP})Al \qquad [\text{kJ}] \tag{9-16}$$

The MEP of a reciprocating engine can be determined by experimentally measuring the work output per cycle Wk_c and the physical measurements of the engine piston area and stroke.

The power output p_o of the Otto cycle engine is

$$p_o = (Wk_{\text{net}})(N) = NM_c(q_i - q_r) = \dot{M}(q_i - q_r) \qquad [\text{kW}] \tag{9-17}$$

where

$$N = \text{number of cycles per second} \qquad [—]$$

When the Otto cycle engine is used to drive a variable load, in addition to variable throttling, a variable speed ratio gear box is often used to adjust both the engine MEP and speed so as to most effectively meet the requirements of the applied load.

The actual open cycle spark-ignition engine today differs from the air standard Otto cycle in several ways including:

1. A combustion process is used instead of a heat transfer process.
2. The specific heat of products of combustion are not the same as that for air and increase with temperature.
3. The mechanical design involves inlet and exhaust valves, which result in a pressure drop.
4. Heat transferred between the combustion gases and the cylinder walls, so the compression and expansion processes are not adiabatic.
5. There are irreversibilities in temperature and pressure gradients.

These factors should be considered when a precise design or performance analysis of an Otto cycle engine is required.

9-5 THE DIESEL ENGINE (DIESEL CYCLE)

The Diesel cycle (named for Rudolf Diesel, who developed the cycle) is a compression-ignition engine. To increase the engine cycle efficiency, Diesel increased the compression ratio. As this could not be accomplished with an air–fuel mixture because of auto ignition during the compression stroke, air alone was compressed and, subsequent to compression, fuel was injected. The high temperature of the air from compression effected immediate ignition of the fuel as it entered the combustion chamber.

Figure 9-11 General Motors' 1979 4.3-Liter V-8 Diesel engine. (Photo courtesy of the Oldsmobile Division of General Motors.)

The typical compression ratio for 1979 automobile Diesel engines is 23 : 1. A typical automobile Diesel engine is shown in Figure 9-11.

The Diesel cycle can be described as follows:

1. A mass of air is compressed.
2. Fuel is injected into the compressed gas resulting in immediate combustion at a rate and time period controlled by the fuel injection.
3. An expansion of the combustion products occurs.
4. The expanded combustion products from the chamber (usually a cylinder) are replaced by a new mass of air.

These steps can be approximated by the air standard processes listed in Table 9-7.

The thermodynamic diagrams of these processes making this cycle are shown in Figure 9-12.

A description of the Diesel cycle can be given by defining the four basic thermodynamic processes.

1. Process 1–2 is an adiabatic compression process that compresses the air, requiring a work input into the system. As with the Otto cycle, the

Table 9-7

Cycle Step	Thermodynamic Process Description
1-2	Adiabatic compression
2-3	Constant pressure (isobaric) heat addition
3-4	Adiabatic expansion
4-1	Constant volume (isometric) heat rejection

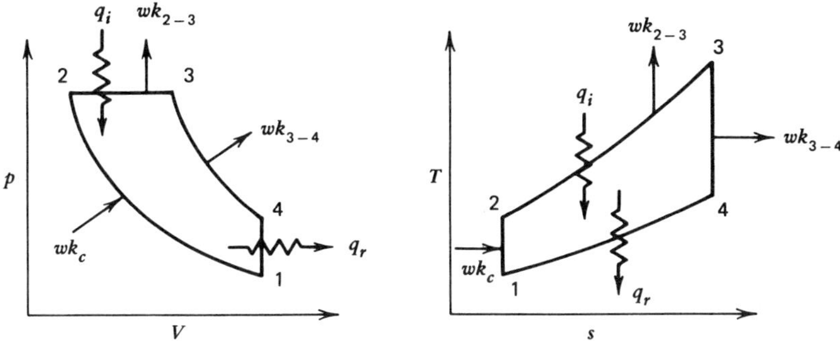

Figure 9-12 Thermodynamic diagrams for the Diesel cycle.

amount of compression is usually expressed as a compression ratio r_c, which is based on volumetric measurements.

$$r_c = \frac{V_1}{V_2} \qquad [-] \qquad \text{(Eq. 9-5)}$$

As with the air standard Otto cycle, the main equations are

$$q_{1-2} = 0$$

$$T_2 = T_1 (r_c)^{\gamma-1} \qquad [\text{K}] \qquad \text{(Eq. 9-6)}$$

$$p_2 = p_1 (r_c)^{\gamma} \qquad [\text{MPa}] \qquad \text{(Eq. 9-7)}$$

$$wk_{1-2} = \frac{R}{1-\gamma}(r_c^{\gamma-1} - 1)T_1 \qquad [\text{kJ/kg}] \qquad \text{(Eq. 9-8)}$$

so for air

$$wk_{1-2} = -0.7176(r_c^{0.4} - 1)T_1 \qquad [\text{kJ/kg}] \qquad \text{(Eq. 9-8a)}$$

2. Process 2–3 is an isobaric heat addition (combustion process) that both increases the air temperature and produces work output.

The amount of heat added will vary according to the length of the isobaric process. The ratio of V_3/V_2 is called the cutoff ratio r_{co}. This defines the length of the fuel injection period in the cycle and the amount of heat input to the cycle.

For this isobaric heat input process, the main equations are

$$p_3 = p_2 \qquad [\text{kPa}] \qquad \text{(process definition)}$$

$$\frac{V_3}{V_2} = r_{co} \qquad [-] \qquad (9\text{-}18)$$

$$T_3 = T_2\left[\frac{V_3}{V_2}\right] \qquad [\text{K}] \qquad \text{(Eq. 7-1x)}$$

so that

$$T_3 = T_2(r_{co}) \qquad [\text{K}] \tag{9-19}$$

$$q_{2-3} = c_p(r_{co} - 1)T_2 \qquad [\text{kJ/kg air}] \tag{9-20}$$

$$wk_{2-3} = p_2(v_3 - v_2) \qquad [\text{kJ/kg}] \qquad (\text{Eq. 7-3})$$

3. Process 3–4 is an adiabatic expansion process that expands the combustion products producing a work output from the system in a similar manner to the Otto cycle expansion. In this process the main equations are

$$T_4 = T_3\left[\frac{r_{co}}{r_c}\right]^{\gamma-1} \qquad [\text{K}] \tag{9-21}$$

$$p_4 = p_3\left[\frac{r_{co}}{r_c}\right]^{\gamma} \qquad [\text{kPa}] \tag{9-22}$$

$$wk = \frac{R}{1-\gamma}\left[\left(\frac{r_{co}}{r_c}\right)^{\gamma-1} - 1\right]T_3 \qquad [\text{kJ/kg}] \tag{9-23}$$

These equations are derived as follows:

Temperature

$$T_4 = T_3\left[\frac{p_4}{p_3}\right]^{\gamma-1/\gamma} \qquad [\text{K}] \qquad (\text{Eq. 7-14x})$$

$$\left[\frac{p_4}{p_3}\right]^{1/\gamma} = \left[\frac{V_3}{V_4}\right] \qquad [-] \qquad (\text{Eq. 7-15x})$$

Therefore,

$$T_4 = T_3\left[\frac{V_3}{V_4}\right]^{\gamma-1}$$

$$V_3 = V_2 r_{co} \qquad (\text{by definition})$$

$$V_4 = V_2 r_c \qquad (\text{by definition})$$

and

$$T_4 = T_3\left[\frac{r_{co}}{r_c}\right]^{\gamma-1} \qquad [\text{K}] \qquad (\text{Eq. 9-21})$$

Pressure

$$\left[\frac{p_4}{p_3}\right]^{1/\gamma}=\left[\frac{V_3}{V_4}\right]=\left[\frac{V_2 r_{co}}{V_2 r_c}\right]=\left[\frac{r_{co}}{r_c}\right]$$

$$\left[\frac{p_4}{p_3}\right]=\left[\frac{r_{co}}{r_c}\right]^{\gamma}$$

so that

$$p_4=p_3\left[\frac{r_{co}}{r_c}\right]^{\gamma} \qquad [\text{kPa}] \qquad (\text{Eq. 9-22})$$

Work

$$wk=\frac{R}{1-\gamma}[T_4-T_3] \qquad [\text{kJ/kg}] \qquad (\text{Eq. 7-18a})$$

$$T_4=T_3\left[\frac{r_{co}}{r_c}\right]^{\gamma-1} \qquad [\text{K}] \qquad (\text{Eq. 9-21})$$

$$=\frac{R}{1-\gamma}\left[T_3\left(\frac{r_{co}}{r_c}\right)^{\gamma-1}-T_3\right]$$

Therefore,

$$wk=\frac{R}{1-\gamma}\left[\left(\frac{r_{co}}{r_c}\right)^{\gamma-1}-1\right]T_3 \qquad [\text{kJ/kg}] \qquad (\text{Eq. 9-23})$$

4. Process 4–1 is a constant volume heat rejection process which, as with the Otto air standard cycle, reduces the gas temperature to T_1 with no work involved. The main equations of this process are

$$q_r=c_v(T_1-T_4) \qquad [\text{kJ/kg}] \qquad (\text{Eq. 6-10a})$$

$$p_1=p_4\left[\frac{T_1}{T_4}\right] \qquad [\text{Pa}] \qquad (\text{Eq. 7-5x})$$

$$wk=0 \qquad [\text{kJ/kg}] \qquad (\text{Eq. 7-7})$$

The two major overall performance characteristics of thermal efficiency and power output are discussed in the following sections.

9-5.1 THERMAL EFFICIENCY

The equation for the thermal efficiency is

$$\eta_{\text{th}} = 1 - \left(\frac{1}{r_c}\right)^{\gamma-1}\left[\frac{(r_{co}^{\gamma}-1)}{\gamma(r_{co}-1)}\right] \quad [—] \qquad (9\text{-}24)$$

This equation can be derived from

$$\eta_{\text{th}} = \frac{q_i - |q_r|}{q_i} = 1 - \frac{|q_r|}{q_i}$$

but

$$q_r = c_v(T_4 - T_1)$$

and

$$q_i = c_p(T_3 - T_2)$$

Evaluating the temperatures in terms of T_1, we

$$T_2 = T_1(r_c)^{\gamma-1}$$

$$T_3 = T_2(r_{co}) = T_1(r_{co})(r_c)^{\gamma-1}$$

$$T_4 = T_3\left[\frac{r_{co}}{r_c}\right]^{\gamma-1} = T_1\left[\frac{r_{co}}{r_c}\right]^{\gamma-1}(r_{co})(r_c)^{\gamma-1} = T_1(r_{co})^{\gamma}$$

$$\eta_{\text{th}} = 1 - \left[\frac{c_v[T_1(r_{co}) - T_1]}{c_p[T_1(r_{co})(r_c)^{\gamma-1} - T_1(r_c)^{\gamma-1}]}\right]$$

$$= 1 - \left[\frac{1}{\gamma}\,\frac{(r_{co}^{\gamma}-1)}{(r_{co}-1)(r_c)^{\gamma-1}}\right]$$

$$= 1 - \left(\frac{1}{r_c}\right)^{\gamma-1}\left[\frac{r_{co}^{\gamma}-1}{\gamma(r_{co}-1)}\right] \quad [—] \qquad (\text{Eq. } 9\text{-}24)$$

This equation indicates that the thermal efficiency increases with an increased compression ratio r_c and decreases with an increased fuel cutoff ratio r_{co}.

9-5.2 POWER OUTPUT

The power output of the Diesel cycle engine is the product of the net work output wk_{net} per cycle and the number of cycles per second (speed of the engine).

The net work output per cycle can be obtained in the same ways as for the Otto cycle, namely:

- Evaluating the q_{net} from the heat input and heat rejection processes.
- Evaluating the net work from the work of the compression and expansion processes.
- Determining the MEP.

The net work output is calculated by the following equations:

$$Wk_{net} = Mc(q_i - q_r) \qquad [\text{kJ}] \qquad (\text{Eq. 3-11x})$$

where

M_c = mass of air in the cycle process [kg]
q_i = heat input [kJ/kg]
q_r = heat rejected [kJ/kg]

Then

$$q_i = q_{2-3} = c_p(T_3 - T_2)$$

$$T_3 = T_1(r_{co})(r_c)^{\gamma-1}$$

$$T_2 = T_1(r_c)^{\gamma-1}$$

$$q_1 = c_p r_c^{\gamma-1}(r_{co} - 1)T_1$$

$$q_r = q_{4-1} = c_v(T_4 - T_1)$$

$$T_4 = T_1(r_{co})^{\gamma}$$

$$q_r = c_v(r_{co}^{\gamma} - 1)T_1$$

$$Wk_{net} = M_c([c_p r_c^{\gamma-1}(r_{co} - 1)] - [c_v(r_{co}^{\gamma} - 1)])T_i$$

as $M_c = \dfrac{P_1 V_1}{RT_1}$

$$Wk_{net} = \frac{p_1 V_1}{R}\left[c_p r_c^{\gamma-1}(r_{co} - 1) - c_v(r_{co}^{\gamma} - 1)\right] \qquad [\text{kJ}] \qquad (9\text{-}25)$$

This equation indicates that for a constant initial cycle volume V_1, the Wk_{net} per cycle increases with

- Initial pressure (p_1) increase.
- Compression ratio (r_c) increase.
- Cutoff ratio (r_{co}) increase.

Consequently, for a fixed engine geometry the work output per cycle can be varied by varying (1) the inlet pressure (throttling), and (2) the cutoff ratio.

The power output p_o of the Diesel cycle engine is the product of the net work per cycle and the engine speed.

9-6 DUAL COMBUSTION OR LIMITED PRESSURE CYCLE

Most of the high speed Diesel engines of today inject fuel slightly in advance of the "top dead center" (end of the compression stroke) resulting in some of the heat being added at constant volume (as in the Otto cycle) and the balance at constant pressure (as in the Diesel cycle). This dual combustion cycle is known as the Sabathé cycle or the *limited pressure* cycle. By placing a limit to the peak pressure from the constant volume combustion, the amount of the total heat added in the constant volume heat addition process is established. The balance of the total heat added per cycle is then added in the following constant pressure heat addition process.

These steps can be approximated by the air standard process in Table 9-8.

The thermodynamic diagrams of these processes making the cycle are shown in Figure 9-13.

A description of the limited pressure cycle can be given by defining the five basic thermodynamic processes.

1. Process 1–2 is an adiabatic compression of the air. As with the Otto and Diesel cycles the compression ratio r_c is based on the volumetric measurements:

$$r_c = \frac{V_1}{V_2} \quad [-] \quad \text{(Eq. 9-5)}$$

2. Process 2–3 is a constant volume, heat addition process as in the Otto cycle. The fuel is ignited by the air temperature as the fuel is injected. In the limited pressure cycle a pressure ratio r_p is introduced and is defined as

$$r_p = \frac{p_3}{p_2} \quad [-] \tag{9-26}$$

Table 9-8

Cycle Step	Thermodynamic Process Description
1–2	Adiabatic compression
2–3	Constant volume heat addition
3–4	Constant pressure heat addition
4–5	Adiabatic expansion
5–1	Constant volume heat rejection

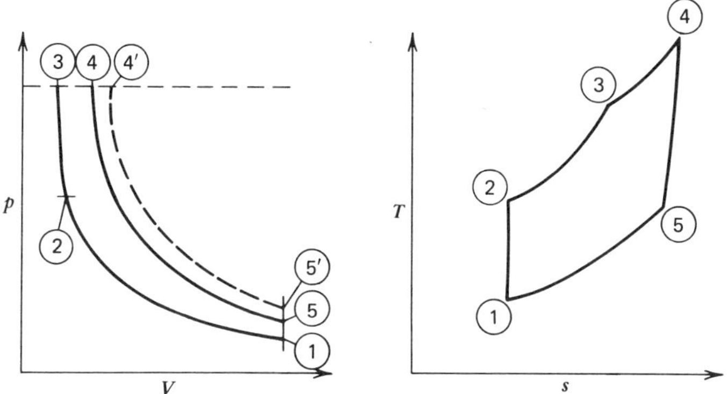

Figure 9-13 Thermodynamic diagrams for the limited pressure cycle.

3. Process 3–4 is a constant pressure, heat addition process as in the Diesel cycle and is described in length by the fuel cutoff ratio (r_{co}), which is defined in the same manner as for the Diesel cycle:

$$r_{co}=\frac{V_4}{V_3} \qquad [—] \qquad \text{(Eq. 9-18)}$$

4. Process 4–5 is an adiabatic expansion process wherein the high pressure hot air is expanded to the initial volume V_1 of the cycle.

5. Process 5–1 is a constant volume, heat rejection process returning the air to the initial starting temperature of the cycle T_1.

The thermal efficiency of the cycle is

$$\eta_{\text{th}}=1-\left(\frac{1}{r_c}\right)^{\gamma-1}\left[\frac{(r_{co}^{\gamma}r_p)-1}{(r_p-1)+\gamma r_p(r_{co}-1)}\right] \qquad \text{(9-27)}$$

When r_{co} (the fuel cutoff ratio) is 1, the cycle is the Otto cycle and Eq. 9-27 becomes Eq. 9-13. When r_p (the pressure ratio) is 1, the cycle is the Diesel cycle and Eq. 9-27 becomes Eq. 9-24.

The net work of the cycle (wk_{net}) is

$$wk_{\text{net}}=q_a-q_r \qquad [\text{kJ/kg}]$$

where

$$q_a=c_v(T_3-T_2)+c_p(T_4-T_3) \qquad [\text{kJ/kg}]$$

$$=c_v(T_5-T_1) \qquad [\text{kJ/kg}]$$

resulting in

$$wk_{\text{net}}=c_v[(T_3-T_2)-(T_5-T_1)]+c_p(T_4-T_3) \qquad [\text{kJ/kg}] \qquad \text{(9-28)}$$

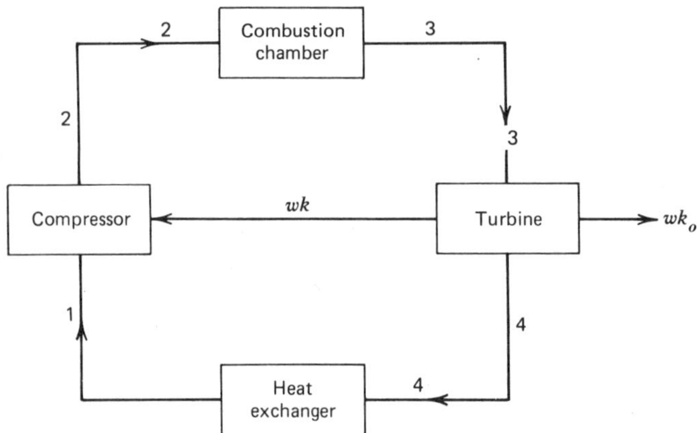

Figure 9-14 Block diagram of the air standard Brayton cycle.

9-7 THE GAS TURBINE (BRAYTON CYCLE)

The gas turbine cycle is characterized by a continuous combustion process, as contrasted with the cyclic combustion process of the Otto and Diesel cycles. The Brayton (or Joule) cycle can be described as follows:

1. A continuous flow of inlet air is compressed.
2. Fuel is injected in a continuous burning process.
3. The combustion products are continuously expanded and exhausted from the system.

These steps can be approximated by the air standard cycle as illustrated in the block diagram in Figure 9-14.

The air standard processes are listed in Table 9-9.

The thermodynamic diagrams of these processes making the cycle are shown in Figure 9-15.

As the gas turbine cycle generally uses turbomachinery (compressor and turbine), a large volume (and mass) of air can be circulated. Because of this, relatively large amounts of power can be produced in a relatively small volume. The combustion temperatures are limited to 925 to 1000°C by the materials and cooling designs of the turbine blades and other parts that are in intimate contact with the combustion products in relatively high

Table 9-9

Cycle Step	Thermodynamic Process Description
1–2	Isentropic compression
2–3	Constant pressure (isobaric) heat addition
3–4	Isentropic expansion
4–1	Constant pressure (isobaric) heat rejection

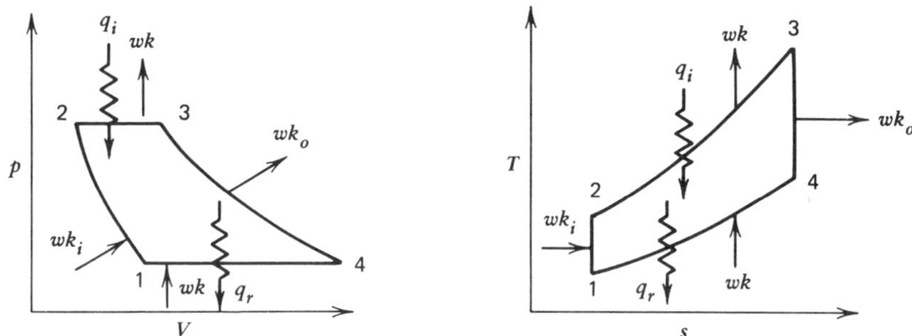

Figure 9-15 Thermodynamic diagrams for the Brayton cycle.

heat transfer rate configurations. A typical gas turbine installation is shown in Figure 9-16.

Process 1–2 is an adiabatic compression that compresses the air with a negligible discharge velocity from the compressor. The compressor pressure ratio r_p is measured in terms of pressure and usually is in the range of 13:1 to 20:1.

$$r_p = \frac{p_2}{p_1} \qquad [-] \tag{9-29}$$

Figure 9-16 Mars 7900 kW Gas Compressor set in Alberta, Canada. (Photo courtesy of Solar Turbines International, an Operating Group of International Harvester.)

The main equations of this process are

$$q_{1-2}=0$$

$$T_2=T_1\left[\frac{p_2}{p_1}\right]^{(\gamma-1)/\gamma}=T_1(r_p)^{(\gamma-1)/\gamma} \qquad [\text{K}] \qquad (9\text{-}30)$$

$$wk_{1-2}=\left[\frac{R}{1-\gamma}\right]\left[r_p^{(\gamma-1)/\gamma}-1\right]T_1 \qquad [\text{kJ/kg}] \qquad (9\text{-}31)$$

For air, then, we obtain

$$wk_{1-2}=0.7176\left[r_p^{0.286}-1\right]T_1 \qquad [\text{kJ/kg}] \qquad (9\text{-}31\text{a})$$

Process 2–3 is an isobaric heat addition (combustion process) that increases the air temperature with work involved. The main equations of this process are

$$p_3=p_2 \qquad \text{(process definition)}$$

The temperature increase is a function of the amount of heat input per kilogram of air. This is an independent variable.

$$\Delta T_{2-3}=\frac{q_{2-3}}{c_p} \qquad [\text{K}] \qquad (\text{Eq. 6-9ax})$$

$$T_3=T_2+\Delta T=T_2+\frac{q_{2-3}}{c_p} \qquad [\text{K}] \qquad (9\text{-}32)$$

$$wk_{2-3}=p_2(v_3-v_2) \qquad [\text{kJ/kg}] \qquad (\text{Eq. 7-3ax})$$

Process 3–4 is an adiabatic expansion process that expands the combustion products producing a work output. A negligible discharge velocity from the expansion is assumed. In this process the main equations are

$$\frac{p_4}{p_3}=\frac{1}{r_p}$$

$$T_4=T_3\left(\frac{1}{r_p}\right)^{(\gamma-1)/\gamma}=T_3(r_p)^{(1-\gamma)/\gamma} \qquad [\text{K}] \qquad (9\text{-}33)$$

$$wk_{3-4}=\frac{R}{1-\gamma}(T_4-T_3)=\left[\frac{R}{1-\gamma}\right]\left(r_p^{(1-\gamma)/\gamma}-1\right)T_3 \qquad [\text{kJ/kg}] \qquad (9\text{-}34)$$

Process 4–1 is an isobaric heat rejection process that decreases the temperature of the expanded combustion products to the compressor inlet temperature T_1. The main equations of this process are

$$p_1 = p_4 \qquad \text{(process definition)}$$

$$q_{4-1} = c_p(T_1 - T_4) \qquad [\text{kJ/kg}] \qquad \text{(Eq. 6-9x)}$$

$$wk_{4-1} = p_4(v_1 - v_4) \qquad [\text{kJ/kg}] \qquad \text{(Eq. 7-3ax)}$$

The two major overall performance characteristics of thermal efficiency and power output are discussed in the following sections.

9-7.1 THERMAL EFFICIENCY

The equation for the thermal efficiency is

$$\eta_{\text{th}} = 1 - \left(\frac{1}{r_p}\right)^{(\gamma-1)/\gamma} = 1 - (r_p)^{(1-\gamma)/\gamma} \qquad [-] \qquad \text{(9-35)}$$

This equation can be derived from

$$\eta_{\text{th}} = \frac{|q_i| - |q_r|}{|q_i|} = 1 - \frac{|q_r|}{|q_i|}$$

$$|q_r| = c_p(T_4 - T_1)$$

$$|q_i| = c_p(T_3 - T_2)$$

$$\eta_{\text{th}} = 1 - \frac{c_p(T_4 - T_1)}{c_p(T_3 - T_2)} = 1 - \left[\frac{T_1 c_p(T_4/T_1 - 1)}{T_2 c_p(T_3/T_2 - 1)}\right] = [-]$$

The adiabatic compression and expansion process have the same compression ratio so that

$$\frac{T_2}{T_1} = (r_p)^{(\gamma-1)/\gamma} \qquad [-] \qquad \text{(Eq. 9-30)}$$

$$\frac{T_3}{T_4} = (r_p)^{(\gamma-1)/\gamma} \qquad [-] \qquad \text{(Eq. 9-33)}$$

Therefore,

$$\frac{T_2}{T_1} = \frac{T_3}{T_4}$$

so that

$$\frac{T_4}{T_1} = \frac{T_3}{T_2}$$

and

$$\left(\frac{T_4}{T_1}-1\right)=\left(\frac{T_3}{T_2}-1\right)$$

Then

$$\eta_{\text{th}}=1-\frac{T_1}{T_2}$$

$$\frac{T_1}{T_2}=\left(\frac{1}{r_p}\right)^{(\gamma-1)/\gamma} \qquad [-] \qquad \text{(Eq. 9-30)}$$

and therefore,

$$\eta_{\text{th}}=1-\left(\frac{1}{r_p}\right)^{(\gamma-1)/\gamma}=1-(r_p)^{(1-\gamma)/\gamma} \qquad [-] \qquad \text{(Eq. 9-35)}$$

The thermal efficiency equation is similar to the Otto cycle, thermal efficiency equations and indicates that the thermal efficiency increases as the compressor pressure ratio r_p increases.

9-7.2 POWER OUTPUT

The power output of the Brayton cycle engine can be determined from the net heat input rate.

$$p_o=\dot{M}q_{\text{net}}=\dot{M}(|q_i|-|q_r|) \qquad [\text{W}]$$

$$\dot{M}=\text{air mass flow rate} \qquad [\text{kg/s}]$$

$$|q_i|=c_p(T_3-T_2)=q_{2-3} \qquad [\text{kJ/kg}]$$

$$|q_r|=c_p(T_4-T_1) \qquad [\text{kJ/kg}]$$

$$|q_r|=q_{2-3}(r_p)^{(1-\gamma)/\gamma} \qquad [\text{kJ/kg}]$$

$$p_o=\dot{M}(q_{2-3})\left(1-r_p^{(1-\gamma)/\gamma}\right) \qquad [\text{W}] \qquad (9\text{-}36)$$

which leads to

$$p_p=\dot{M}(q_{2-3})\eta_{\text{th}} \qquad [\text{W}] \qquad (9\text{-}37)$$

This equation indicates that the power output increases with an increased heat input q_{2-3} and an increased compressor pressure ratio r_p. In practice the temperature of the hot gases to the expansion turbine is usually limited to approximately 1100°C with the latest materials.

9-8 ENGINES POWERED BY EXTERNAL HEAT SOURCES

In this book the authors have categorized power-producing thermodynamic cycles as *engines powered by external heat sources* when the heat input and heat rejection is accomplished by heat transfer through the boundary of the system, and the thermodynamic substance remains in one phase throughout the cycle. The two major cycles, discussed in the following sections, are

- Stirling cycle.
- Ericsson cycle.

In general, Stirling and Ericsson engines did not successfully compete with steam and internal combustion engines due to limitations in the high temperature life and strength of materials available in the nineteenth century and the relatively low power output for the engine size. Power outputs ranged up to 50 horsepower, which limited the scope of applications. In the nineteenth century steam was the only way to obtain large power outputs. When the gasoline engine was developed, it provided a more compact and efficient source of power in the low horsepower sizes of the Stirling and Ericsson engines. Consequently, these engines fell into disuse.

Currently, with the availability of high temperature materials, some interest has developed in these cycles for special power-producing or refrigeration cycles. Power-producing cycles can be heated by a wide range of fuels and concentrated sunlight (solar) or nuclear heat sources. Refrigeration temperatures using helium can achieve temperatures of 80 K and lower.

9-8.1 THE STIRLING CYCLE

The Stirling cycle was invented by Robert Stirling in 1816. The engine was basically a closed cycle system using regenerators to aid in the transfer of heat into, and out of, the working fluid. Initially, air at a nominal atmospheric pressure was used, and later at substantially higher pressures. The reliability of the engine was poor because the available materials for the hot portions were subject to repetitive failure due to oxidation and low strength characteristics at the operating temperatures. Consequently, these engines did not achieve general acceptance.

Modern developments make use of high system pressures (for greater output for the engine size) and modern high temperature materials to eliminate material failures.

The thermodynamic diagrams of the air standard Stirling cycle are shown in Figure 9-17.

A thermodynamic description of the Stirling cycle can be given by defining the four basic thermodynamic processes as in Table 9-10.

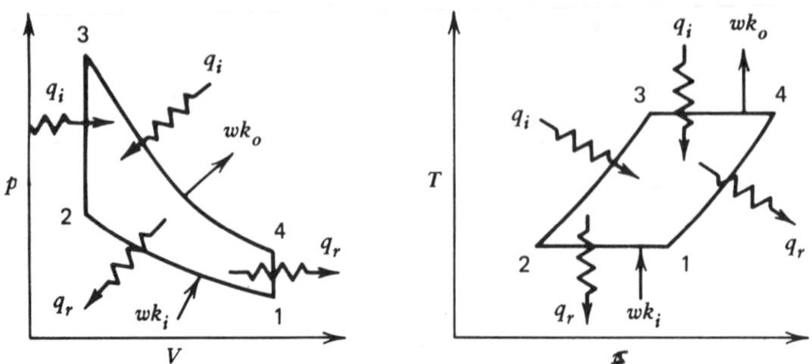

Figure 9-17 Thermodynamic diagrams for the Stirling cycle.

Table 9-10

Cycle Step	Thermodynamic Process Description
1–2	Isothermal compression with heat rejection
2–3	Constant volume (isometric) heat addition
3–4	Isothermal expansion with heat addition
4–1	Constant volume (isometric) heat rejection

A regenerator absorbing the rejected heat $q_{r_{4-1}}$ and inputting the heat back into the system as $q_{i_{2-3}}$ can significantly increase the thermal efficiency in an ideal reversible system $q_{i_{3-4}} = q_{r_{4-1}}$. With an ideal regenerator the system heat input would be $q_{i_{3-4}}$ at the constant high temperature level, and the system heat rejection would be $q_{r_{1-2}}$ at the constant low temperature level. The thermal efficiency would then equal that of a Carnot cycle operating between the same temperature levels.

9-8.2 THE ERICSSON CYCLE

John Ericsson promoted the Ericsson cycle in an open cycle form, and referred to these engines as "Caloric" engines. In 1883 Ericsson demonstrated a 5-horsepower engine in London. By 1860 a one-half-horsepower engine was being produced in quantity for light industrial users.

The Ericsson cycle differs from the Stirling cycle in that a constant pressure heat addition and heat rejection processes are used instead of constant volume processes. The use of regenerators is important for these cycles as shown in Figure 9-18. The $q_{r_{4-1}}$ rejected from the system can be stored in the regenerator to be inputted into the system as $q_{i_{2-3}}$. In an ideal cycle $q_{r_{4-1}} = q_{i_{2-3}}$. Hence, with an ideal regenerative system, the thermal efficiency of the Carnot cycle is approached. The thermodynamic diagrams of this cycle are shown in Figure 9-18.

A thermodynamic description of the Ericsson cycle can be given by defining the four basic thermodynamic processes as in Table 9-11.

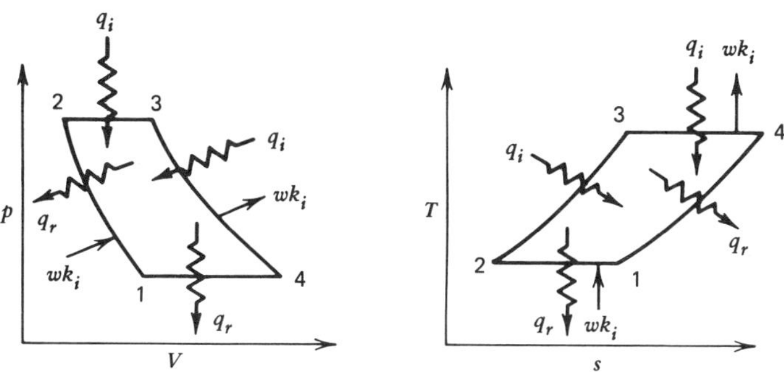

Figure 9-18 Thermodynamic diagrams for the Ericsson cycle.

Table 9-11

Cycle Step	Thermodynamic Process Description
1–2	Isothermal compression with heat rejection
2–3	Constant pressure (isobaric) with heat addition
3–4	Isothermal expansion with heat addition
4–1	Constant pressure (isobaric) with heat rejection

9-9 PROBLEMS

9-1. Draw the p–V and T–s diagrams of the Carnot cycle. List the processes involved. Indicate the work and heat flow involved in each process.

9-2. Draw the p–V and T–s diagrams of the Rankine cycle. List the processes involved. Indicate the work and heat flow involved in each process.

9-3. A Rankine cycle powerplant operates on steam at a boiler pressure of 4.00 MPa and a condenser pressure of 20 kPa. The steam enters the turbine at 450°C. Calculate the thermal efficiency of this simple Rankine cycle.

9-4. A Rankine cycle powerplant operates on steam at a boiler pressure of 3.50 MPa, a turbine inlet temperature of 500°C and a turbine exhaust pressure of 10 kPa. Calculate the thermal efficiency of this simple Rankine cycle.

9-5. What is the state of the exhaust steam from the turbine in Problem 9-4?

9-6. How does the reheat cycle modify the simple Rankine cycle?

9-7. Superheated steam enters the high pressure turbine of a Rankine reheat cycle powerplant at 3.5 MPa pressure and 500°C temperature. After expansion to 0.40 MPa the steam is reheated to 500°C and then expands in the low pressure turbine to 10 kPa. Calculate the thermal efficiency of this reheat cycle powerplant.

9-8. What is the state of the exhaust steam from: (a) the high pressure turbine, and (b) the low pressure turbine in Problem 9-7.

9-9. Compare the thermal efficiency of the simple Rankine cycle in Problem 9-4 with the thermal efficiency of the reheat cycle in Problem 9-7.

9-10. Name four assumptions required for use of an air standard cycle as the basis for comparison of internal combustion engine cycles.

9-11. Draw the p–V and T–s diagrams of the air standard Otto cycle engine. List the processes involved. Indicate the work and heat involved in each process.

9-12. Why is the thermal efficiency of the ideal air standard Otto cycle engine a function only of the compression ratio? Why is the thermal efficiency independent of the heating value of the fuel?

9-13. What aspect(s) of the performance of the Otto cycle does the heating value of the fuel affect?

9-14. Calculate the thermal efficiency of an air standard Otto cycle engine operating at a compression ratio of 8.0 : 1.

9-15. If the compression ratio were increased to 8.5 : 1, what is the thermal efficiency of the air standard Otto cycle?

9-16. An air standard Otto cycle engine has a compression ratio of 8.0 : 1. What is the air temperature at the end of the compression process for an air intake temperature of (a) 25°C, and (b) 40°C?

9-17. If the compression ratio of the engine in Problem 9-16 were increased to 8.5 : 1, what would be the corresponding air temperatures at the end of the compression process?

9-18. What considerations limit the compression ratio of an actual automobile engine?

9-19. How is the work per cycle output of an Otto cycle engine controlled? Does the thermal efficiency change with changes in work output per cycle? Explain the basis for your answer.

9-20. In an Otto cycle engine, if a modified fuel with a standard air cycle heat input value of $Q = 1200$ kJ/kg of air were substituted for gasoline (with a $Q = 1500$ kJ/kg) and the compression ratio r_c of the engine were 8.0, what would be the change in (a) thermal efficiency and (b) work output per kg of air?

9-21. If the modified fuel of Problem 9-20 would permit increasing the Otto cycle engine's compression ratio r_c to 12.0 with the same initial cylinder intake volume, what would be the comparison between the performance of the modified fuel at a compression ratio r_c of 12 and gasoline at a compression ratio r_c of 8.0 with respect to the (a) thermal efficiency, and (b) the work output per cycle?

9-22. To evaluate the degrading effect of delayed combustion in an air standard Otto cycle engine (operating with an inlet pressure of 100 kPa, inlet temperature of 25°C, compression ratio of 8, and a heat input of 1500 kJ/kg) compare the specific work output of the expansion process for the following two cases:

(a) Full heat release at constant volume (end of compression stroke) with an isentropic expansion.

(b) Three-quarters of the full heat release at constant volume (end of compression stroke) and the balance of the heat released during the expansion (delayed combustion). Treat the expansion as a polytropic expansion with a constant value of $n = 1.33$.

9-23. If you were to consider water injection to improve the performance of an Otto cycle engine, in what process would you add water and with what objective in mind? Select an amount of injected water and calculate the thermal efficiency and work output (net) for your conditions. Compare with the air standard cycle for the same compression ratio, and heat input.

9-24. Draw the p–V and T–s diagrams of the air standard Diesel cycle. List the processes involved. Indicate the work and heat flow involved in each process.

9-25. Calculate the thermal efficiency of an air standard Diesel cycle engine operating at a volumetric compression ratio of 23 : 1 with a cutoff ratio of 2.25.

9-26. If the Diesel engine of Problem 9-25 were operated at a cutoff ratio of 2.0, what would be (a) the thermal efficiency, and (b) the specific work output if $T_1 = 25°C$?

9-27. What difference (if any) is there in the thermal efficiency and specific work output between the operating conditions in Problems 9-25 and 9-26. Explain why these differences occur.

9-28. A Diesel engine has a compression ratio of 23 : 1. If the intake air temperature is 0°C, what is the air temperature at the end of the compression process?

9-29. If the diesel engine described in Problem 9-28 had a heat input of 1500 kJ/kg of air, what would be the cutoff ratio? What would be the thermal efficiency?

9-30. How is the work per cycle output of a Diesel cycle engine controlled? Does the thermal efficiency change with changes in the work output cycle? Explain the basis of your answer.

9-31. Compare the exhaust temperature of the air standard cycles with a 25°C "inlet" air temperature between:

(a) An Otto cycle engine with a compression ratio 8.0 : 1 and a heat input of 1500 kJ/kg of air.

(b) A Diesel cycle engine with a compression ratio of 23 : 1 and a cutoff ratio of 2.25.

9-32. If you were to consider water injection to improve the performance of a Diesel cycle engine, in what process would you add water and with what objective in mind? Select an amount of injected water and calculate the thermal efficiency and work output (net) for your conditions. Compare with the air standard cycle for the same compression ratio and heat input.

9-33. Draw the p–V and T–s diagrams of the limited pressure cycle modification of the air standard Diesel cycle. List the processes involved. Indicate the work and heat flow involved in each process.

9-34. Calculate the thermal efficiency of a limited pressure cycle with a compression of 23 : 1, an initial pressure p_1 of 100 kPa, initial temperature T_1 of 25°C, a pressure limit of 12 MPa, and a total heat input of 1500 kJ/kg of air.

9-35. How does the thermal efficiency of the limited pressure cycle of Problem 9-34 compare with the thermal efficiency of a Diesel cycle with the same initial conditions, compression ratio, and total heat input?

9-36. How could an engine running on the Diesel cycle be altered to run on the limited pressure cycle?

9-37. Draw the p–V and T–s diagrams of the air standard Brayton cycle. List the processes involved. Indicate the work and heat flow involved in each process.

9-38. Calculate the thermal efficiency of an air standard Brayton cycle engine that has a compressor pressure ratio of 13.5 : 1.

9-39. What would the specific work output of the Brayton cycle in Problem 9-38 with a compressor inlet temperature of 25°C and pressure of 100 kPa, if the temperature of the expansion turbine inlet is (a) 1000°C and (b) 900°C.

9-40. In order to obtain more power from the Brayton cycle in Problem 3-39, more heat is added raising the air temperature to 1250°C, followed by cooling the air by injecting liquid water saturated at 25°C to reduce the turbine inlet temperature to 1000°C.

(a) Calculate the net specific work output considering the additional mass of water vapor expanding through the turbine.

(b) How does the thermal efficiency compare with that of Problem 9-39 for 1000°C turbine inlet temperature?

9-41. Draw the p–V and T–s diagrams of the Stirling cycle. List the processes involved. Indicate the work and heat flow involved in each process. Explain how a regenerator can improve the cycle thermal efficiency. What heat flows and processes are involved with the regenerator?

9-42. A regenerating Stirling cycle engine using nitrogen as the working fluid operates with a high temperature of 400°C and a low temperature of 100°C. If helium were used instead of nitrogen as the thermodynamic substance, would there be any change in the amount of work obtained per cycle per kilogram of thermodynamic substance using the same volumetric compression ratio?

9.43. If the high temperature in Problem 9-42 were increased to 450°C and the low temperature remained at 100°C, what would be the work obtained per kilogram of nitrogen?

9-44. If the pressure level of the gas in the Stirling cycle engine of Problem 9-42 were doubled, what effect would the increased pressure have on the following?

(a) The work obtained per cycle per kg of thermodynamic substance.

(b) The volume of the engine for the same work output per cycle.

9-45. Draw the p–V and T–s diagrams of the Ericsson cycle. List the processes involved. Indicate the work and heat flow involved in each process. Explain how a regenerator can improve the cycle thermal efficiency? What heat flows and processes are involved with the regenerator?

9-46. How does the Ericsson cycle differ thermodynamically from the Stirling cycle?

9-7 $\eta_{RANKINE} = \dfrac{Wk_{TURA} + Wk_{TURB} - Wk_{PUMP}}{q(BOILER) + q(REHEAT)} = \dfrac{Wk_{NET}}{q_{in}}$

$Wk_A = h_1 - h_2 \quad Wk_B = h_3 - h_4 \quad h_6 = Wk_{pum} - h_5$

9-16 – $T_2 = T_1 (R_c)^{\gamma-1}$

9-20 – $\dfrac{\eta_1}{\eta_2} = \dfrac{1 - \frac{1}{(R_c)^{\gamma-1}}}{1 - \frac{1}{(R_c)^{\gamma-1}}} ; \dfrac{\eta_1}{\eta_2} = 1.0 ; Wk \propto q \Rightarrow \dfrac{W_{o1}}{W_{o2}} = \dfrac{q_1}{q_2}$

9-25 – $T_2 = T_1 (K_c)^{\gamma-1}$

9-42 – $Wk_c = C \ln \left[\dfrac{V_2}{V_1}\right] ; C = RT ; \dfrac{Wko_{He}}{Wko_{N_2}} = \dfrac{C_{He} \ln \left[\frac{V_4}{V_3}\right] \cdot \eta_{TH}}{C_{N_2} \ln \left[\frac{V_4}{V_3}\right] \eta_{TH}}$

WORK RATIO = $\dfrac{C_{He}}{C_{N_2}} = \dfrac{RT_{He}}{RT_{N_2}} = \dfrac{Wk_{He}}{Wk_{N_2}}$

CHAPTER 10
HEAT-PUMPING CYCLES

Thermodynamic cycles can be made to receive work input and to absorb heat at lower temperatures and reject heat at higher temperatures. Such systems can be used to absorb heat (refrigeration systems) or to reject heat (heat pump systems). The heat-pumping and refrigeration systems are discussed in the following sections.

10-1 THE HEAT PUMP

A diagram of the elements of the heat pump cycle was given in Figure 4-2. Such a cycle is usually implemented by a vapor compression cycle illustrated by the block diagram in Figure 10-1. The heat output (q_{out}) is obtained by the condensation of the compressed vapor of the thermodynamic substance. The condensed saturate liquid is then expanded into a lower pressure evaporator by passing through a throttling valve. Heat input at a lower temperature is obtained by boiling the thermodynamic substance to form saturate vapor. This vapor is then compressed to a higher pressure with the desired saturation (condensing) temperature.

This cycle can be represented by the processes listed in Table 10-1, page 260.

The thermodynamic diagrams of the cycle with these processes are shown in Figure 10-2, page 260. The thermodynamic substance for such a cycle is usually selected on the basis of characteristics including the saturation pressures at the desired heat input and rejection temperature, the vapor density, and the latent heat of evaporation.

The adiabatic vapor compression process (1–2) compresses the vapor, requiring a work input into the system. As a vapor near saturation conditions (initially) is involved, the perfect gas laws do not apply without a compressibility factor (as discussed in Section 6-6). Consequently, the compressed state is usually obtained from thermodynamic tables of the

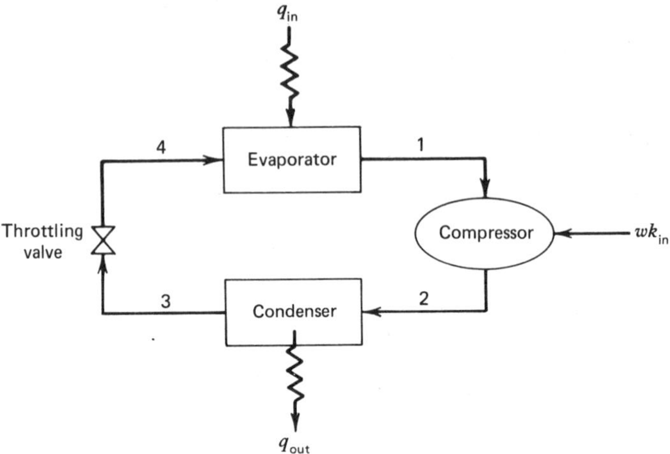

Figure 10-1 Block diagram of a vapor compression heat pump cycle.

Table 10-1

Cycle Step	Thermodynamic Process Description
1–2	Adiabatic (isentropic) compression
2–3	Constant pressure (isobaric) heat rejection
3–4	Constant enthalpy (isenthalpic) expansion
4–1	Constant pressure (isobaric) heat addition

substance based on an isentropic compression ($\Delta s = 0$). The input is

$$wk_{in} = (h_1 - h_2) \qquad [\text{kJ/kg}] \tag{10-1}$$

The constant pressure (isobaric) heat rejection process 2–3 removes the vapor superheat and the latent heat of vaporization from the thermodynamic substance. The temperature of the compressed superheated vapor is reduced to saturation temperature by the removal of the sensible superheat (2–2′). The latent heat of vaporization is then removed at the saturation

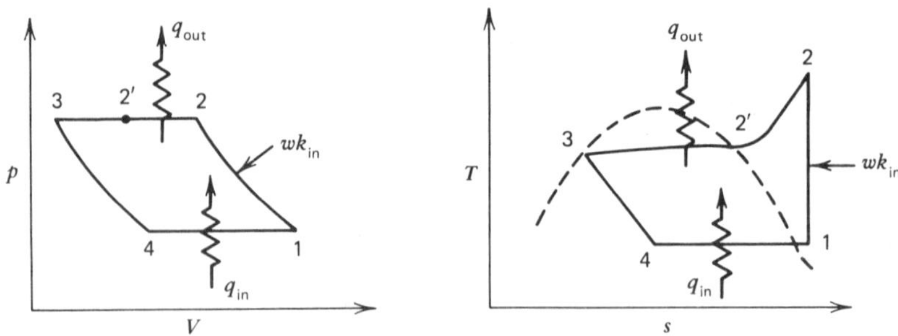

Figure 10-2 Thermodynamic diagrams for the ideal vapor compression cycle.

temperature (2′–3), resulting in a saturated liquid. The heat rejected is

$$q_{\text{out}} = (h_3 - h_2) \qquad [\text{kJ/kg}] \tag{10-2}$$

The constant enthalpy (isenthalpic) expansion process 3–4 to the evaporator pressure was discussed in Section 8-5. During the flow through the throttling value (expansion), some of the high pressure saturated liquid becomes saturated vapor at the lower (evaporator) pressure, since $h_3 = h_4$, and the substance temperature drops to the saturation temperature at the evaporator pressure.

The constant pressure (isobaric) heat addition process 4–1 absorbs heat by vaporizing the saturate liquid. This is accomplished at the saturation temperature corresponding to the evaporator pressure. The heat added q_{in} is

$$q_{\text{in}} = (h_1 - h_4) \qquad [\text{kJ/kg}] \tag{10-3}$$

The major overall performance parameters for heat-pumping cycles are the coefficient of performance (COP) and the capacity C_r as discussed in Sections 4-3.1 and 4-3.3.

Thus

$$\text{COP} = \frac{q_{\text{out}}}{wk_{\text{in}}} \qquad [-] \tag{10-4}$$

and

$$C_r = \dot{M} q_{\text{out}} \qquad [\text{kW}] \tag{10-5}$$

where

q_{out} = the net heat rejected by the heat pump cycle [kJ/kg]

wk_{in} = the net work input to the heat pump cycle [kJ/kg]

$\dot{M}$ = refrigerant flow rate [kg/s]

10-2 REFRIGERATION CYCLES

The refrigeration cycle is used to transfer energy (heat) from a cold chamber, which is at a temperature lower than its surroundings. In order to accomplish this thermodynamic cycle, energy must be supplied to the system so as not to violate the first and second laws of thermodynamics. The refrigeration cycle then is simply a reversed heat engine cycle. (See Section 4-3.2.)

The basic refrigeration cycle consists of a sequence of processes utilizing a working fluid, called the refrigerant, usually in continuous circulation within a closed system. The refrigerant receives energy in the evaporator (cold chamber) at a temperature below that of the surroundings, and then rejects this energy in the condenser (hot chamber) prior to returning to its initial state. The high temperature energy can be used for heating purposes (i.e., a heat pump); an example is the warm air being circulated from the common kitchen refrigerator.

The type of refrigerant depends on the specific type of refrigeration cycle used (e.g., ammonia, Freon, gas, or air). In the following sections, the vapor compression, absorption, gas, and vacuum refrigeration cycles will be discussed briefly. For a more detailed discussion the reader should refer to a book on refrigeration system.

10-2.1 THE VAPOR COMPRESSION CYCLE

The vapor compression cycle for refrigeration is basically the same as the heat pump vapor compression cycle. In the refrigeration system the evaporator pressure is controlled to give the saturation temperature desired for the refrigeration application. The heat rejection should be made at the lowest practical temperature. In contrast, in the heat pump system the condenser pressure is controlled to give the saturation temperature desired for the heating application and the heat input is made at the highest practical temperature.

The major overall performance parameters for refrigeration cycles are the coefficient of refrigeration (COR) and the capacity C_r are discussed in Sections 4-3.2 and 4-3.3.

Thus

$$\mathrm{COR}=\frac{-q_{\mathrm{in}}}{wk_{\mathrm{in}}} \qquad [—] \tag{10-6}$$

and

$$C_r=\dot{M}q_{\mathrm{in}} \qquad [\mathrm{kW}] \tag{10-7}$$

where

q_{in} = net heat absorbed by the refrigeration cycle [kJ/kg]

wk_{in} = net work input to the refrigeration cycle [kJ/kg]

$\dot{M}$ = refrigerant flow rate [kg/s]

Example Problem 10-1

A vapor compression refrigeration cycle utilizes Freon-12 as the working fluid. The temperature of the refrigerant in the evaporator is $-15°\mathrm{C}$ and in the condenser is

35°C. If the mass flow rate of the refrigerant is 0.03 kg/s, determine (1) the capacity in terms of rate of refrigeration, and (2) the coefficient of refrigeration (COR). See Figure 10-3 and Tables 10-2 and 10-3.

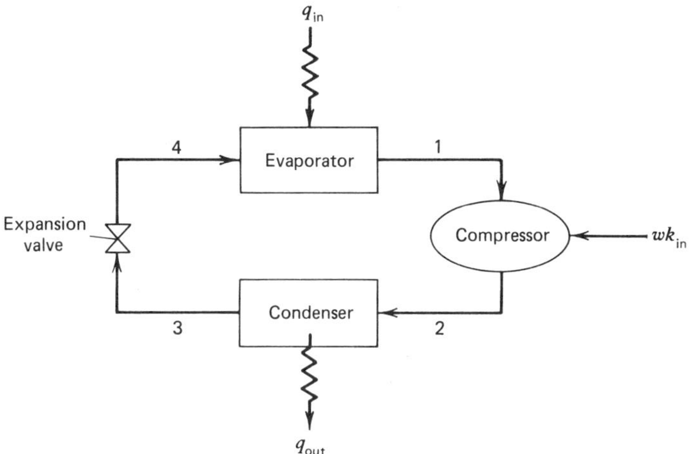

Figure 10-3 Diagram of a vapor compression refrigeration cycle.

Solution (a)

Determine C_r using Equation 10-7 with $\dot{M}$ given and q_{in} obtained from h_1 and h_4.
Equation

$$C_r = \dot{M} q_{in} \qquad [\text{kW}] \qquad (\text{Eq. 10-7})$$

Table 10-2

Processes	Thermodynamic Process Description
1–2	Adiabatic (isentropic) vapor compression
2–3	Isobaric heat rejection
3–4	Isenthalpic expansion
4–1	Isobaric heat input

Table 10-3 Table of State

Substance: F12	1	2	3	4
Condition	Saturated Vapor	Superheated Vapor	Saturated Liquid	2ϕ Liquid/Vapor
Mass [kg]	1 (a)	→	→	1
Temperature [°C]	−15 (g)		35 (g)	−15 (g)
Pressure [Pa]				
Quality [—]	1.00 (a)		0.00 (a)	
Specific enthalpy [kJ/kg]	180.846(tt)	207.146(C-1)	69.494(tt)	69.494(tt)
Specific entropy [kJ/kg·K]	0.7046(tt)	0.7046		

Note: (a) assumed; (g) given; (tt) thermodynamic tables for Freon-12 (Appendix A, Table A-5).

Parameters

$$\dot{M}=0.03 \text{ kg/s} \qquad \text{(given)}$$

$$q_{\text{in}}=h_1-h_4 \text{ [kJ/kg]}$$

$$h_1=180.846 \text{ kJ/kg} \qquad \text{(Table of state)}$$

$$h_4=69.494 \text{ kJ/kg} \qquad \text{(Table of state)}$$

$$q_{\text{in}}=180.846-69.494 \qquad \text{[kJ/k]}-\text{[kJ/kg]}=\text{[kJ/kg]}$$

$$=111.354 \text{ kJ/kg}$$

Substitution

$$C_r=(0.03)(111.354) \qquad \text{[kJ/s][kJ/kg]}=\text{[kJ/s]}=\text{[kW]}$$

Answer

$$\mathbf{C_r=3.34 \text{ kW}}$$

Solution (b)

Determine COR using Equation 10-6 with q_{in} from Solution (a) above and wk_{in} from h_1 and h_2. To determine h_2, treat as an isentropic compression from state 1 ($s_2=s_1$). Then from the thermodynamic tables of superheated F-12, determine h_2 from s_2 and p_2.

Equation

$$\text{COR}=\frac{-q_{\text{in}}}{wk_{\text{in}}} \qquad [\text{—}] \qquad \text{(Eq. 10-6)}$$

Parameters

$$q_{\text{in}}=111.354 \text{ kJ/kg} \qquad \text{[Solution (a)]}$$

$$wk_{\text{in}}=h_1-h_2 \text{ [kJ/kg]} \qquad \text{(Eq. 3-29a)}$$

$$s_2=s_1=0.7046 \text{ kJ/kg}\cdot\Delta_1\text{K}$$

$$p_2=p_3=0.8477 \text{ MPa} \qquad \text{(Appendix A, Table A-5 at 35°C)}$$

(C–1). To obtain h_2 from $s_2=0.7046$ kJ/kg·Δ_1K and $p_2=0.8477$ MPa, a double interpolation of the thermodynamic table data is required.

1. Determine the h at 0.80 MPa and then at 0.90 MPa by interpolating between the data for 40 and 50°C.
2. Determine the h at 0.8477 MPa by interpolating between the two values for h at 0.80 MPa and 0.90 MPa.

First Iteration at 0.80 MPa

$$h = 205.924 + \frac{\Delta s}{\Delta s'}(213.290 - 205.924) \qquad [\text{kJ/kg}]$$

$$\Delta s = 0.7046 - 0.7016 = 0.0030 \text{ kJ/kg}\cdot\Delta_1\text{K}$$

$$\Delta s' = 0.7248 - 0.7016 = 0.0236 \text{ kJ/kg}\cdot\Delta_1\text{K}$$

$$h = 205.924 + \frac{0.0030}{0.0232}(7.366) \qquad \text{kJ/kg} + \left[\frac{\text{kJ/kg}\cdot\Delta_1\text{K}}{\text{kJ/kg}\cdot\Delta_1\text{K}}\right] = [\text{kJ/kg}]$$

$$h = 206.882 \text{ kJ/kg}$$

At 0.90 MPa, we obtain

$$h = 204.170 + \frac{\Delta s}{\Delta s'}(211.765 - 204.170) \; [\text{kJ/kg}]$$

$$\Delta s = 0.7046 - 0.6982 = 0.0064 \text{ kJ/kg}\cdot\Delta_1\text{K}$$

$$\Delta s' = 0.7131 - 0.6982 = 0.0149 \text{ kJ/kg}\cdot\Delta_1\text{K}$$

$$h = 204.170 + \frac{0.0064}{0.0149}(7.595) \qquad \text{kJ/kg} + \left[\frac{\text{kJ/kg}\cdot\Delta_1\text{K}}{\text{kJ/kg}\cdot\Delta_1\text{K}}\right][\text{kJ/kg}] = [\text{kJ/kg}]$$

$$h = 207.436 \text{ kJ/kg}$$

Second Iteration at 0.8477 MPa

$$h_2 = 206.882 + \frac{\Delta p}{\Delta p'}(207.436 - 206.882) \qquad [\text{kJ/kg}]$$

$$\Delta p = 0.8477 - 0.8000 = 0.0477 \text{ MPa}$$

$$\Delta p' = 0.9000 - 0.8000 = 0.1000 \text{ MPa}$$

$$h_2 = 206.882 + \frac{0.0477}{0.1000}(0.554) \qquad [\text{kJ/kg}] + \frac{[\text{MPa}]}{[\text{MPa}]}[\text{kJ/kg}] = [\text{kJ/kg}]$$

$$h_2 = 207.146 \text{ kJ/kg}$$

$$h_1 = 180.846 \text{ kJ/kg} \qquad (\text{Table of state})$$

$$wk_i = 180.846 - 207.146 \qquad [\text{kJ/kg}] - [\text{kJ/kg}] = [\text{kJ/kg}]$$

$$wk_i = -26.300 \text{ kJ/kg}$$

Substitution

$$\text{COR} = \frac{-111.354}{-26.300} \qquad \frac{[\text{kJ/kg}]}{[\text{kJ/kg}]} = [-]$$

Answer

$$\mathbf{COR = 4.23[-]}$$

A wide range of substances can be used as the refrigerant in the vapor compression systems. An initial requirement is that the freezing point be well below the lowest temperature in the system. Then there are considerations of the saturation pressures at the high and low temperatures as well as the specific volumes. High pressures and large specific volumes tend to increase the cost of the required equipment to implement the refrigeration cycle. The thermodynamic characteristics of the substance, such as the heat of evaporation and heat of condensation, must be included in the considerations to select an optimum refrigerant for the refrigeration application. The use of a number of potential refrigerants has been prohibited because of safety and health hazards. Commercially available refrigeration equipment has been standardized for the use primarily of Freon-12, Freon-22, and ammonia.

The COR of the vapor compression refrigeration cycle can be improved by adding a heat exchanger to reduce the temperature of the liquid condensate before it passes through the expansion valve. This modification is illustrated by the block diagram in Figure 10-4.

By subcooling the saturate liquid condensate in the added heat exchanger, the quality of the discharge from the expansion valve is lower (less vapor). This results in an increased refrigeration effect per unit mass of condensed refrigerant.

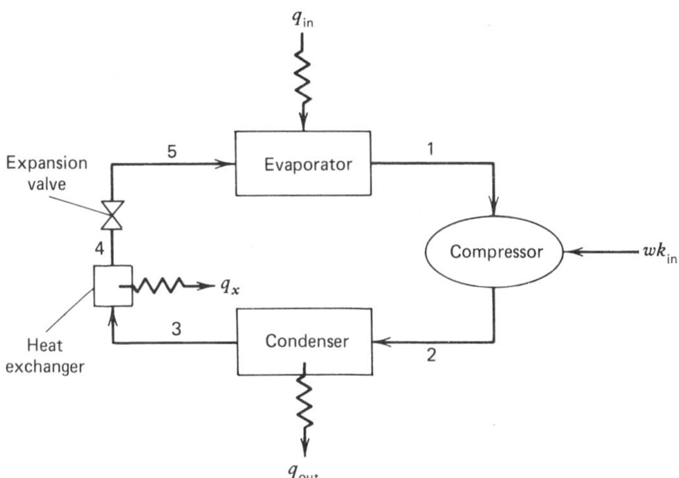

Figure 10-4 Block diagram of a vapor compression refrigeration cycle with heat exchanger.

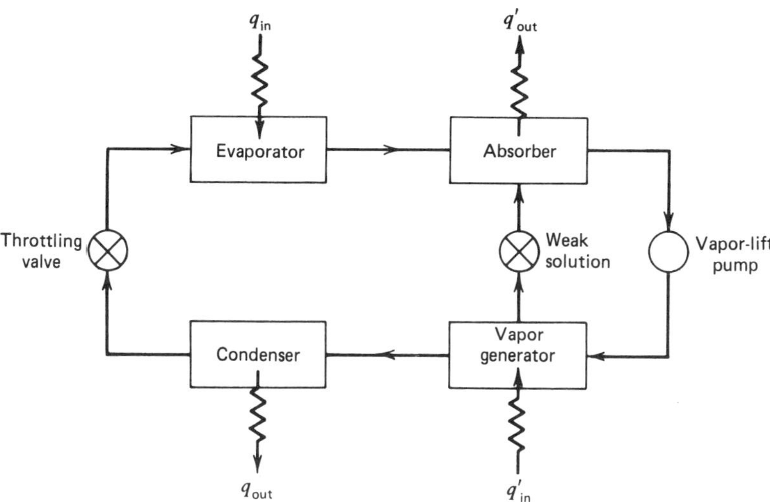

Figure 10-5 Block diagram of an absorption refrigeration cycle.

10-2.2 THE ABSORPTION REFRIGERATION CYCLE

The absorption refrigeration cycle differs from the vapor compression refrigeration cycle in the manner in which compression is achieved. The absorption refrigeration cycle uses an equivalent heat source in place of the mechanical compressor. The use of this additional heat input to effect the vapor compression results in significant additional heat rejection requirements for the system. Auxiliary heat rejection systems, such as cooling towers, with coolant circulation pumping requirements may be required to support the absorption refrigeration system. A block diagram of the elements of the absorption refrigeration cycle is shown in Figure 10-5.

In the 1930–1950 period the absorption refrigeration system, which used a gas flame and had no moving parts, became widely used in household refrigerators. When the home storage of frozen foods became popular (which required significantly lower refrigerator temperatures) the household absorption refrigerator was supplanted by vapor compression refrigerators. Today many air conditioning systems make use of absorption refrigeration systems.

As the details of the absorption refrigeration system and its performance characteristics are relatively complex and are not of primary importance in this text, the interested student is referred to suitable textbooks on absorption refrigeration systems.

10-2.3 GAS REFRIGERATION CYCLE

Among the many gas cycles that have been tried, the reversed Brayton cycle has been used most successfully to provide refrigeration. The COR is

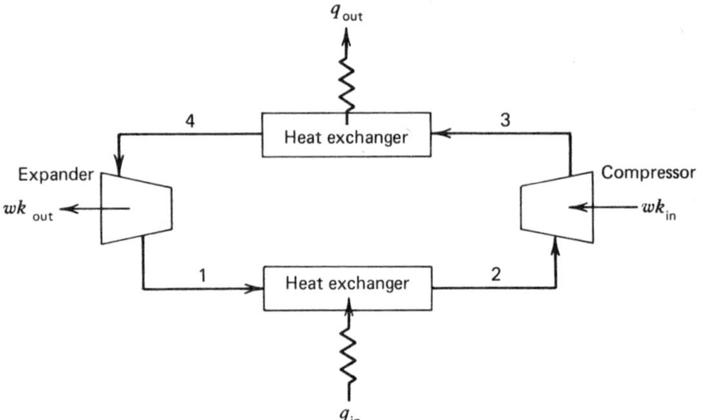

Figure 10-6 Schematic diagram of a simple reversed Brayton cycle.

smaller than for vapor compression cycles, so the input power requirement is larger. The gas refrigeration cycle can be applied to very low temperature refrigeration applications where no suitable two-phase refrigerant for the vapor compression cycle is available. One example is the refrigeration of air to the lower temperatures in the air liquefaction process. Also, there are applications where other considerations favor the use of the gas refrigeration cycle. One such application is for refrigeration in pressurized commercial aircraft. Because air compression is required for the cabin pressurization, providing for refrigeration requires adding only small-size and lightweight components with essentially no increase in the overall power requirements. Hence an open system air refrigeration cycle is often used.

A schematic of the simple, reversed Brayton cycle refrigeration system is shown in Figure 10-6. The p–V and T–s diagrams for the reversed Brayton cycle are shown in Figure 10-7.

A thermodynamic description of the reversed Brayton cycle can be given by defining the four basic thermodynamic processes listed in Table 10-4.

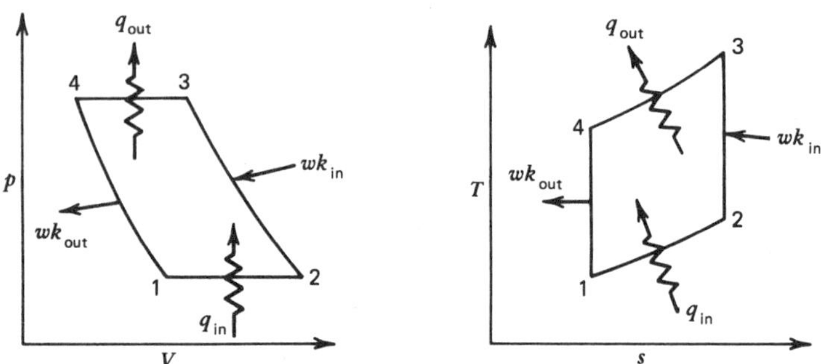

Figure 10-7 Thermodynamic diagrams for reversed Brayton cycle.

Table 10-4

Cycle Step	Thermodynamic Process Description
1–2	Constant pressure (isobaric) heat addition from refrigerated space
2–3	Isentropic compression
3–4	Constant pressure (isobaric) heat rejection to
4–1	Isentropic expansion

A performance parameter of this type of refrigerator cycle can be obtained from the definition of coefficient of refrigeration (COR) and assuming an ideal gas with constant specific heat. Thus

$$\mathrm{COR}_{\text{rev Brayton}} = \left[\frac{\text{heat absorbed}}{\text{net work required}}\right]$$

In terms of temperature we find that

$$\mathrm{COR}_{\text{rev Brayton}} = \left\{\frac{1}{(T_3 - T_4/T_2 - T_1) - 1}\right\}[-] \qquad \text{(10-8)}$$

In terms of pressure we obtain

$$\mathrm{COR}_{\text{rev Brayton}} = \left[\frac{1}{(p_3/p_2)^{(\gamma-1)/\gamma} - 1}\right][-] \qquad \text{(10-9)}$$

The development of Equations 10-8 and 10-9 from the definition of COR follows:

$$\mathrm{COR}_{\text{rev Brayton}} = \left[\frac{-q_{\text{in}}}{wk_{2-3} + wk_{4-1}}\right]$$

Since

$$q_{\text{in}} = c_p \Delta T \qquad \text{(Eq. 6-9a)}$$

$$\mathrm{COR}_{\text{rev Brayton}} = \left\{\frac{-c_p(T_1 - T_2)}{[\gamma R/(1-\gamma)](T - T) + [\gamma R/(1-\gamma)](T_4 - T_1)}\right\}$$

$$= \left\{\frac{c_p}{[\gamma R/(1-\gamma)]}\right\}\left[\frac{(T_1 - T_2)}{(T_2 - T_3) + (T_4 - T_1)}\right]$$

However,

$$c_p = \frac{\gamma R}{1-\gamma} \qquad \text{(Eq. 6-18)}$$

$$= [-1]\left[\frac{(T_2-T_1)}{(T_2-T_3)+(T_4-T_1)}\right] = \left[\frac{-(T_2-T_1)}{(T_2-T_1)-(T_3-T_4)}\right]$$

Thus

$$\text{COR}_{\text{rev Brayton}} = \left\{\frac{1}{(T_3-T_4)/(T_2-T_1)-1}\right\} \qquad \text{(Eq. 10-8)}$$

Introducing Eq. 7-14 into Eq. 10-8, and noting that $p_4 = p_3$, we obtain

$$\text{COR}_{\text{rev Brayton}} = \frac{-(T_2-T_1)}{T_2\left[1-(p_3/p_2)^{(\gamma-1)/\gamma}\right]-T_1\left[1-(p_3/p_2)^{(\gamma-1)/\gamma}\right]}$$

$$= \frac{-(T_2-T_1)}{(T_2-T_1)\left[1-(p_3/p_2)^{(\gamma-1)/\gamma}\right]}$$

Thus

$$\text{COR}_{\text{rev Brayton}} = \left[\frac{1}{(p_3/p_2)^{(\gamma-1)/\gamma}-1}\right] \qquad \text{(Eq. 10-9)}$$

10-2.4 VACUUM REFRIGERATION CYCLE

The vacuum refrigeration cycle effects refrigeration by reducing the pressure on a liquid, causing the liquid to vaporize to the extent necessary to lower the liquid temperature to saturation. Hence the refrigeration temperature can be controlled by the applied vacuum pressure. This refrigeration cycle can be used to cool water or aqueous solutions directly, as illustrated in Figure 10-8, or solid substances can be cooled by applying enough water or other fluid to the solid surface to effect refrigeration of the solid substance by the evaporation of the applied fluid.

In the cycle shown, water is being supplied and withdrawn with the difference in the mass of water vapor going into the pumping system. Often steam jet ejectors are used instead of vacuum pumps.

Now applying the continuity principle (conservation of mass), we obtain:

$$\dot{M}_1 = \dot{M}_2 + \dot{M}_3 \qquad \text{and} \qquad \dot{M}h_1 = \dot{M}_2 h_2 + \dot{M}_3 h_3$$

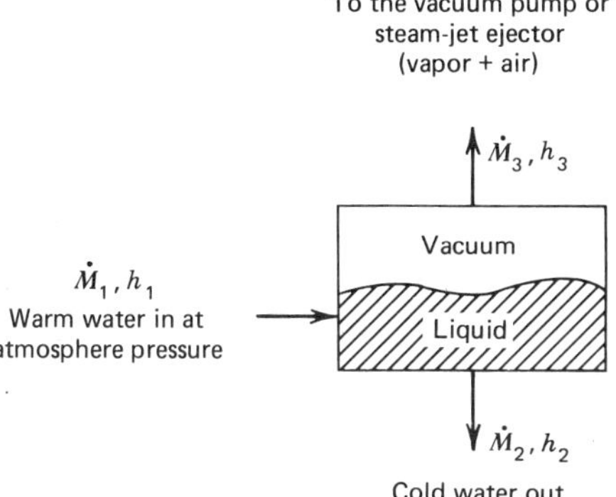

Figure 10-8 Block diagram of a vacuum refrigeration cycle.

The refrigerating effect (heat removed from the system) can be determined from the vapor as follows:

$$\dot{Q} = \dot{M}_3(h_3 - h_1) \qquad [\text{kW}]$$

or from the liquid as follows:

$$\dot{Q} = \dot{M}_2(h_1 - h_2) \qquad [\text{kW}]$$

The vapor refrigeration cycle using water is currently being used extensively by the produce industry for precooling produce for subsequent refrigerated transport to market. After processing from the field, the produce is appropriately wet with water, packaged in bags or boxes with holes to permit vapor flow, and then placed in a vacuum chamber to evaporate the added water and precool the produce to the desired temperature for shipment.

10-3 PROBLEMS

10-1. Draw a block diagram of a vapor compression heat pump cycle and describe the thermodynamic processes of the cycle.

10-2. Draw the p–V and T–s diagrams of the vapor compression heat pump cycle. Indicate the processes wherein heat and/or work cross the system boundary.

10-3. A heat pump using F-12 is being designed for obtained heat from a well at 10°C (evaporator temperature) and rejecting the heat into the heating system at 50°C (condenser temperature). To supply 5 kW of heat to the heating system, how much power input does the heat pump require? What is its COP? What mass flow rate of the Freon-12 will be required? Describe the state of the Freon-12 at the beginning and end of each process in the cycle.

10-4. If ammonia were used as the heat pump substance for the conditions of Problem 10-3, how much power would be required, and what is the COP? Describe the state of the ammonia at the beginning and end of each process in the cycle.

10-5. List four common types of refrigeration cycle.

10-6. Draw a block diagram of a vapor compression refrigeration cycle and describe the thermodynamic processes of the cycle.

10-7. Draw the p–V and T–s diagrams of the vapor compression refrigeration cycle. Indicate the processes wherein heat and/or work cross the system boundary.

10-8. Do the vapor compression heat pump and refrigeration cycles differ? If so, in what manner?

10-9. A refrigerator is to operate with a $-10°C$ evaporator and a $+50°C$ condenser. What would be the COR for the following systems?
(a) A F-12 system.
(b) An ammonia system.

10-10. To provide 1 kW of refrigeration by the two systems in Problem 10-9, determine for each of the two systems:
(a) The power input required.
(b) The mass flow rate of the refrigerant.
(c) The volumetric flow rate of the saturated vapor from the evaporator.

10-11. If a heat exchanger, such as shown in Figure 10-4, were added to the vapor compression cycle described in Problem 10-9, what would be the change in the COR for the F-12 system if the condensate were subcooled to $+30°C$? The c_p of a Freon-12 subcooled liquid is a variable. For this problem assume an average value of 1.0 kJ/kg·Δ_1K.

10-12. Repeat Problem 10-11 for ammonia, assuming an average value of c_p for the liquid ammonia of 4.9 kJ/kg·Δ_1K.

10-13. If you were to select a refrigerant for a vapor compression refrigeration cycle what factors would you consider and why?

10-14. Draw a block diagram of the reverse Brayton (refrigeration) cycle and describe the thermodynamic processes of the cycle.

10-15. Draw the p–V and T–s diagrams of the reverse Brayton cycle. Indicate the processes wherein heat and/or work cross the system boundary.

10-16. Identify a refrigeration application for which you would consider the use of a gas refrigeration cycle and explain why.

10-17. To cool an aircraft cabin, compressed air from the aircraft engines is cooled and expanded isentropically to a temperature lower than the cabin. The cold air enters the cabin, picking up heat as its temperature rises to the cabin temperature. The amount of heat it absorbs rising to the cabin temperature is the refrigerating effect. How much refrigerating effect can be obtained from 1 kg of compressed air at 150°C when it is expanded isentropically through a pressure ratio of 8:1 and the cabin temperature is 25°C?

10-18. If the compressed air in Problem 10-17 were cooled to 90°C before expansion, what would be the refrigerating effect per kilogram of air?

10-19. Describe or explain what vacuum refrigeration is and how it works.

10-20. If 1 metric ton (1000 kg) of oranges is to be precooled from a temperature of 25 to 7°C, what mass of water saturated at 25°C must be evaporated to effect the cooling? Assume that the specific heat of the oranges is 3.5 kJ/kgΔ_1 K; and use an effective average temperature of the saturated water vapor of 16°C.

10-21. To effect the cooling in Problem 10-20, what final vacuum is required? What would be the volume of the mass of evaporated water at the final vacuum pressure?

CHAPTER 11

THRUST-PRODUCING CYCLES

Thrust-producing heat engine cycles differ from power-producing cycles because their objective is to produce a thrust (force) by changing the velocity of a mass flow rate ($\dot{M}$). According to Newton's law, we obtain

$$F = \dot{M}(\mathbf{V}_j - \mathbf{V}_i) \qquad \left[\frac{\text{kg}}{\text{s}}\right]\left[\frac{\text{m}}{\text{s}}\right] = \left[\frac{\text{kg}\cdot\text{m}}{\text{s}^2}\right] = [\text{N}] \tag{11-1}$$

where

$\dot{M}$ = mass flow rate of gas [kg/s]

$\mathbf{V}_j$ = exhaust jet velocity [m/s]

$\mathbf{V}_i$ = inlet velocity of gas [m/s]

The common thrust-producing cycle burns fuel at an elevated temperature and pressure and then expands the heated combustion products ($\dot{M}$) to obtain a high jet velocity ($\mathbf{V}_j$).

The two major thrust-producing heat engine cycles are

- The rocket thrust chamber.
- The aircraft jet engine.

These cycles are discussed in the following sections, primarily in the form of an overview. The reader is referred to appropriate textbooks for detailed discussions of these propulsion systems.

11-1 THE ROCKET THRUST CHAMBER

The basic rocket thrust chamber includes a combustion chamber with a de Laval (convergent–divergent) exhaust nozzle, as illustrated in Figure 11-1.

A propellant is burned at pressure p_c in the combustion chamber producing gaseous combustion products with a mass flow rate of $\dot{M}$. The

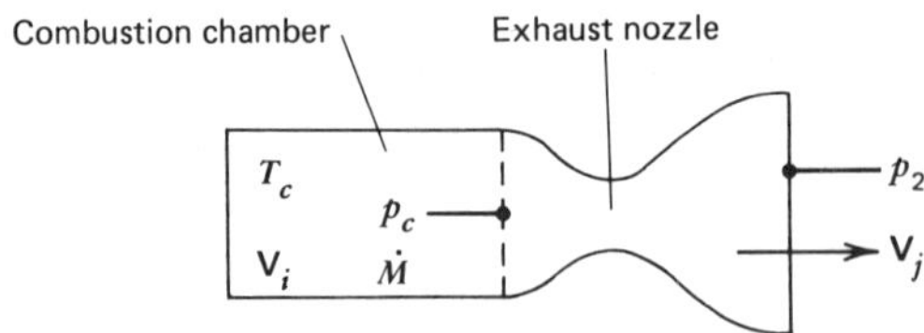

Figure 11-1 A simple rocket thrust chamber.

two basic types of rocket thrust chambers are

- Solid propellant.
- Liquid propellant.

In solid propellant rockets, the solid propellant is placed in the combustion chamber and burns continuously when ignited. The combustion occurs on the exposed surface of the solid propellant, which is designed to give the desired burning rate $\dot{M}$.

In liquid propellant rockets, the propellant (separate fuel and oxidizer fluids, or a monopropellant fluid) are pumped into the combustion chamber at the desired mass flow rate $\dot{M}$. Combustion of the liquid propellants occurs immediately after entrance to the combustion chamber. A schematic diagram of a typical liquid propellant rocket engine is shown in Figure 11-2.

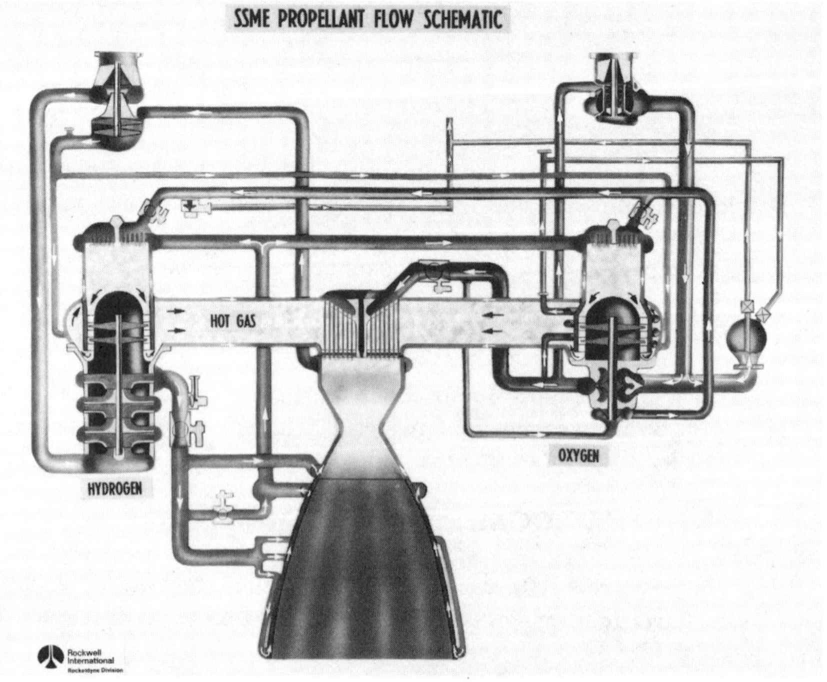

Figure 11-2 Liquid propellant rocket engine flow schematic. (Photo courtesy of Rockwell International Corporation, Rocketdyne Division.)

The two major performance parameters for rocket thrust chambers are

- Thrust.
- Specific impulse.

11-1.1 THRUST

The thrust produced by a rocket thrust chamber is usually based upon the intake and exhaust jet velocities measured with respect to the thrust chamber. In most cases the intake velocity can be neglected. In solid propellant rockets the solid propellant is an integral part of the combustion chamber so that its intake velocity is zero. The injection velocities of liquid propellants are relatively low (usually less than 30 m/s) and can be neglected.

The general equation for thrust (Equation 11-1) then becomes

$$F = \dot{M}\mathbf{V}_j \qquad [\mathrm{N}] \tag{11-2}$$

where

F = thrust [N]

$\dot{M}$ = propellant combustion rate [kg/s]

$\mathbf{V}_j$ = exhaust jet velocity [m/s]

The propellant combustion mass rate $\dot{M}$ is a design parameter that is controlled by the solid propellant design or the injection rate of the liquid propellant.

The exhaust velocity is obtained by the conversion of heat energy Q into kinetic energy E_k. From the general open system equation, the following equation can be written:

$$\tfrac{1}{2}\dot{M}\mathbf{V}_j^2 = \Delta Q + (H_1 - H_2) \qquad [\mathrm{J}] \qquad (\text{Eq. 3-26x})$$

The two main processes involved in the rocket thrust chamber are

1–2. An irreversible constant pressure heat addition (combustion).
2–3. A reversible adiabatic expansion.

The thermodynamic diagrams for the combustion and expansion processes are shown in Figure 11-3.

The exhaust jet velocity $\mathbf{V}_j$ can be calculated for a contemplated design by assuming an adiabatic expansion of the combustion products through an ideal nozzle with a constant molecular weight and specific heats of the combustion gas throughout the expansion process. The exhaust jet velocity equation based upon the above assumptions is

$$\mathbf{V}_j = \sqrt{\frac{2\gamma\bar{R}}{(\gamma-1)\mathfrak{M}} T_c\left[1 - \frac{p_j}{p_c}^{(\gamma-1)/\gamma}\right]} \qquad [\mathrm{m/s}] \tag{11-3}$$

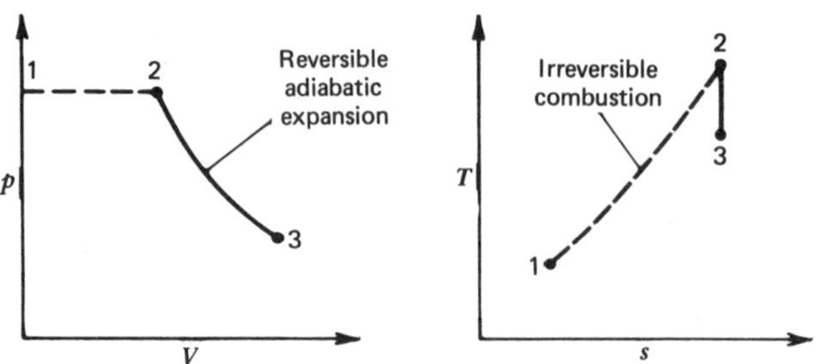

Figure 11-3 Thermodynamic diagrams for the rocket thrust chamber.

where

$\mathbf{V}_j$ = exhaust jet velocity [m/s]

γ = ratio of specific heats [—]

$\bar{R}$ = universal gas constant $[\mathrm{J/kg \cdot mol^{-1} \cdot \Delta_1 K}]$

$\mathfrak{M}$ = molecular weight of the exhaust gas $[\mathrm{mol^{-1}}]$

T_c = combustion chamber temperature [K]

p_j = nozzle exit pressure [Pa]

p_c = combustion chamber pressure [Pa]

From this equation it is seen that the exhaust velocity $\mathbf{V}_j$ increases with

- A smaller value of γ.
- An increased pressure ratio p_c/p_j.
- A smaller value of the molecular weight of the combustion products.
- A higher value of T_c.

The derivation of this equation as well as the various relationships between the mass flow rate, nozzle areas, and pressures are presented by authors, such as Sutton.[11] Because real nozzles and expanding combustion gases deviate significantly from ideal performance, the exhaust gas velocity of Equation 11-3 is not used as much in actual rocket design as other parameters such as C_F and c^*.

***C_F*.** The *thrust coefficient* C_F is a figure of merit of the exhaust nozzle design and the pressure expansion ratio. The higher the numerical value, the better the rocket performance. The equation for C_F is

$$C_F = \sqrt{\frac{2\gamma^2}{\gamma-1}\left(\frac{2}{\gamma+1}\right)^{(\gamma+1)/(\gamma-1)}\left[1-\left(\frac{p_2}{p_c}\right)^{(\gamma-1)/\gamma}\right]} + \frac{p_2-p_3}{p_c}\left(\frac{A_2}{A_t}\right)[—] \tag{11-4}$$

where

Subscript 2 is for the nozzle exit condition

Subscript 3 is for ambient condition

Subscript t is for nozzle throat condition

c*. The *characteristic exhaust velocity* (called cee star) is a figure of merit of the propellant/combustion performance, and is essentially independent of the nozzle and nozzle expansion characteristics. The higher the numerical value of c^*, the better the rocket performance. The equation for c^* is

$$c^* = \frac{p_c A_t}{\dot{M}} \qquad [\mathrm{m/s}] \tag{11-5}$$

where

p_c = the combustion chamber pressure [Pa]

A_t = the cross-sectional area of the nozzle throat (normal to the flow of the combustion gases) [m^2]

g = the gravitational constant [m/s^2]

$\dot{M}$ = the combustion mass flow rate [kg/s]

An equivalent equation for c^* is

$$c^* = \frac{I_s}{gC_F} = \frac{\sqrt{g\gamma RT_c}}{\gamma\sqrt{[2/(\gamma+1)]^{(\gamma+1)/(\gamma-1)}}} \qquad [\mathrm{m/s}] \tag{11-6}$$

A further discussion can be found in Sutton.[11]

11-1.2 SPECIFIC IMPULSE PARAMETER

The specific impulse I_s is one of the most important performance parameters for evaluating and comparing rocket thrust chamber performance; it is defined as

$$I_s = \frac{F}{\dot{M}} \qquad [\mathrm{kN \cdot s/kg}] = [\mathrm{km/s}] \tag{11-7}$$

where

F = force (thrust) [kN]

$\dot{M}$ = propellant combustion rate [kg/s]

Typical specific impulse values for solid and liquid propellant thrust chambers are

Solid: 1.67 – 2.45 kN·s/kg or km/s

Liquid: 2.05 – 3.95 kN·s/kg or km/s

11-2 THE AIRCRAFT JET ENGINE

One of the main reasons for the growth of the aircraft industry during the past 25 years has been the application of the jet engine as its powerplant. The jet engine of today embodies the most advanced applications of aerodynamics, thermodynamics, materials and mechanical design. The aircraft gas turbine (jet engine) was developed to efficiently and reliably propel aircraft in the high subsonic ($0.7 < \text{Mach number} < 1.0$) and low supersonic ($1.0 < \text{Mach number} < 3.0$) regimes. A photograph of a Pratt & Whitney JT9D turbofan jet engine currently used on the Boeing 747 aircraft is shown in Figure 11-4.

All jet engines contain five basic elements: the inlet diffuser, the compressor, the combustion chamber, the turbine, and the exhaust nozzle, as illustrated in the block diagram of Figure 11-5, page 279.

Although the mechanical design of these components varies with each manufacturer, the basic function remains the same. The useful thrust acting on an aircraft by the jet engine is derived from the change of momentum of atmospheric air $\dot{M}_a$ and fuel $\dot{M}_f$ passing through the engine. If the velocity of the exhaust discharged $\mathbf{V}_{ex}$ to the rear is greater than the

Figure 11-4 JT9D aircraft jet engine. (Photo courtesy of Pratt & Whitney Aircraft Group of United Technologies Corporation.)

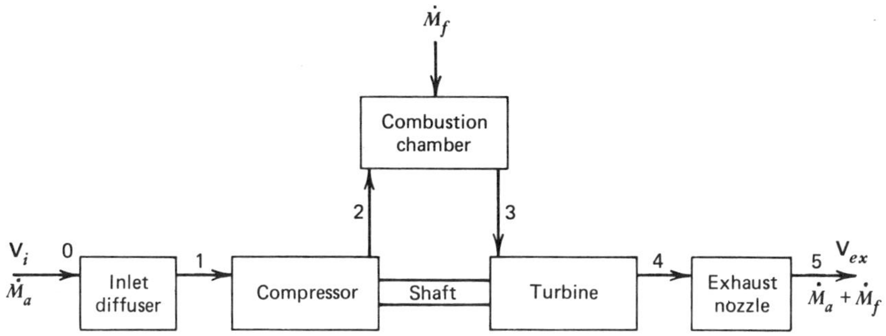

Figure 11-5 Diagram of the aircraft "standard" jet engine.

intake air velocity $\mathbf{V}_i$ (both considered relative to the aircraft), the momentum of the air (and fuel) has been increased, and consequently a reactive propulsive force acts on the aircraft.

$$F=\left(\dot{M}_a+\dot{M}_f\right)\mathbf{V}_{ex}-\dot{M}_a\mathbf{V}_i \qquad [\mathrm{N}] \tag{11-8}$$

The p–V and T–s diagrams for the aircraft standard jet engine are given in Figure 11-6.

The performance of the aircraft standard jet engine is primarily limited by the maximum temperature T_3 to the expansion turbine blades and the optimum compression rato (p_2/p_1) dictated by the mechanical design (compressor and turbine efficiencies, weight, etc.). The turbine inlet temperature T_3 in Figure 11-6 sets the amount of thermal energy input to the system per kg of input air. The compressor and turbine efficiencies and the resultant net power then establishes the remaining energy in the turbine exhaust. The exhaust velocity leaving the exhaust nozzle at the atmospheric (altitude) pressure is effectively a nominal basic fixed value that varies inversely with the atmospheric (altitude) pressure.

A typical mechanical configuration of the aircraft standard jet engine (often called the turbojet engine) is shown in Figure 11-7, page 280.

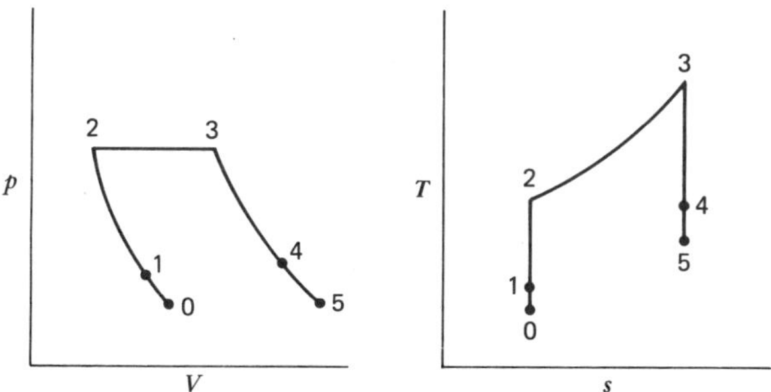

Figure 11-6 Thermodynamic diagrams for the aircraft standard jet engine.

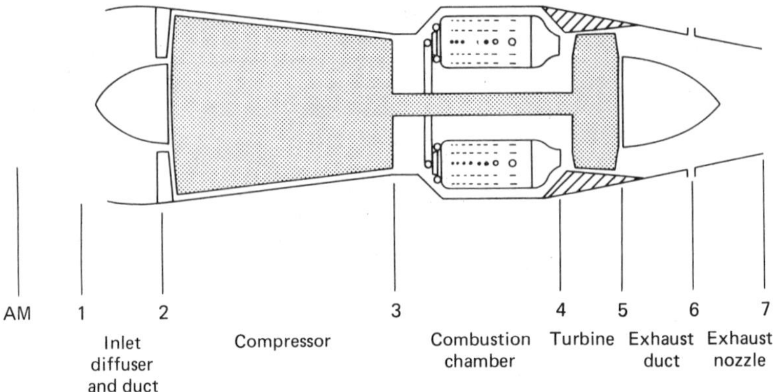

Figure 11-7 Typical turbojet configuration.

To increase the thrust, making the best use of the available kinetic energy in the exhaust gases, Newtonian mechanics indicates that if the mass of air flow can be increased (with the reduction in velocity required to maintain constant kinetic energy), an increased momentum change and hence an increased thrust can be obtained. This is illustrated by the following equations.

For constant kinetic energy, we obtain

$$\frac{\dot{M}_1\mathbf{V}_1^2}{2}=\frac{\dot{M}_2\mathbf{V}_2^2}{2} \qquad [\mathrm{J}]$$

Solving for $\mathbf{V}_2$ yields

$$\mathbf{V}_2=\left[\left(\frac{\dot{M}_1}{\dot{M}_2}\right)^{1/2}\mathbf{V}_1\right] \qquad [\mathrm{m/s}]$$

Using the momentum equation ($F=\dot{M}\mathbf{V}$), we find that

$$F_2=\dot{M}_2\mathbf{V}_2 \qquad [\mathrm{N}]$$

Then substituting the constant kinetic energy value for $\mathbf{V}_2$ gives

$$F_2=\left[\left(\dot{M}_1\dot{M}_2\right)^{1/2}\mathbf{V}_1\right] \qquad [\mathrm{N}]$$

This shows, for example, if the air mass flow were doubled (i.e., $\dot{M}_2=2\dot{M}_1$), that

$$F_2=1.4\dot{M}_1\mathbf{V}_1 \qquad [\mathrm{N}]$$

which is a 40 percent increase in thrust.

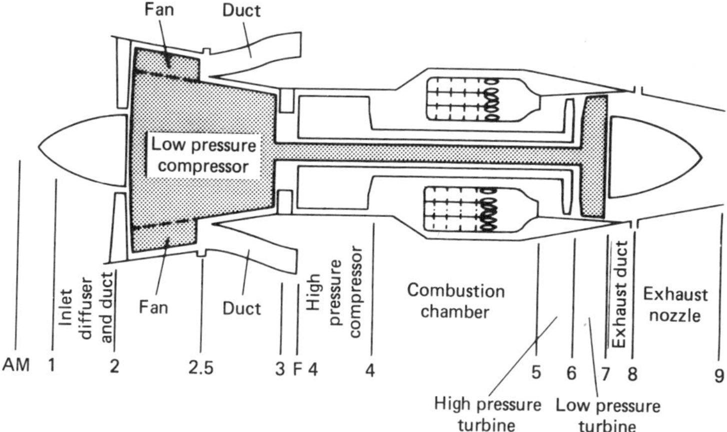

Figure 11-8 Typical turbofan configuration.

This concept for increasing the thrust for the same fuel consumption has been implemented in the turbofan engine. A typical design configuration is shown in Figure 11-8.

In this jet engine design, more kinetic energy is absorbed in the turbine and transmitted to a fan, which imparts the kinetic energy to an additional mass flow of air external to the standard jet engine system. This reduces the kinetic energy in the exhaust gases from the standard jet engine system and adds the kinetic energy removed from the exhaust gases to an added air mass flow external to the standard jet engine system. The net effect is an increase in thrust for the same fuel consumption.

Various design configurations have been used for implementing the ducting and discharge of the added external air flow. The term *ducted fan jet* may be encountered. This term is usually applied to a turbofan jet engine configuration ducting the added external air flow along the length of the engine for discharge in the rear of the engine.

11-2.1 MAJOR PERFORMANCE PARAMETERS

The two major performance parameters for aircraft jet engines are

- Thrust.
- Specific fuel consumption.

Thrust

As indicated earlier in this chaper, the thrust (force) produced by an aircraft jet engine results from the momentum change of the mass of air and fuel flowing through the jet engine between the engine intake and exhaust. If we designate the initial velocity of air entering the engine as $\mathbf{V}_i$

(equal to air speed of vehicle), and the velocity of the exhaust leaving the jet nozzle as $\mathbf{V}_j$, then the change in velocity will be $(\mathbf{V}_j - \mathbf{V}_i)$. Then from Newton's law of motion, we can develop the thrust equations as follows:

$$F = Ma = \dot{M}\,\Delta\mathbf{V} \qquad [\mathrm{N}] \qquad (\text{Eq. } 11\text{-}1)$$

Based on air flow we obtain

$$F_{\mathrm{air}} = \dot{M}\,\Delta\mathbf{V} = \dot{M}_{\mathrm{air}}(\mathbf{V}_j - \mathbf{V}_i) \qquad [\mathrm{N}] \tag{11-9}$$

and based on fuel flow we obtain

$$F_{\mathrm{fuel}} = \dot{M}_{\mathrm{fuel}}(\mathbf{V}_j - 0) \qquad [\mathrm{N}] \tag{11-10}$$

The total thrust based on air and fuel mass flow rates is

$$F = \dot{M}_{\mathrm{air}}(\mathbf{V}_j - \mathbf{V}_i) + \dot{M}_{\mathrm{fuel}}(\mathbf{V}_j) \qquad [\mathrm{N}] \tag{11-11}$$

Sometimes, there is an additional (minor) effect on thrust due to the static pressure in the jet nozzle (overpressure). This term is given by the following equation:

$$F_{\mathrm{nozzle}} = (p_{s\,\mathrm{nozzle}} - p_{\mathrm{ambient}})(A_{\mathrm{nozzle}}) \qquad [\mathrm{N}] \tag{11-12}$$

This term becomes effective only under certain conditions and is often omitted in general calculations. The most general thrust equation for the aircraft jet engine is obtained by combining Equations 11-11 and 11-12. The resultant equation is

$$F_{\mathrm{total}} = \dot{M}_{\mathrm{air}}(\mathbf{V}_j - \mathbf{V}_i) + \dot{M}_{\mathrm{fuel}}(\mathbf{V}_j) + (p_{s_{\mathrm{jet}}} - p_{\mathrm{ambient}})A_{\mathrm{jet}} \qquad [\mathrm{N}] \tag{11-13}$$

At sea level conditions the thrust produced by a jet engine will be larger (for the same jet nozzle area) than at any higher altitude because of the variation in mass flow rate through the engine. Therefore, it is desirable to have a variable area nozzle that can be adjusted to give optimum thrust at any given altitude. In most cases the velocity of the exhaust gas expanded to the ambient altitude pressure is subsonic relative to the aircraft. Therefore, the exhaust nozzle is convergent only. In extreme cases, where a convergent–divergent nozzle could be used to ideally optimize thrust, reduced nozzle efficiency due to increased friction caused by the device is encountered.(12) Consequently, a convergent–divergent nozzle is not generally used.

Table 11-1 Typical Jet Engine Performance Data

Manufacturer	Model Number	Thrust[a], kN	TSFC[b], kg/kN·h	Compression Ratio
General Electric	CF6-6D	174.8	36.06	26.6:1
Pratt & Whitney	JT9D	200.2	36.71	22:1
Rolls Royce	RB211	186.8	36.71	25:1
Pratt & Whitney	JT8D	64.5	58.13	16.9:1

[a]Values at takeoff rating (from manufacturer's data).
[b]Thrust specific fuel consumption.

Specific Fuel Consumption Parameter

The ratio of fuel flow rate to thrust produced is called the thrust specific fuel consumption (TSFC) and is expressed in kg/kN·h. Thus

$$\text{TSFC} = \frac{\dot{M}_{\text{fuel}}}{F} \qquad [\text{kg/kN}\cdot\text{h}] \tag{11-14}$$

where

$$\dot{M}_{\text{fuel}} = \text{mass flow rate of fuel} \qquad [\text{kg/h}]$$

$$F = \text{thrust} \qquad [\text{kN}]$$

Typical performance values for the most widely used commercial aircraft jet engines are given in Table 11-1.

11-2.2 THRUST AUGMENTATION

In aircraft operation there are times when thrust augmentation for a relatively short time duration period is essential. One example is in military aircraft combat operations, wherein a short time significant increase in engine thrust could make the difference between victory or defeat. Aircraft takeoff procedure (military or commercial) is a time when maximum thrust is essential. Building the capability of significantly larger thrusts into the aircraft engine results in a size and weight penalty that seriously degrades the aircraft's performance.

Because of the temperature limitation on the turbine inlet, the overall air–fuel mixture ratio is very lean (a large excess of unreacted air). Consequently, the engine system energy can be increased simply by injecting additional fuel into (1) the turbine exhaust gases, or (2) into the combustion chamber.

The burning of additional fuel in the turbine exhaust is called *afterburning*. It is discussed in the following section.

The burning of additional fuel in the combustion chamber increases the gas temperature. To reduce the increased combustion temperature to the turbine inlet limit, water (liquid) is injected. Hence this method is called "water injection." The water injection method of thrust augmentation is discussed in a later section.

Afterburner (Reheat)

An afterburner can usually provide up to a 50 percent increase in thrust for a given jet engine; it has its main applications in high speed (supersonic) military aircraft. The Concorde supersonic transport aircraft also uses jet engines with afterburners to provide the necessary thrust for its flight envelope. The afterburner increases the turbine exhaust gas temperature, which results in a significant increase in the specific volume and jet velocity of the exhaust gases. Since an afterburner is very high in fuel consumption, it is activated mainly for short periods of operation that require high thrust. A typical afterburner configuration added to a standard jet engine is shown in Figure 11-9.

A variable area jet nozzle is required when afterburning in order to keep the mass flow through the jet engine constant, and thus allow the compressor and turbine components to operate efficiently. Since the velocity of sound in a gas is a function of the square root of the temperature, and the volume (at constant pressure) varies directly as the temperature, sonic velocities may be reached in the afterburner nozzle. If this happens, a variable area covergent–divergent nozzle may be required to optimize the afterburner performance.

The thermodynamic diagrams for a jet engine with afterburner are shown in Figure 11-10, page 285.

Water Injection

Water injection is not used to a very large extent today. The relative magnitude of the thrust augmentation realized by water injection is degraded because of the practical aspects in added system requirements, such as assurance that the injection water would not freeze and additional controls, drains, tanks, and other hardware components.

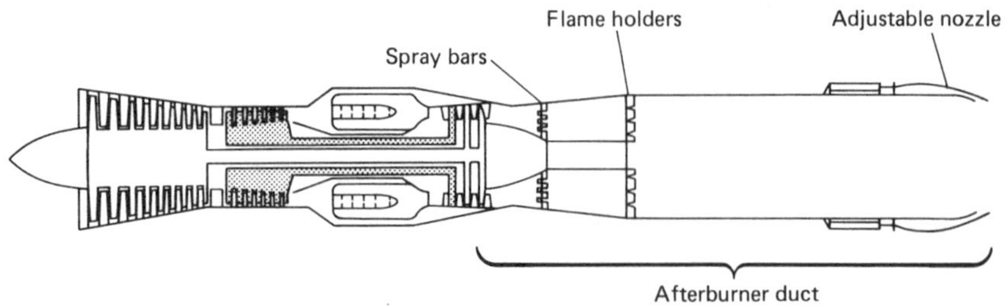

Figure 11-9 Typical afterburning turbojet configuration.

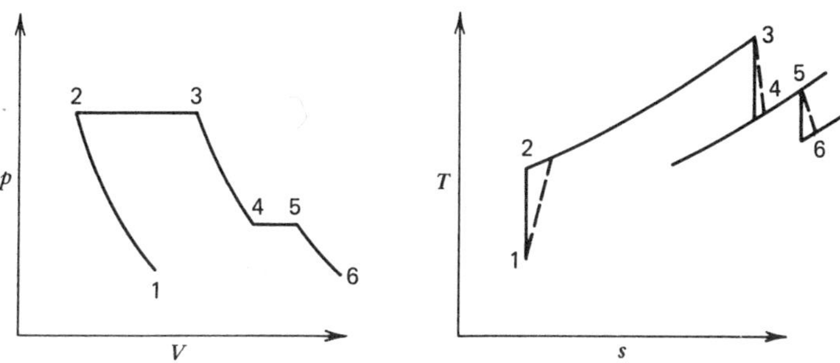

Figure 11-10 Thermodynamic diagrams for the aircraft gas turbine with reheat (afterburner).

Water injection into the compressor inlet would reduce the compressor power required, since the evaporation of the liquid water during the compression process would change the compression process from adiabatic to a polytropic compression with an exponent less than γ. The injection of distilled water or a mixture of water and methanol (to prevent freezing) is discussed in Reference 13.

Water injection into the combustion products without an increase in heat input degrades the performance of the jet engine as the heat of evaporation of the liquid water comes from the hot combustion products. The net result is that the product of the total gas mass flow rate and the nozzle exhaust specific volume is lower after the water injection. Therefore, the resultant thrust of the engine is lower.

11-3 PROBLEMS

11-1. How does a thrust-producing heat engine cycle differ from a power-producing cycle?

11-2. List two major thrust-producing heat engine cycles.

11-3. Draw a block diagram of a rocket thrust chamber and describe the thermodynamic processes involved.

11-4. Draw the p–V and T–s diagrams of a rocket thrust chamber. Indicate the processes wherein heat and/or work cross the system boundary.

11-5. A solid propellant rocket engine has a propellant combustion rate of 125 kg/s. If the exhaust jet velocity is 2120 m/s, calculate the thrust produced.

11-6. A liquid propellant rocket engine has a total propellant mass flow rate of 1405 kg/s. If the thrust rating is given as 4075 kN, determine
(a) The exhaust jet velocity.
(b) The specific impulse.

11-7. What pumping power would be required to inject the liquid propellant into the thrust chamber in Problem 11-6 if the liquid propellant density were 1300 kg/m^3, the propellants were stored in a tank at 200 kPa, and the injection pressure were 6 MPa?

11-8. What parameter of a rocket thrust chamber is an indicator of the following conditions?

(a) The combustion efficiency with a given propellant.
(b) The effectiveness of the exhaust nozzle.
(c) The chemical/energy characteristics of a propellant.

11-9. Is the thrust coefficient affected by the rocket combustion gas expansion ratio (p_c/p_3)? If so, what is the effect?

11-10. For a given rocket exhaust nozzle configuration and combustion chamber pressure, what change in the thrust coefficient from the design point (the nozzle exit pressure p_2 equals the ambient pressure p_3) would be expected if the ambient pressure changes as follows?
(a) Increases.
(b) Decreases.

11-11. For the same ratio change in the ambient pressure p_3 with respect to the nozzle exit pressure p_2, would the change in the thrust coefficient be the same for the + and − variations in p_3? If not, would the + or − variation in p_3 cause a greater change from the "design point" thrust coefficient value? Do these respective changes in p_3 increase or decrease the thrust coefficient from the design point value?

11-12. Draw a block diagram of the aircraft standard jet engine, and describe the thermodynamic processes involved.

11-13. Draw the p–V and T–s diagrams of the aircraft standard jet engine. Indicate the process wherein heat and/or work cross the system boundary.

11-14. How does the aircraft turbofan engine differ from the standard turbojet engine? Why was this configuration developed?

11-15. Calculate the total thrust produced by an aircraft jet engine that is rated for an air mass flow rate of 520 kg/s and TSFC=35.2 kg/kN·h. The aircraft is moving at a speed of 100 m/s and has an exhaust jet velocity of 350 m/s. (Assume that $p_{s_{\text{nozzle}}} = p_{\text{ambient}}$).

11-16. Describe two methods of short-time-period thrust augmentation.

11-17. Draw the p–V and T–s diagrams of a standard jet engine with an afterburner.

11-18. If an afterburner were added to the aircraft jet engine in Problem 11-15 and it increased the exhaust jet velocity to 700 m/s, what would be the increase in thrust realized from the use of the afterburner?

11-19. Assuming that the exhaust gas temperature of the engine in Problem 11-15 is 600 K, and the c_p of the exhaust gas is 1.1 kJ/kg Δ_1K, how much heat (kJ/kg) must be added in the afterburner in Problem 11-18? If the heat obtained from the fuel were equivalent to 18,000 kJ/kg (for this analysis) what would be the fuel consumption rate for the afterburner in kg/hr? What would be the TSFC for the afterburner?

11-5 − $F = \dot{m}(V_j)$ $F = 125\ \frac{kg}{s}\left(2120\ \frac{m}{s}\right)$ $\{TH. = 2.65\times10^5\ N\}$

11-15 − ① $F_j = \dot{m}(\Delta V)$ $F_j = 520\ \frac{kg}{s}\left(350\ \frac{m}{s} - 100\ \frac{m}{s}\right)$

② $\dot{m}_{FUEL} = F_j(TSFC)$ ③ $THRUST_{TOT} = F_{J\,TOT} = (\dot{m}\,\Delta V)_{AIR} + \dot{m}(V_j)_{FUEL}$

11-18 − $F_{J\,AB} = \dot{m}_{TOT}(\Delta V)$ $\dot{m}_{TOT} = \dot{m}_{AIR} + \dot{m}_{FUEL}$

11-19 − ① $T_2 = \left[\frac{V_2}{V_1}\right] T_1$ (°K) $\Delta q = C_p\,\Delta T$

② $\dot{m}_{FUEL}\, q_{FUEL} = \dot{m}_{AIR}\,\Delta q \Rightarrow \dot{m}_{FUEL} = \frac{\dot{m}_{AIR}\,\Delta q}{q_{FUEL}}$

③ $TSFC = \frac{\dot{m}_{FUEL}}{F}$

CHAPTER 12
PSYCHROMETRY

Psychrometry is concerned with mixtures of condensible vapors in a noncondensible gas. Although the major application is to water vapor in air (for air conditioning or moisture control in compressed air systems), the principles can be applied to air–fuel carburetion and to water injection in heat engine cycles. The discussion in this chapter is primarily concerned with water vapor in air.

12-1 COMMON PARAMETERS

In air–water vapor mixtures solutions to problems can be most easily obtained by relating the state of the system to the mass of dry air. The fundamental parameters defining the state of the system are listed in Table 12-1.

Other common parameters, which can be derived from the key parameters describing the state, are listed in Table 12-2.

These parameters are discussed in the following sections.

Table 12-1 Psychrometric Table of State

Substance	Moist Air	
Mass of dry air	M_a	[kg]
Pressure (barometric)	p_b	[Pa]
Temperature (dry bulb)	T_{db}	[°C]
Humidity ratio	ω	[—]

Table 12-2 Other Common Psychrometric Parameters

Substance	Moist Air	
Relative humidity	ϕ	[—]
Wet bulb temperature	T_{wb}	[°C]
Dew point	T_{dew}	[°C]
Enthalpy	h	[kJ/kg dry air]

12-1.1 MASS OF DRY AIR

In most psychrometric systems, such as air conditioning, compressed air, and internal combustion engine carburetion systems, the mass of dry air entering and leaving remains unchanged, but the content of the condensible vapor does not. Consequently, the mass of dry air is usually an invariant, making a convenient reference base or datum.[14]

12-1.2 PRESSURE (BAROMETRIC)

The term *barometric pressure* is applied to the total absolute pressure of the system. The total pressure of the system is the sum of the partial pressures of the constituents of the system. In air conditioning and compressed air systems it is composed of the pressure of the dry air plus the vapor pressure of the condensible substance such as H_2O (water vapor) or fuel.

$$p_b = p_a + p_{vp} \qquad [\mathrm{kPa}] \tag{12-1}$$

where

p_b = barometric pressure (absolute) [kPa]

p_a = pressure of the dry air [kPa]

p_{vp} = pressure of the condensible substance [kPa]

The pressure of the dry air can be calculated from the perfect gas laws by

$$p_a = \frac{0.287 M_a T}{V} \qquad [\mathrm{kPa}] \tag{12-2}$$

where

M_a = mass of dry air [kg]

T = temperature (dry bulb) [K]

V = volume [m^3]

The vapor pressure of the condensible substance usually must be determined from the conditions of the problem such as the humidity ratio ω or the relative humidity ϕ. The vapor pressure determination for each is discussed in the respective sections.

12-1.3 DRY BULB TEMPERATURE

The dry bulb temperature (T_{db}) is the temperature of the air condensible vapor mixture as measured by a thermometer or sensor that is dry. This is the ordinary temperature. However, to minimize confusion between this temperature and the wet bulb temperature, the term *dry bulb* is often attached to the ordinary temperature to eliminate potential uncertainty.

12-1.4 HUMIDITY RATIO

The humidity ratio ω of an air–vapor mixture is the ratio of the mass of vapor in the mixture to the mass of dry air in the mixture. Occasionally, other authors use the term *specific humidity* for the humidity ratio.

$$\omega=\frac{M_{wv}}{M_a} \qquad [-] \tag{12-3}$$

where

ω = humidity ratio [kg vapor/kg dry air]

M_{wv} = mass of vapor [kg]

M_a = mass of dry air [kg]

It is important to note that the humidity ratio in a system remains unchanged during changes in temperature and/or pressure until (a) condensation occurs, or (b) more vapor mass is added. Hence the humidity ratio parameter is used as one of the fundamental parameters defining the state.

As a common application of psychrometry is to air conditioning wherein dry air and water vapor are involved, the following discussion uses that application as a basis. To simplify the presentation of the concepts involved, the perfect gas laws can be applied to the water vapor as well as to the air. Since the water vapor is not a perfect gas ($Z\neq 1.0$), some error is introduced by this assumption; however, this is not a significant factor in the introduction of the student to the primary factors in psychrometry.

By applying the perfect gas laws, Equation 12-3 can be expressed in terms of the partial pressure of the water vapor and the partial pressure of dry air or the barometric pressure, as indicated in the following equation.

$$\omega=0.622\left(\frac{p_{wv}}{p_a}\right)=0.622\left(\frac{p_{wv}}{p_b-p_{wv}}\right) \qquad [-] \tag{12-4}$$

where

p_{wv} = partial pressure of water vapor [kPa]

p_a = partial pressure of dry air [kPa]

p_b = barometric pressure [kPa]

Equation 12-3 is derived from Equation 12-2 using the perfect gas laws as follows:

$$M_{wv} = \left[\frac{(p_{wv})(V)(\mathfrak{M}_{wv})}{\bar{R}T}\right] \qquad [\text{kg}] \qquad (\text{Eq. 6-2})$$

$$M_a = \left[\frac{(p_a)(V)(\mathfrak{M}_a)}{\bar{R}T}\right] \qquad [\text{kg}] \qquad (\text{Eq. 6-2})$$

$$\frac{M_{wv}}{M_a} = \left\{\frac{\dfrac{(p_{wv})(V)(\mathfrak{M}_{wv})}{\bar{R}T}}{\dfrac{(p_a)(V)(\mathfrak{M}_a)}{\bar{R}T}}\right\} = \left[\frac{(p_{wv})(\mathfrak{M}_{wv})}{(p_a)(\mathfrak{M}_a)}\right] \qquad [-]$$

From Appendix A, Table A-8, we can obtain the molecular weights:

$$\mathfrak{M}_{wv} = 18.016\,\text{mol}^{-1} \qquad \text{and} \qquad \mathfrak{M}_a = 28.966\,\text{mol}^{-1}$$

Now, substituting the values for molecular weight into the previous equation, we obtain

$$\frac{M_{wv}}{M_a} = \left[\frac{(p_{wv})(18.016)}{(p_a)(28.966)}\right] = 0.622\left(\frac{p_{wv}}{p_a}\right)$$

but

$$p_a = p_b - p_{wv}$$

and therefore

$$\frac{M_{wv}}{M_a} = 0.622\left[\frac{p_{wv}}{(p_b - p_{wv})}\right]$$

Example Problem 12-1

If an air–water vapor mixture is saturate at 15°C with a barometric pressure of 101.3 kPa, what is the humidity ratio ω?

Equation

$$\omega = 0.622\left(\frac{p_{wv}}{p_b - p_{wv}}\right) \qquad [-] \qquad (\text{Eq. 12-4})$$

Parameters

$p_{wv} = 1.705$ kPa (Steam tables A-3.1, saturation pressure at 15°C)

$p_b = 101.3$ kPa (given)

Substitution

$$\omega = 0.622\left(\frac{1.705}{101.3-1.7}\right) \qquad [—]\frac{[\text{kPa}]}{[\text{kPa}]-[\text{kPa}]}=[—]$$

$$= 0.622\frac{1.705}{99.6}$$

Answer

$$\boldsymbol{\omega = 0.0106} \qquad [—]$$

12-1.5 RELATIVE HUMIDITY

The relative humidity ϕ of an air–water vapor mixture is the ratio of the partial pressure of water vapor in the mixture to the partial pressure of water vapor in saturated air at the same dry bulb temperature and barometric pressure.

$$\phi = \frac{p_{wv}}{p_{vs}} \qquad [—] \tag{12-5}$$

where

ϕ = relative humidity [—]

p_{wv} = partial pressure of water vapor [kPa]

p_{vs} = partial pressure of saturated water vapor at dry bulb temperature [kPa]

The pressure of the condensible vapor (p_{vp}) for Equation 12-1 is

$$p_{vp} = p_{wv} = \phi p_{vs} \qquad [\text{kPa}] \tag{12-5a}$$

The relative humidity ϕ of an air–water vapor mixture with a constant humidity ratio ω will naturally vary as the temperature or pressure of the mixture changes. As the temperature changes at constant pressure, the saturation pressure p_{vs} of the water changes, but the partial pressure p_{wv} of the water pressure does not. As the pressure p_b changes at constant temperature, the saturation pressure p_{vs} remains constant, but the partial pressure p_{wv} changes in direct relationship with the change in p_b.

To measure the relative humidity, a "sling psychrometer" is often used. This device makes use of a standard dry bulb thermometer and a wet bulb thermometer, a prescribed procedure for swirling the thermometers around, and an appropriate table giving the relationship between the measured T_{db}, T_{wb}, and ϕ.

Example Problem 12-2

If the air–water vapor mixture saturate at 15°C is heated to 25°C with a constant barometric pressure of 101.3 kPa, what is the relative humidity of the air–water vapor mixture at 25°C?

Equation

$$\phi=\frac{p_v}{p_{vs}} \quad [-] \qquad \text{(Eq. 12-5)}$$

Parameters

$p_v=1.705$ kPa (Steam tables A-3.1, saturation pressure at 15°C)

$p_{vs}=3.169$ kPa (Steam tables A-3.1, saturation pressure at 25°C)

Substitution

$$\phi=\frac{1.705}{3.169} \quad \frac{[\text{kPa}]}{[\text{kPa}]}=[-]$$

Answer

$$\boldsymbol{\phi=0.538} \quad [-]$$

12-1.6 WET BULB TEMPERATURE

The wet bulb temperature T_{wb} is the temperature of the air–water vapor mixture as measured by a thermometer with a film of water covering the bulb. Because of the evaporation of the water, the wet bulb temperature is lower than the dry bulb temperature. This temperature difference is a function of the dryness of the air as well as the velocity of the air–water vapor mixture across the thermometer bulb. In order to use the wet bulb temperature in combination with the dry bulb temperature to determine the relative humidity, a standard thermometer design and measuring technique has been established, incorporating both a wet bulb thermometer and a dry bulb thermometer in a sling for swirling around in a prescribed manner. This device is known as the "Sling psychrometer." The wet bulb temperature obtained with the sling psychrometer is the value usually used as the wet bulb temperature (T_{wb}).

Since water vapor deviates from the perfect gas laws, W. H. Carrier developed an empirical equation to increase the accuracy of the determination of the partial pressure of water vapor in a mixture with dry air using wet bulb and dry bulb temperatures.[(15)] His equation for calculating the pressure of the water vapor for a set of wet bulb and dry bulb temperatures is usually referred to as the "Carrier equation," which is

$$p_{wv}=p'_{vs}-\frac{(T_{db}-T_{wb})(p_b-p_{vs})}{1500-1.44T_{wb}} \quad [\text{kPa}] \qquad (12\text{-}6)$$

where

p'_{vs} = saturation vapor pressure at the wet bulb temperature [kPa]

p_b = barometric pressure [kPa]

T_{db} = dry bulb temperature [°C]

T_{wb} = wet bulb temperature [°C]

12-1.7 DEW POINT (TEMPERATURE)

The dew point (T_{dew}) of an air–water vapor mixture is the dry bulb temperature to which an unsaturated mixture must be cooled to reach the water vapor saturation conditions ($\phi = 1.00$). For each humidity ratio there is a corresponding dew point temperature. The plot of this relationship on the psychrometric chart defines the "saturation line" on the chart.

If the air–water vapor mixture is cooled below the dew point, moisture (condensate) precipitates from the mixture. As the mass of water vapor per unit mass of dry air is ω, the mass of water vapor condensing out between two states of the air–water vapor mixture is

$$M_c = M_a(\omega_1 - \omega_2) \quad [\text{kg}] \tag{12-7}$$

where

M_c = mass of condensed water vapor [kg]

M_a = mass of dry air [kg]

ω_1 = humidity ratio at state 1 [—]

ω_2 = humidity ratio at state 2 [—]

Example Problem 12-1

The air supplied to a room is to be at 20°C and have a relative humidity of 50 percent. Calculate (a) the humidity ratio, and (b) the dew point when the barometric pressure is 0.100 MPa.

Solution (a)

Calculate the humidity ratio using Equation 12-3.

Equation

$$\omega = 0.622 \frac{p_{wv}}{(p_b - p_{wv})} \quad [—] \qquad \text{(Eq. 12-3)}$$

Parameters

$$p_{wv}=\phi p_{vs} \qquad [\text{kPa}] \qquad (\text{Eq. 12-5a})$$

$$\phi=0.5 \qquad [—] \qquad (\text{given})$$

$$p_{vs}=2.339\ \text{kPa} \qquad (\text{Appendix A, Table A-3.1 at } 20°\text{C})$$

$$p_{wv}=(0.5)(2.339) \qquad [—] \qquad [\text{kPa}]=[\text{kPa}]$$

$$=1.1695\ \text{kPa}$$

$$p_b=100.0 \qquad (\text{given})$$

Substitution

$$\omega=0.622\frac{1.1695}{(100.0-1.17)} \qquad [—]\frac{[\text{kPa}]}{[\text{kPa}]}=[—]$$

Answer

$$\boldsymbol{\omega=0.00736\ \textbf{kg moisture/kg dry air}}$$

Solution (b)

The dew point temperature T_{dew} is obtained by interpolation of the steam tables at the saturation temperature corresponding to $p=1.169$ kPa. From Table A-3.1, Appendix A, it will be between 5 and 10°C.

Equation

$$T_{\text{dew}}=5+\Delta T \qquad [°\text{C}]$$

Parameters

(See Table 12-3.)

$$\Delta T=\frac{\Delta p}{\Delta p'}\Delta T' \qquad [°\text{C}]$$

$$\Delta T'=10-5=5\Delta°\text{C}$$

$$\Delta p'=1.2276-0.8721=0.3555\ \text{kPa}$$

$$\Delta p=(1.1695-0.8721)=0.2974\ \text{kPa}$$

$$\Delta T=\frac{0.2974}{0.3555}(5)=4.2\Delta°\text{C}$$

Substitution

$$T_{\text{dew}}=5+\Delta T=(5+4.2) \qquad [°\text{C}]+[\Delta°\text{C}]=[°\text{C}]$$

Table 12-3 Interpolation Data

T [°C]	p [kPa]
5	0.8721
T_{dew}	1.169
10	1.2276

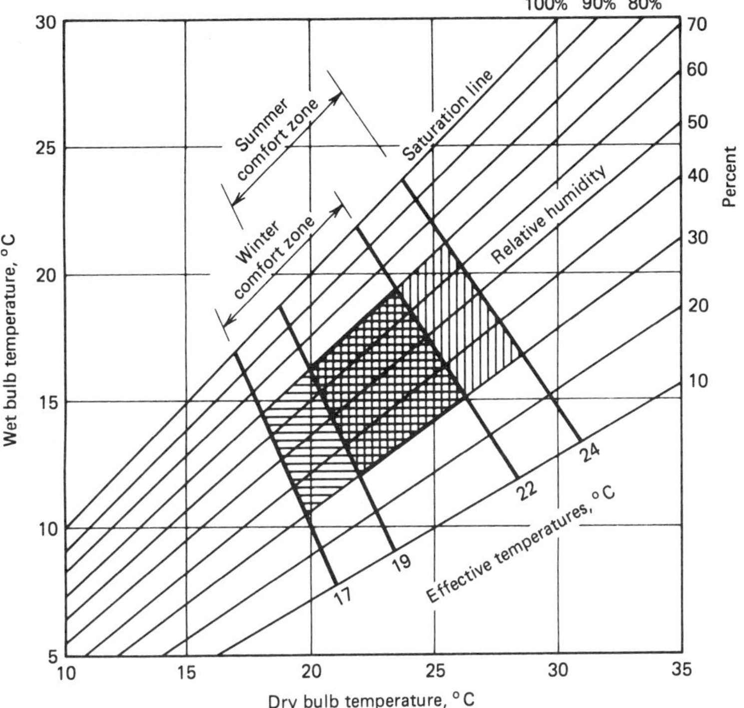

Figure 12-1 Comfort zone chart. (Adapted from the Frigidaire division, General Motors.)

Answer

$$T_{\mathbf{dew}} = 9.2°\mathrm{C}$$

Example Problem 12-2

If the air in Example Problem 12-1 flows at the rate of 0.5 m³/s over a cooling coil that is at a temperature of 5°C, determine the amount of water vapor (kg/h) that will be condensed. Assume that the air leaving the coil is saturated and the barometric pressure is the same. (See Figure 12-2 and Table 12-4.)

Solution

Use Equation 12-7 to calculate $\dot{M}_c$

$$\dot{M}_c = \dot{M}_a(\omega_1 - \omega_2) \qquad [\mathrm{kg/s}] \qquad (\text{Eq. 12-7})$$

Obtain the mass flow rate $\dot{M}_a$ of dry air from state 1 (given) and the given volumetric flow rate. Also, obtain ω_1 and ω_2 using Equation 12-2 and the steam tables. If ω_2 for saturation is less than ω_1, water will precipitate. If ω_2 for saturation is greater than ω_1, then $\omega_2 = \omega_1$ and no water will precipitate.

Equation

$$\dot{M}_a = \rho_a \dot{V} \qquad [\mathrm{kg/s}]$$

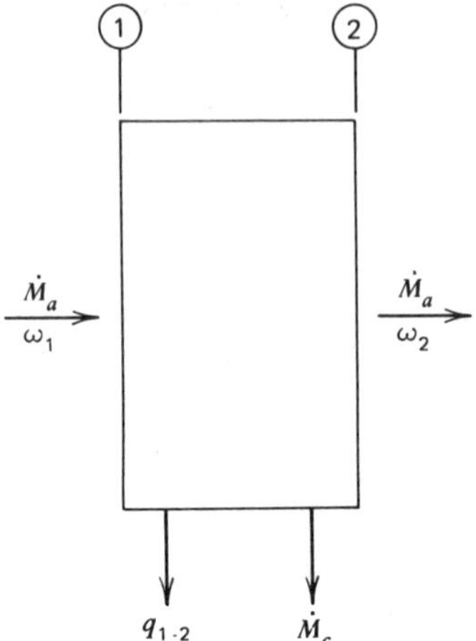

Figure 12-2 Open system, constant pressure process sketch for Example Problem 12-1.

Parameters

From the perfect gas law

$$\rho_a = \frac{p_{a_1}}{(R_a)(T_1)} \qquad [\text{kg/m}^3]$$

$$p_{a_1} = p_{b_1} - p_{wv_1} \qquad [\text{kPa}]$$

$$p_{b_1} = 0.100\ \text{MPa} = 100\ \text{kPa}$$

$$p_{wv_1} = \phi p_{vs} \qquad [\text{kPa}] \qquad (\text{Eq. 12-4})$$

$$\phi = 0.50 \qquad [—] \qquad (\text{given})$$

$$p_{vs} = 2.339\ \text{kPa} \qquad (\text{Appendix A, Table A-3.1, at } 20°\text{C})$$

$$p_{wv_1} = (0.50)(2.339) \qquad [—] \qquad [\text{kPa}] = [\text{kPa}]$$

$$= 1.1695\ \text{kPa}$$

$$p_{a_1} = (100) - (1.1695) \qquad [\text{kPa} - \text{kPa}] = [\text{kPa}]$$

$$= 98.831\ \text{kPa}$$

$$R_a = 0.287\ \text{kJ/kg}\cdot\Delta_1\text{K} \qquad (\text{Appendix A, Table A-9})$$

Table 12-4 Table of State.

Substance: Moist Air		①	②
$\dot{M}_a$	[kg/s]	0.5876	→
p_b	[MPa]	0.100 (g)	→
T_{db}	[°C]	20 (g)	5 (g)
ω	[—]	0.00736	0.005472

Note: g (given).

$$T_1 = 20°\text{C} = 293\ \text{K} \qquad \text{(given)}$$

$$\rho_a = \frac{(98.831)}{(0.287)(293)} \qquad \frac{[\text{kPa}]}{[\text{kJ/kg}\cdot\Delta_1\text{K}][\text{K}]} = \frac{[\text{kg}]}{[\text{m}^3]}$$

$$\rho_a = 1.1753\ \text{kg/m}^3$$

$$\dot{V} = 0.5\ \text{m}^3/\text{s} \qquad \text{(given)}$$

Substitution

$$\dot{M}_a = (1.1753)(0.5) \qquad [\text{kg/m}^3] \quad [\text{m}^3/\text{s}] = [\text{kg/s}]$$

Answer

$$\mathbf{\dot{M}_a = 0.5876\ kg/s}$$

Equation

$$\omega = 0.622\left(\frac{p_{wv_1}}{p_{a_1}}\right) \qquad [\text{—}] \qquad \text{(Eq. 12-2)}$$

Parameters

$$p_{wv_1} = 1.1695\ \text{kPa} \qquad \text{(from previous calculation)}$$

$$p_{a_1} = 98.831\ \text{kPa} \qquad \text{(from previous calculation)}$$

Substitution

$$(0.622)\frac{(1.1695)}{(98.831)} \qquad \frac{[\text{kPa}]}{[\text{kPa}]} \qquad [\text{—}]$$

Answer

$$\mathbf{\omega_1 = 0.00736} \qquad [\text{—}]$$

Equation

$$\omega_2 = 0.622\left(\frac{p_{wv_2}}{p_{a_2}}\right) \qquad [\text{—}] \qquad \text{(Eq. 12-2)}$$

Parameters

$$p_{wv_2} = p_{vs_2} = 0.8721\ \text{kPa} \quad \text{(Appendix A, Table A-3.1, at 5°C) saturated condition}$$

$$p_{a_2} = p_{b_2} - p_{wv_2} \qquad [\text{kPa}]$$

$$p_{b_2} = 0.10\ \text{MPa} = 100\ \text{kPa} \qquad \text{(given)}$$

$$p_{a_2} = 100 - 0.8721 \qquad [\text{kPa} - \text{kPa}] = [\text{kPa}]$$

$$= 99.1279\ \text{kPa}$$

Substitution

$$\omega_2 = 0.622 \frac{(0.8721)}{(99.1279)} \qquad \frac{[\text{kPa}]}{[\text{kPa}]} = [\text{kPa}]$$

Answer

$$\boldsymbol{\omega_2 = 0.005472} \qquad [—]$$

Substituting in previous answers into Equation 12-7, we obtain

$$\dot{M}_c = (0.5876)(0.00736 - 0.00547) \qquad [\text{kg/s}] \ \big[[—]-[—]\big] = [\text{kg/s}]$$

Final Answer

$$\boldsymbol{\dot{M}_c = 0.0011106 \text{ kg/s}}$$

or

$$\boldsymbol{\dot{M}_c = 3.998 \text{ kg/hr}}$$

12-1.8 ENTHALPY

The enthalpy of the air–water mixture is the sum of the component substance enthalpies, and the specific enthalpy of the mixture is usually based on 1 kg of dry air. Thus

$$h_m = h_a + \omega h_{wv} \qquad [\text{kJ/kg}] \tag{12-8}$$

where

h_m = enthalpy of the mixture [kJ/kg dry air]
h_a = enthalpy of the dry air [kJ/kg]
h_{wv} = enthalpy of water vapor [kJ/kg]
ω = humidity ratio [—]

The enthalpy of the air–water vapor mixtures contains both sensible and latent heat components. These mixtures are discussed in the following sections.

Sensible Heat for Air–Water Vapor Mixtures

The process of sensible heating or cooling is characterized by constant moisture content, dew point temperature and vapor pressure (partial pressure of moisture in the air), and corresponding changes in the relative humidity, dry bulb temperature, wet bulb temperature, and specific volume.

The sensible heat can be calculated considering the dry air and the water vapor separately, using appropriate gas and steam tables, or from the psychrometric chart.

Latent Heat for Air–Water Vapor Mixtures

The process of latent heat input or output is involved when the moisture content (humidity ratio) is changed. When the humidity ratio is increased, energy (heat) is required to change any increase in the mass of the water vapor from liquid into vapor. The amount of heat required is that of the latent heat of vaporization of the water for the conditions involved. Conversely, if the moisture content (mass) is reduced, the corresponding latent heat of vaporization must be removed. Often in air conditioning the latent heat involved in dehumidification or humidification can be the dominant heat transfer requirement, and consequently must be considered.

Example Problem 12-3

Calculate the latent heat required to vaporize 30 kg of water in the saturated liquid phase to saturated vapor with a dew point of 10°C.

Equation

$$H_1 = Mh_{fg} \qquad [\text{kJ}] \qquad (\text{Eq. 3-23}x)$$

Parameters

$$M = 20 \text{ kg} \qquad (\text{given})$$

$$h_{fg} = 2477.7 \text{ kJ/kg} \qquad (\text{Appendix A, Table A-3.1, at } 10°\text{C})$$

Substitution

$$H_1 = (20)(2477.7) \qquad [\text{kg}][\text{kJ/kg}] = [\text{kJ}]$$

Answer

$$\mathbf{H_1 = 49{,}554 \text{ kJ} = 49.55 \text{ MJ}}$$

12-2 THE PSYCHROMETRIC CHART

The psychrometric chart is constructed to avoid many detailed calculations relating

- Dry bulb temperature T_{db}.
- Humidity ratio ω.
- Relative humidity ϕ.
- Dew point T_{dew}.
- Wet bulb temperature T_{wb}.
- Specific enthalpy h_m.

The common air–water vapor psychrometric chart (Appendix A-12) is generally used in solving air conditioning problems and is based on a barometric pressure of 1 atmosphere. For barometric pressures deviating significantly from 1 atm, corrections must be made if the common psychrometric chart is used. Such corrections would affect the relative humidity, dew point, and enthalpy. Normal wet bulb temperatures would not be applicable.

The psychrometric chart is constructed on an orthogonal grid of humidity ratio ω versus temperature T_{db}. This temperature is the actual temperature of the air–water vapor mixture and is called the "dry bulb" temperature to differentiate it from the "wet bulb" temperature used to determine the relative humidity. (See Figure 12-3.)

As the air–water vapor mixture is cooled (temperature reduced) for a constant humidity ratio ω, the relative humidity ϕ will increase until the temperature is reached at which the vapor attains saturation. This saturation line has been drawn across the basic grid of the humidity ratio and temperature, as well as a family of lines of constant relative humidity. Thus the relative humidity ϕ for any humidity ratio ω at any temperature above the saturation temperature can be directly determined by the position of the intersection of the temperature and humidity ratio with respect to the relative humidity lines. The dry bulb temperature at the saturation line is the *dew point*. Any further cooling will result in the precipitation of moisture from the air and a reduction in the humidity ratio ω.

The wet bulb temperature grid is provided as the common method for measuring (or determining) the relative humidity. This procedure requires taking simultaneous readings of wet bulb temperature (thermometer covered with a thin film of water) and the normal air temperature (called the dry bulb temperature to differentiate the two temperatures). The lines of constant wet bulb temperature are carried throughout the area covered by the relative humidity lines. Figure 12-4 shows lines of constant wet bulb temperature on a basic psychrometric chart.

A family of constant enthalpy lines is superimposed on this grid. In the variation of the basic psychrometric chart (Figure 12-5), the constant enthalpy lines are defined by the scale crossing the chart at approximately

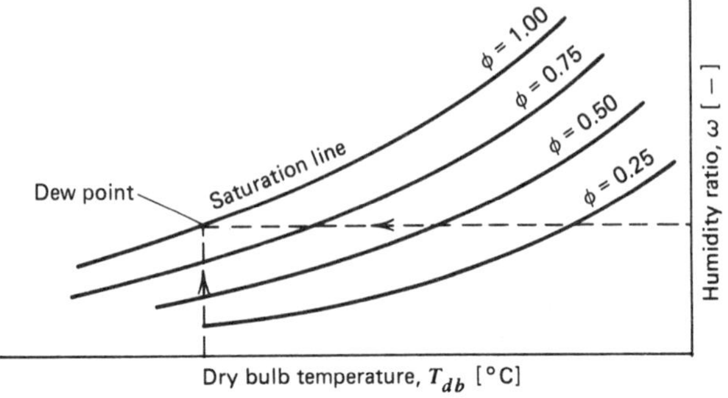

Figure 12-3 Basic psychrometric chart.

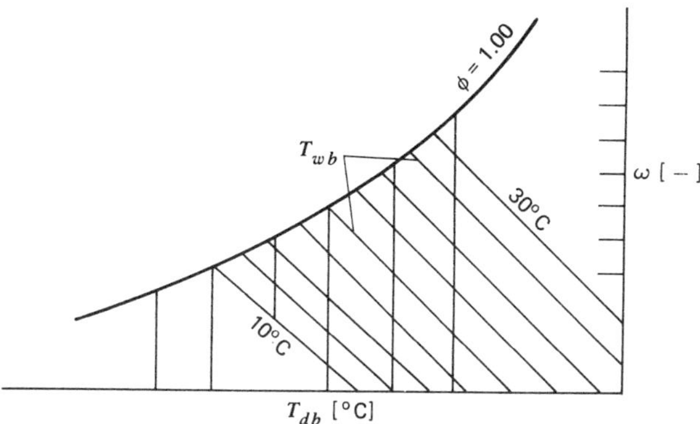

Figure 12-4 Lines of constant wet bulb temperature (T_{wb}) on a psychrometric chart.

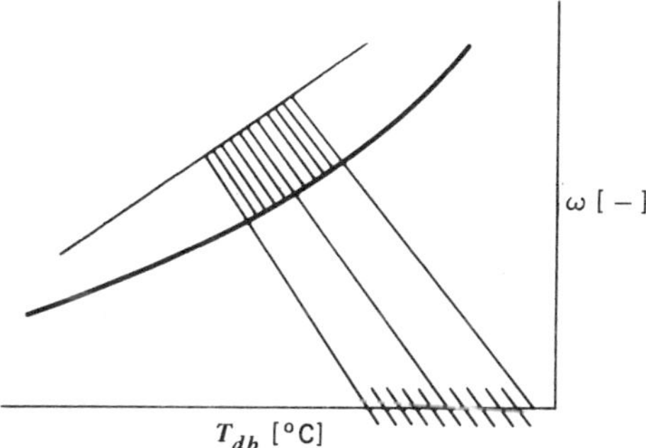

Figure 12-5 Lines of constant enthalpy on a psychrometric chart.

a 45° angle with scale marks on the outside temperature and humidity ratio grid edge. Only the major lines are drawn through the grid of the sample chart presented in this book. Consequently, for reading a value of enthalpy for a particular humidity ratio/temperature point, a straight edge is necessary. This technique is used to permit the superimposing of a wet bulb temperature network through the grid.

Example Problem 12-4

Solve Example Problem 12-1 using the psychrometric chart to obtain the humidity ratio and dew point. (See Figure 12-6, page 302.)

Answer

Locate the dry bulb temperature (20°C) on the horizontal scale, and then proceed vertically until the 50 percent relative humidity curve is intersected. This gives the humidity ratio ω, which is approximately 0.0073. Next, move horizontally along the ω line until the intersection with the saturation line is reached. At this point, read the dew point temperature on the dry bulb temperature scale. In this case it will be approximately 9°C.

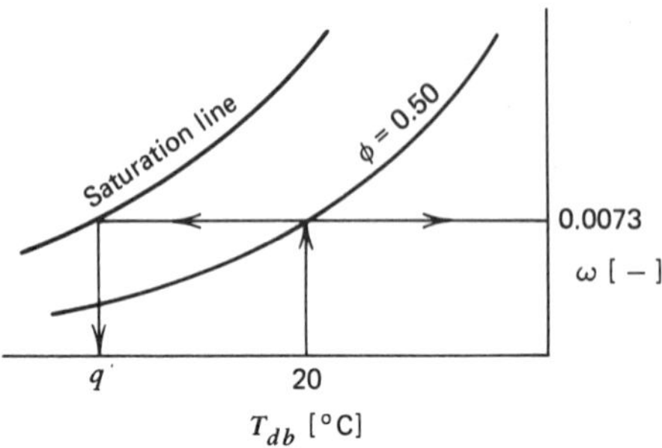

Figure 12-6 Sketch for Example Problem 12-4.

Example Problem 12-5

Solve Example Problem 12-2 using the psychrometric chart to obtain the condensate flow rate. (See Figure 12-7.)

Answer

Locate the initial dry bulb temperature (20°C) on the horizontal scale, and then proceed vertically until the 50 percent relative humidity curve is intersected. At this point read the humidity ratio on the vertical scale to the right. For the initial conditions this is approximately 0.00735 kg moisture/kg dry air. At the final

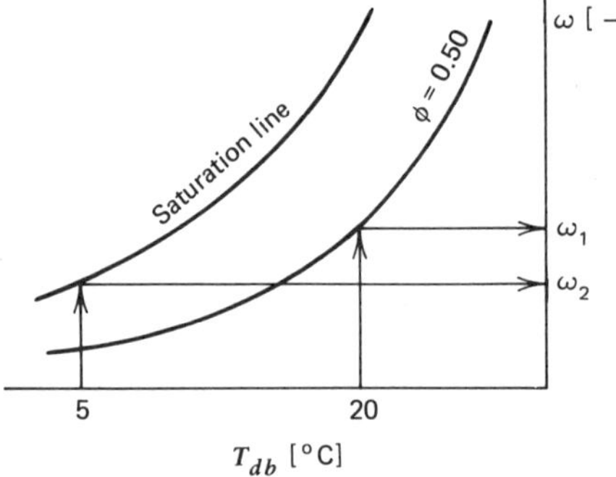

Figure 12-7 Sketch for Example Problem 12-5.

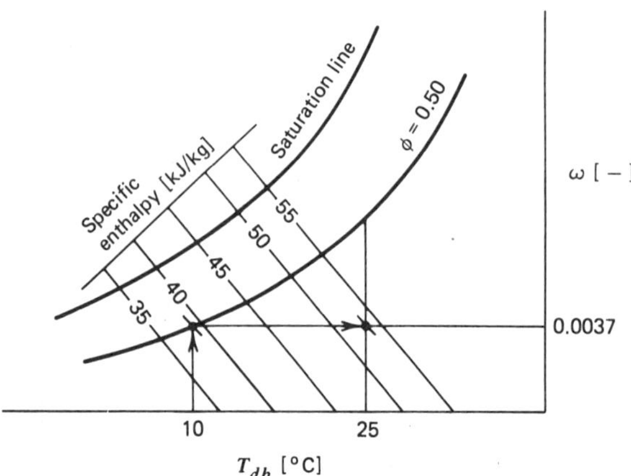

Figure 12-8 Sketch for Example Problem 12-6.

condition of 5°C and 100 percent relative humidity (on the surface of the cooling coil), the humidity ratio is approximately 0.00550 kg moisture/kg dry air. Thus the amount of condensate will be the difference between these two humidity ratios, which is 0.00185. For a mass flow rate of 0.59 kg/s, the condensate flow is 3.96 kg/h.

Next for h_2, move horizontally across the chart (constant humidity ratio) to the 25°C dry bulb temperature line. At this point the enthalpy $h_2 = 54.7$ kJ/kg dry air.

Equation

$$q = h_2 - h_1 \qquad [\text{kJ/kg dry air}]$$

Substitution

$$q = (54.7 - 39.4) \qquad [\text{kJ/kg dry air} - \text{kJ/kg dry air}]$$

Answer

$$\mathbf{q = 15.3\ kJ/kg\ dry\ air}$$

Example Problem 12-6

Calculate the amount of sensible heat required to heat air with a dry bulb temperature of 10°C and relative humidity of 50 percent to 25°C using the psychrometric chart. (See Figure 12-8.)

Plan

Use the equation $q = h_2 - h_1$ kJ/kg dry air. Obtain the values of h_2 and h_1 by using the psychrometric chart (Appendix A-12). Locate h_1 at the point where the 10°C dry bulb temperature intersects the 50 percent relative humidity line. At this point we can obtain the following values:

$$\omega_1 = 0.0037 \text{ kg moisture/kg dry air}$$

$$h_1 = 39.4 \text{ kJ/kg dry air}$$

12-3 PROBLEMS

12-1. Define the humidity ratio ω and relative humidity ϕ. Which one remains constant with changes in pressure and temperature? Why is this so?

12-2. Explain the difference between the dry bulb temperature T_{db} and the wet bulb temperature T_{wb}. How are measurements of each obtained? If you had a wet bulb temperature, a dry bulb temperature, and the barometric pressure, could you obtain the relative humidity? If so, by what means could this be done?

12-3. Define or describe the dew point.

12-4. Air is to be supplied to a room at 18°C with a relative humidity of 45 percent. Calculate the humidity ratio and dew point when the barometric pressure is 0.10 MPa.

12-5. If air has a 10°C dew point, what is the humidity ratio and relative humidity when the barometric pressure is 0.10 MPa and the dry bulb temperature is 25°C?

12-4 ① $\omega_{HUM. RAT.} = \frac{0.622(P_{WV})}{P_b - P_{WV}}$; $P_{WV} = \phi P_{VS}$; P_{VS} from table A3.1

② $T_{DEW} = \Delta T$ from table A3.1 $\left(\frac{P_{VS} - L}{H - L}\right)$

12-6. If the air in Problem 12-4 flows at the rate of 0.45 m^3/s over a cooling coil, which is at a temperature of 5°C, determine the amount of water vapor in kg/h that will be condensed.

12-7. If the cooling coil temperature is lowered to 3°C in Problem 12-6, what amount of water vapor in kg/h will be condensed?

12-8. If 20°C air at atmospheric pressure with a relative humidity of 50 percent is heated to 30°C, what is its relative humidity at 30°C?

12-9. If 20°C air at atmospheric pressure is saturated, what is its relative humidity when it is heated to 30°C?

12-10. The air at sea level (0.1013 MPa) is 20°C with a relative humidity of 0.60. It rises to an altitude of 3000 m, where the pressure is 0.0695 MPa and the temperature is 5°C.
(a) How much moisture precipitates per kg of dry air?
(b) What relative humidity at sea level would result if the air just reaches saturation at the 3000 m altitude conditions?

12-11. The air in a compressed air system at 900 kPa pressure is saturated at 25°C. What is the relative humidity when it is expanded to 100 kPa pressure at 25°C?

12-12. Atmospheric air at 100 kPa, 25°C, and $\phi=0.15$ is compressed to 900 kPa. When it has cooled to 25°C in the compressed air system receiver, how much water condenses (if any) in terms of kilograms of water per kilogram of dry air?

12-13. If the atmospheric air in Problem 12-12 had a $\phi=0.50$, how much water would condense (if any)?

12-14. Atmospheric air at 25°C and 60 percent relative humidity is compressed to 1 MPa and stored in a tank called a receiver at a temperature of 25°C. Subsequently, the air is expanded isothermally to the atmospheric pressure. What is the relative humidity of the expanded air at 25°C and 1 atm pressure?

12-15. In a manufacturing plant compressed air system, outside air at 0.103 MPa, 28°C, and relative humidity ϕ of 0.5 is compressed to 1 MPa and is stored in a receiver (tank) in room A, where the temperature is 28°C. The compressed air in the receiver (tank) cools to the room A temperature (28°C). The air is drawn from the receiver to a regulator in a temperature-controlled room, at 25°C temperature. The air pressure is dropped to 0.5 MPa; its temperature is 25°C.
(a) What is the relative humidity of the regulated air (0.5 MPa at 25°C)? It is important that the relative humidity in the regulated air does not exceed 0.90.
(b) What temperature of room A would result in a relative humidity of 0.90 in the regulated air?

12-16. Describe the psychrometric chart and indicate how to obtain
(a) The value of the humidity ratio ω at a set of dry bulb temperature T_{db} and relative humidity ϕ.
(b) The dew point T_{dew} for any humidity ratio ω.

12-17. Using the psychrometric chart, solve Problems 12-4 and 12-6.

12-18. Outside air at 30°C and 50 percent relative humidity is processed through an air conditioning system to provide conditioned air at 25°C and 60 percent relative humidity. To what dew point must the outside air be cooled in the air conditioning system to remove the excess water (if any)?

If the outside air warms up to 35°C and the relative humidity increases to 75 percent, how much lower must the dew point be to compensate?

12-19. In Problem 12-18, how much heat per kg of dry air is removed to bring the outside air down to the required dew point temperature in both cases? How much heat must be added to reheat the cooled air to 25°C?

12-20. Outside air at 35°C and 30 percent relative humidity is processed through an air conditioning system to provide conditioned air at 25°C and 50 percent relative humidity. To what dew point must the outside air be cooled in the air conditioning system to remove the excess water (if any)?

12-21. In Problem 12-20, how much heat per kilogram of dry air is removed to bring the outside air down to the required dew point temperature? Then how much heat must be added to reheat the cooled air to 25°C?

12-22. If the outside air relative humidity in Problem 12-20 dropped, below what value would the addition of water be required to obtain 50 percent relative humidity at 25°C?

12-23. If, in Problem 12-20, the outside air had a relative humidity of 15 percent, how much water is required per kg of dry air to obtain 50 percent relative humidity at 25°C?

12-24. In Problem 12-23, if the water added is saturated liquid at 25°C, how much heat (if any) must be added to obtain 50 percent relative humidity at 25°C?

12-25. In an engine carburetor, 0.1 kg of liquid butane (saturated at 23.9°C) is injected into 1 kg of dry atmospheric air at 100 kPa barometric pressure at 25°C. If no heat is added, what will be the final equilibrium temperature of the air–fuel mixture?

12-26. In an engine carburetor, 0.08 kg of liquid butane (saturated at 18.3°C) is injected into 1 kg of dry atmospheric air at 70 kPa barometric pressure at 10°C. If no heat is added, what will be the final equilibrium temperature of the air–fuel mixture?

12-27. In Problem 12-25, if the atmospheric air has a relative humidity of 40 percent will there be a danger of carburetor icing? If so, what is the minimum amount of heat required to prevent icing?

12-28. If 0.2 kg of water saturated at 25°C were injected into 1 kg of dry air at 7.4 MPa and 2500°C, what would be the final equilibrium temperature and pressure with no change in volume?

12-18 – $P_{wv} = \phi P_{vs}$ T_{DEW} INTERPOLATED (TABLE A 3.1)

b.) since dewpoint is independent of outside air, there is no change in temperature.

CHAPTER 13
HEAT TRANSFER PRINCIPLES

Whenever a temperature gradient exists, thermal energy (heat) is transferred. The free flow of heat or thermal energy is always along a temperature gradient from a higher temperature to a lower temperature, in accordance with the second law of thermodynamics temperature gradients exist throughout the world around us. Temperature gradients are inherent in material aspects of our everyday life such as power generation, propulsion and transportation, chemical and metallurgical processing, and heating and refrigeration.

An understanding of the laws of the flow of thermal energy (heat transfer) is important to all fields of engineering because these laws are of controlling importance in the design, construction, and operation of the many diverse forms of heat exchange apparatus required in our scientific and industrial technology. In thermodynamic cycles, heat transfer processes are involved whenever heat crosses the system boundary (heat input or heat rejection). Also, in the mechanical implementation (design) of a thermodynamic system, heat transfer devices are usually required to cool mechanical parts in order to maintain safe operating temperatures. Every engineer can expect to be confronted, from time to time, with problems relating to the efficient transmission or the control and regulation of thermal energy.

To facilitate the solution of heat transfer problems, much effort has been expended by investigators to obtain test data and to develop basic theories and equations that will provide relationships for the solution of whatever heat transfer problem is encountered. Test data have been correlated in terms of dimensionless parameters so as to provide useful empirical equations for general application. Theoretical considerations and models have been used to develop analytical equations. Such analytical equations are usually limited in the scope of their application because real material properties and fluid flow deviate from the theoretical models used in the development of the equations. To extend the scope of application of theoretical equations, empirical factors and coefficients are often applied to the basic equations.

13-1 BASIC HEAT TRANSFER MECHANISMS

Heat is transferred by three basic mechanisms. In two of these mechanisms the heat is transferred through material, while in the third mechanism it is transmitted through empty space and vacuum as well as through certain materials transparent to thermal radiation. The physical laws for heat transfer are different for each of these basic mechanisms, which are designated

- Conduction.
- Convection.
- Radiation.

CONDUCTION

In conduction heat flow, the thermal energy is transmitted by direct molecular communication without appreciable displacement of the molecules. Conduction occurs through bodies of solids, liquids, and gases, and from one body to another, when they are in physical contact with each other. A common example of heat transfer by conduction is when one picks up a hot plate and burns one's fingers. The heat energy is conducted from the hot plate to the colder fingers, where there is physical contact between the two.

CONVECTION

In convection heat flow, the thermal energy is primarily transmitted through mass motion (physical displacement) of the molecules within the convective body. Convection occurs only in fluids (liquids or gases), and convective heat transfer takes place between a solid surface and a fluid. The fluid molecules in contact with the solid surface are heated or cooled by conduction, and are swept away from the solid surface by convective flow patterns or flow turbulence, with subsequent thermal energy transfer by conduction with adjacent fluid molecules occurring. Or conversely, molecules at the temperature of the fluid body are swept toward a solid surface, and upon physical contact thermal energy transfer is effected. An example of heat transferred by convection is the chilling effect of a cold wind on a warm body. Without the flow of cold air past a warm body, the primary mode of energy transfer would be conduction, which would be much less.

When there is no phase change in the fluid (a liquid remains a liquid or a gas remains a gas), there are two categories of convection. When the convective flow patterns in the fluid are a result of the effect of the thermal gradient upon the density of the fluid, the resulting convective heat transfer process is called *free convection*. If, on the other hand, the convective flow patterns in the fluid are a result of mechanical pumping, the resulting heat transfer process is called *forced convection*.

When there is a phase change in the fluid (a liquid becomes a gas, or a gas becomes a liquid), there are two categories of convection, namely *boiling* (evaporation) and *condensation*. Boiling may occur under conditions of either free or forced convection. A commonly observed example of free convection boiling heat transfer comes from an open pan of water on a stove. As energy is progressively added to the water from the surface of the hot pan, first a vapor comes from the surface, followed by the formation of bubbles on the bottom of the pan that rise up through the liquid and discharge as steam at the surface when these bubbles break. On the other hand, condensation is a forced convective process only. The phase change of the gas to a liquid on the condensation surface always causes a forced gas flow toward the condensation surface; however, a special case of condensing heat transfer occurs when moisture (water vapor) in the air condenses on a cold surface such as a window or a water pipe.

RADIATION

In radiation heat transfer, the thermal energy is transmitted by electromagnetic waves, differing from light (the visible spectrum) only by their respective wavelengths. These electromagnetic waves (radiant energy) pass readily through a vacuum and empty space, and can pass through gases, liquids, and some solids. The emitted radiant energy is not a continuous variable but is formed by finite batches (or "quanta") of energy associated with step changes in the energy level of the emitting molecule of the body. The rate of this radiant energy emission, and hence the rate of radiant heat transfer, is proportional to the fourth power of the absolute temperature. This is in contrast to the direct proportionality of conductive and convective heat transfer to the temperature difference or temperature gradient. The transmitted radiant energy is absorbed in varying amounts on or near the surface of bodies upon which it impinges, depending upon the absorptivity of the body. Sunshine is a common example of radiant heat transfer. The radiant energy of the sun is transmitted through space and the earth's atmosphere. This transmitted energy is absorbed according to the characteristics of the surface upon which it is incident. Dark objects usually absorb a major portion of this radiant energy.

13-2 CONDUCTION

Conduction is fundamental to almost all facets of heat transfer. Consequently, the beginning student should acquire a good understanding of the basic conduction process. Three key terms that may be new to the reader are

- Thermal conductivity k.
- Thermal conductance $\mathcal{K}$ or $\overline{\mathcal{K}}$.

- Thermal resistivity r.
- Thermal resistance $\mathcal{R}$ or $\overline{\mathcal{R}}$.

Thermal conductivity (k) is a measurement of the capacity of a material to conduct heat. Thermal conductivity can be considered to be an index of the material indicating the amount of heat conducted per unit conduction path area per unit temperature gradient.

$$k = \frac{\dot{Q}}{A\ \Delta T/\mathrm{m}} \qquad \left[\frac{\mathrm{W}}{\mathrm{m}\cdot\Delta_1{}^{\circ}\mathrm{C}}\right] \tag{13-1}$$

where

$\dot{Q}$ = heat flow rate [W]

A = area of the heat transfer path $[\mathrm{m}^2]$

$\Delta T/m$ = temperature gradient along the path $[\Delta_1{}^{\circ}\mathrm{C/m}]$

Thermal conductance $\mathcal{K}$ or $\overline{\mathcal{K}}$ is a measurement of the capacity of a *heat transfer path* to conduct heat. The thermal conductance $\mathcal{K}$ is for a unit area of the conduction path.

$$\mathcal{K} = \frac{\dot{Q}}{A\Delta T} \qquad \left[\frac{\mathrm{W}}{\mathrm{m}^2\cdot\Delta_1{}^{\circ}\mathrm{C}}\right] \tag{13-2}$$

where

$\dot{Q}$ = heat flow rate [W]

A = area of the heat transfer path $[\mathrm{m}^2]$

T = temperature difference between ends of the path $[\Delta_1{}^{\circ}\mathrm{C}]$

The thermal conductance $\overline{\mathcal{K}}$ is for the conduction path.

$$\overline{\mathcal{K}} = A\mathcal{K} = \frac{\dot{Q}}{\Delta T} \qquad \frac{\mathrm{W}}{\Delta_1{}^{\circ}\mathrm{C}} \tag{13-3}$$

Thermal resistivity r and thermal resistance $\mathcal{R}$ and $\overline{\mathcal{R}}$ are the inverse of the thermal conductivity k and thermal conductance $\mathcal{K}$ and $\overline{\mathcal{K}}$.

$$\mathcal{R} = \frac{1}{\mathcal{K}} \qquad [\mathrm{m}^2\,\Delta_1{}^{\circ}\mathrm{C/W}] \tag{13-4}$$

$$\overline{\mathcal{R}} = \frac{1}{\overline{\mathcal{K}}} = \frac{1}{A\mathcal{K}} = \frac{\mathcal{R}}{A} \qquad [\Delta_1{}^{\circ}\mathrm{C/W}] \tag{13-5}$$

The following sections will first describe the thermal conductivity characteristics of various materials and then describe the heat transfer equations for various conductive heat transfer paths.

13-2.1 THERMAL CONDUCTIVITY

Thermal conductivity is the property of a material that indicates its ability to conduct heat. A material with a low thermal conductivity is termed a *good insulator*. Materials have a wide range of thermal conductivities as indicated in Figure 13-1, which is a plot of thermal conductivities versus temperature for an illustrative group of gases, liquids, and solids. In general, gases have the lowest conductivities, liquids are in the midrange, and solids have the highest conductivities. Pure silver has the highest thermal conductivity of any material.

For a free flow of heat through a material there must be a temperature gradient, with the heat flowing from the hotter region to the colder region. According to the kinetic theory of heat, the temperature of a molecule is proportional to the kinetic energy of the molecule. The basic concept of thermal conductivity is that of the rapidity of the diffusion of the molecular kinetic energy through the material. The diffusion or transfer of the molecular kinetic energy takes place by molecular collisions and/or by diffusion of faster-moving electrons from higher to lower temperature molecules, and/or by vibrations along crystal lattice bonds. The theoretical models for thermal conductivities and the magnitude of the values for thermal conductivities differ for gases, liquids, and solids.

Although the mechanism of the propagation of molecular motion in heat transfer has similarities to the propagation of sound and to the electrical properties of the material, only limited theoretical correlations have been achieved. Consequently, the evaluation of thermal conductivities has been primarily experimental; such data are available from many published sources expanding on the information given in Appendix B of this book.

13-2.2 CONDUCTION HEAT TRANSFER PATHS

Any heat transfer path can be characterized by a "thermal conductance" and a "thermal resistance," which is the reciprocal of the thermal conductance. The heat flow through the conductive heat transfer path is described by the following equation:

$$\dot{Q}=\overline{\mathcal{K}}(T_1-T_2)=\overline{\mathcal{K}}\,\Delta T=\frac{\Delta T}{\overline{\mathcal{R}}} \qquad [\mathrm{W}] \tag{13-6}$$

where

$\dot{Q}$ = heat flow rate [W]

$\overline{\mathcal{K}}$ = thermal conductance [W/Δ_1°C]

$\overline{\mathcal{R}}$ = thermal resistance [Δ°C/W]

T_1 = temperature at one end of the path [°C]

T_2 = temperature at the other end of the path [°C]

ΔT = temperature drop along the path [Δ°C]

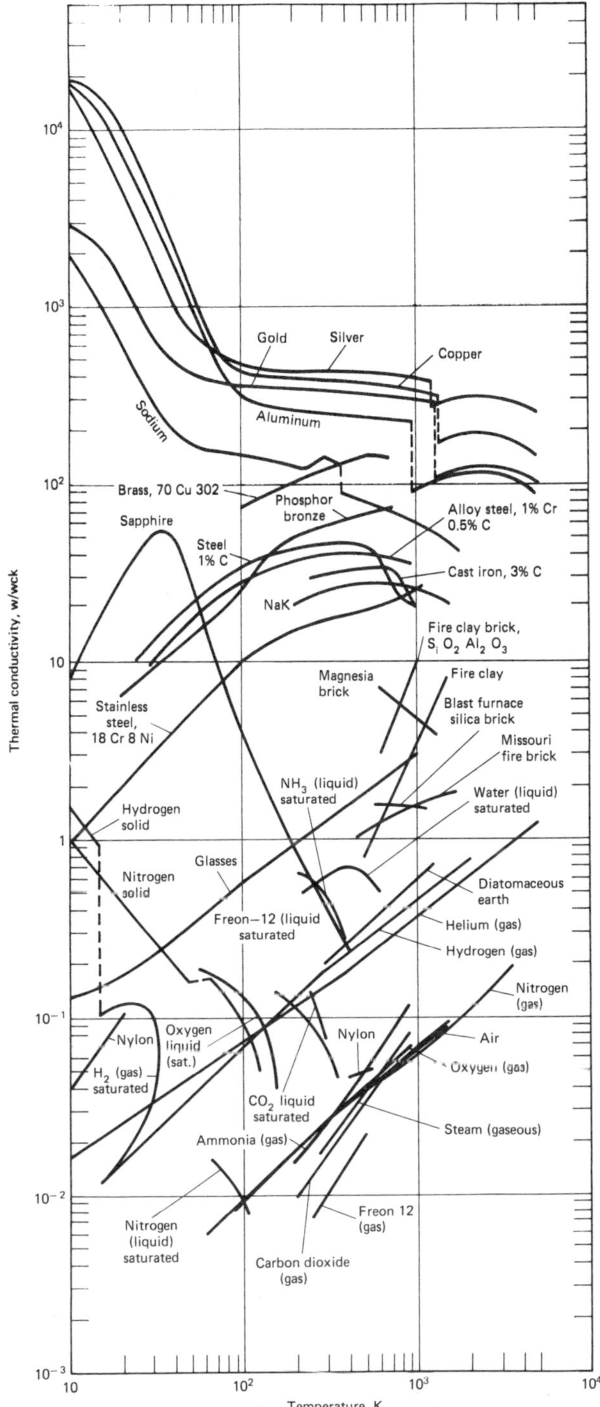

Figure 13-1 Thermal conductivity of gases, liquids, and solids.

Sometimes a heat transfer path thermal conductance ($\mathcal{K}$) per unit path cross-sectional area A is used. Then

$$\dot{Q}=A\mathcal{K}(T_1-T_2)=A\mathcal{K}\,\Delta T=\frac{A\,\Delta T}{\mathcal{R}} \qquad [\mathrm{W}] \tag{13-7}$$

where

$\dot{Q}$, T_1, T_2, and ΔT are defined as for Equation 13-6

$A=$ heat transfer path, cross-sectional area normal to the heat flow $[\mathrm{m}^2]$

$\mathcal{K}=$ thermal conductance per path unit cross-sectional area $[\mathrm{W/m^2\,\Delta_1{}^\circ C}]$

$\mathcal{R}=$ thermal resistance per path unit cross-sectional area $\left[\dfrac{\mathrm{m^2\,\Delta^\circ C}}{\mathrm{W}}\right]$

This equation is similar in form to Ohm's law for the flow of electrical current, which is usually expressed as follows:

$$I=\frac{E_v}{R_e} \qquad [\mathrm{A}] \tag{13-8}$$

where

$I=$ current flow rate $[\mathrm{A}]$

$E_v=$ voltage drop $[\mathrm{V}]$

$R_e=$ electrical resistance $[\Omega]$

The conduction of heat (energy) can be shown to be analogous to the flow of electrical current as $\dot{Q}=\Delta T/\mathcal{R}$. A further discussion of an electrical analogy in heat transfer and its application in solving complex heat transfer problems can be found in references such as Kreith(16).

In most engineering applications, conduction heat transfer paths are varied in form. The basic forms include:

- Linear conduction paths.
- Two-dimensional conduction paths.
- Three-dimensional conduction paths.
- Extended surfaces (rods and fins).
- Bodies with internal heat sources.

The following discussion is limited to the linear conduction paths. For discussions of the other forms the reader is referred to appropriate textbooks.

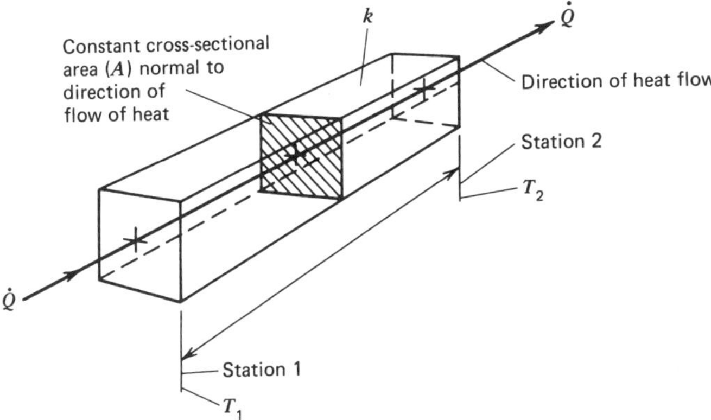

Figure 13-2 The linear conduction, heat transfer path element.

Linear conduction heat transfer path elements can be used singly or arranged in the form of:

- A number of paths with the same temperature drop forming a *parallel* system.
- A single path across the temperature drop forming a *series* system.

The characteristics of the path element and the parallel and series systems are discussed in the following sections.

The Linear Conduction Heat Transfer Path Element

The classical linear conduction heat transfer path element has a constant cross section perpendicular to the heat flow and a constant value of thermal conductivity of its material throughout its length, as illustrated in Figure 13-2.

The arrangements of the basic equations for a simple linear conduction path element are given in Table 13-1.

Example Problem 13-1

What is the thermal conductance $\overline{\mathcal{K}}$ of the heat transfer path along the length of a square metal bar 2 cm×2 cm×1 m long if the thermal conductivity of the metal were 3.70 W/cm Δ_1°C? (See Figure 13-3, page 314.)

Equation

$$\overline{\mathcal{K}}=\frac{Ak}{x} \qquad [\mathrm{W}/\Delta_1{}^\circ\mathrm{C}] \qquad (\text{Eq. 13-6})$$

Parameters

$$A=(2\text{ cm})(2\text{ cm})=4\text{ cm}^2 \qquad (\text{given})$$

$$k=3.70\text{ W/cm }\Delta_1{}^\circ\mathrm{C} \qquad (\text{given})$$

$$x=(1\text{ m})(100\text{ cm/m})=100\text{ cm} \qquad (\text{given})$$

Table 13-1 Arrangements of the Basic Equation (13-6) for Simple Linear Conduction

Rate of heat flow	$\dot{Q}=A\frac{k}{x}(T_1-T_2)=A\frac{k}{x}\Delta T=\overline{\mathcal{K}}\,\Delta T=\frac{\Delta T}{\overline{\mathcal{R}}}$	[W]
Path cross-sectional area	$A=\dot{Q}\frac{x}{k}\frac{1}{(T_1-T_2)}=\dot{Q}\frac{x}{k}\frac{1}{\Delta T}$	$[m^2]$
Path length	$x=A\frac{k}{\dot{Q}}(T_1-T_2)=\frac{Ak}{\dot{Q}}\Delta T$	[m]
Path material conductivity	$k=\frac{\dot{Q}x}{A(T_1-T_2)}=\frac{\dot{Q}x}{A}\cdot\frac{1}{\Delta T}$	[W/m Δ_1°C]
Path conductance	$\overline{\mathcal{K}}=A\frac{k}{x}$	[W/Δ_1°C]
	$\mathcal{K}=\frac{k}{x}$	[W/m^2 Δ_1°C]
Path resistance	$\overline{\mathcal{R}}=\frac{x}{Ak}$	[Δ°C/W]
	$\mathcal{R}=\frac{x}{k}$	[m^2 Δ°C/W]
Path temperature drop	$\Delta T=(T_1-T_2)=\frac{\dot{Q}x}{Ak}=\frac{\dot{Q}}{\overline{\mathcal{K}}}=\dot{Q}\overline{\mathcal{R}}$	[Δ°C]
Temperature at one end of path T_1	$T_1=T_2+\frac{\dot{Q}x}{Ak}=T_2+\frac{\dot{Q}}{\overline{\mathcal{K}}}=T_2+\dot{Q}\overline{\mathcal{R}}$	[°C]
Temperature at other end of path T_2	$T_2=T_1-\frac{\dot{Q}x}{Ak}=T_1-\frac{\dot{Q}}{\overline{\mathcal{K}}}=T_1-\dot{Q}\overline{\mathcal{R}}$	[°C]

Substitution

$$\overline{\mathcal{K}}=\frac{(4)(3.70)}{100}\qquad\frac{[cm^2][W/cm\,\Delta_1°C]}{[cm]}=[W/\Delta_1°C]$$

Answer

$$\overline{\mathcal{K}}=\mathbf{0.148\ W/\Delta_1°C}$$

Example Problem 13-2

How much heat would be conducted by the metal bar in Example Problem 13-1 if the temperature drop between the ends of the bar were 100 Δ°C? (See Figure 13-4.)

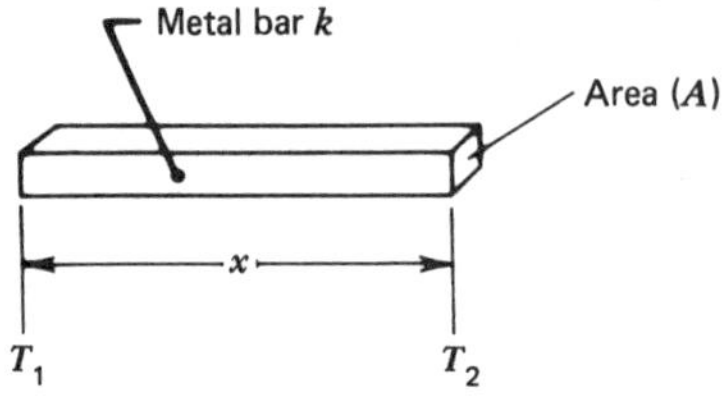

Figure 13-3 Sketch for Example Problem 13-1. Single one-dimensional conduction path.

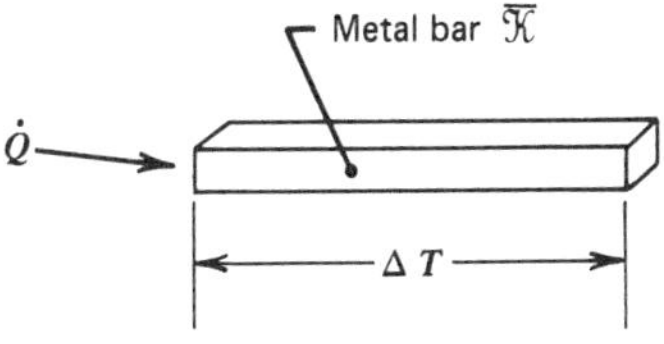

Figure 13-4 Sketch for Example Problem 13-2. Single one-dimensional conduction path.

Equation

$$\dot{Q}=\overline{\mathcal{K}}\,\Delta T \qquad [\mathrm{W}] \qquad (\text{Eq. 13-6})$$

Parameters

$$\overline{\mathcal{K}}=0.148[\mathrm{W}/\Delta_1{}^\circ\mathrm{C}] \qquad (\text{from Example Problem 13-1})$$

$$\Delta T=100[\Delta{}^\circ\mathrm{C}] \qquad (\text{given})$$

Substitution

$$\dot{Q}=(0.148)(100) \qquad [\mathrm{W}/\Delta_1{}^\circ\mathrm{C}][\Delta{}^\circ\mathrm{C}]=[\mathrm{W}]$$

Answer

$$\dot{Q}=\mathbf{14.8\ W}$$

The Parallel Conduction Path System

When a number of heat transfer path elements are arranged so as to provide multiple heat transfer paths between two temperature levels (T_1 and T_2), as illustrated in Figure 13-5, the conduction system is called a *parallel* path system.

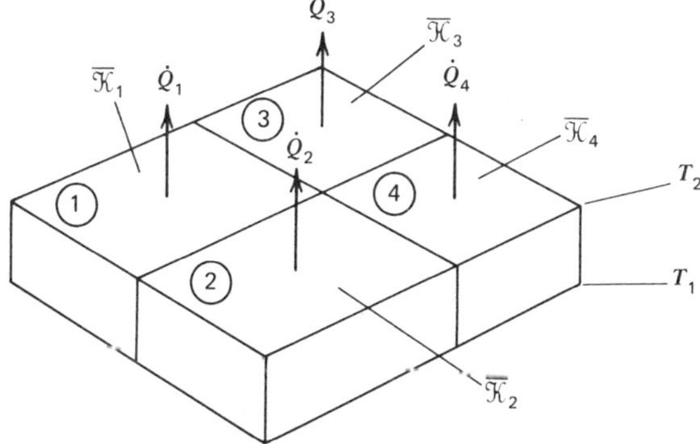

Figure 13-5 The parallel conduction path system.

The parallel conduction path system illustrated in Figure 13-5 consists of four separate conduction paths. Each path has the same temperature drop (T_1-T_2). Each conduction path has its characteristic thermal conductance $(\overline{\mathcal{K}}_1, \overline{\mathcal{K}}_2$, etc.). The total heat flow through the parallel condition system is the sum of the heat flows of the separate conduction paths.

$$\dot{Q}=(\overline{\mathcal{K}}_1+\overline{\mathcal{K}}_2+\overline{\mathcal{K}}_3+\overline{\mathcal{K}}_4)\Delta T \qquad (13\text{-}9)$$

This equation is derived as follows: The total heat flow $\dot{Q}_t$ is the sum of the heat flow through each element:

$$\dot{Q}_t=\dot{Q}_1+\dot{Q}_2+\dot{Q}_3+\dot{Q}_4$$

The heat flow rate through any element is given by Equation 13-4 as

$$\dot{Q}_i=\overline{\mathcal{K}}_i\,\Delta T$$

Then

$$\dot{Q}_t=\overline{\mathcal{K}}_1\,\Delta T_1+\overline{\mathcal{K}}_2\,\Delta T_2+\overline{\mathcal{K}}_3\Delta T_3\cdots+\overline{\mathcal{K}}_n\,\Delta T_n$$

Since ΔT is the same for all paths in a parallel conduction path system (by definition of a parallel conduction path system)

$$\dot{Q}_t=(\overline{\mathcal{K}}_1+\overline{\mathcal{K}}_2+\overline{\mathcal{K}}_3\cdots+\overline{\mathcal{K}}_n)\Delta T \qquad [\mathrm{W}] \qquad (\text{Eq. 13-7})$$

Since

$$\dot{Q}_t=\overline{\mathcal{K}}_t\Delta T \qquad [\mathrm{W}] \qquad (\text{Eq. 13-4})$$

$$\overline{\mathcal{K}}_t=\overline{\mathcal{K}}_1+\overline{\mathcal{K}}_2+\overline{\mathcal{K}}_3\cdots+\overline{\mathcal{K}}_n \qquad [\mathrm{W}/\Delta_1{}^\circ\mathrm{C}] \qquad (13\text{-}10)$$

The basic equations for parallel conduction path systems are listed in Table 13-2.

Table 13-2 Basic Equations for Parallel Conduction Path Systems

Rate of heat flow $\dot{Q}_t=\overline{\mathcal{K}}_t\,\Delta T$	[W]	(Eq. 13-6)
Total thermal conductance $\overline{\mathcal{K}}_t=\overline{\mathcal{K}}_1+\overline{\mathcal{K}}_2+\overline{\mathcal{K}}_3\cdots+\overline{\mathcal{K}}_n$	[W/Δ_1°C]	(Eq. 13-10)
Temperature drop $\Delta T=\dfrac{\dot{Q}_t}{\overline{\mathcal{K}}_t}$	[°C]	(Eq. 13-6x)

In a parallel conduction path system, a single high thermal conductance path can exercise a dominant control of the amount of heat flow. By way of an illustration, if the thermal conductance of one path is 80 percent of the total parallel system conductance, relatively large changes in the thermal conductances of another path, such as doubling or halving its thermal conductance, would have relatively little effect upon the thermal conductance of the total path system.

Example Problem 13-3

A molybdenum plate structure 5 cm thick is attached to a supporting structure with carbon steel rivets. The total cross-sectional area of the rivets is equal to 0.5 percent of the total surface area. (See Figure 13-6.) At 25°C, what is the thermal conductance of the following parts of the structure?

1. The molybdenum material per unit area of structure.
2. The rivets per unit area of structure.
3. The combination of the molybdenum plate structure with steel rivets per unit area.

C-1. For the Molybdenum Plate:

Equation

$$\bar{\mathcal{K}}_m = \frac{k_m A_m}{x_m} \qquad [\mathrm{W}/\Delta_1{}^\circ\mathrm{C}] \qquad \text{(Eq. 13-6x)}$$

Parameters

$$k_m = 123\ \mathrm{W/m}\ \Delta_1{}^\circ\mathrm{C} \qquad \text{(Appendix B-5)}$$

$$A_m = (1\ \mathrm{m})(0.995\ \mathrm{m}) = 0.995\ \mathrm{m}^2 \qquad \text{(given)}$$

$$x = 0.05\ \mathrm{m} \qquad \text{(given)}$$

Substitution

$$\bar{\mathcal{K}}_m = \frac{(123)(0.995)}{0.05} \qquad \frac{[\mathrm{W/m}\ \Delta_1{}^\circ\mathrm{C}][\mathrm{m}^2]}{[\mathrm{m}]} = [\mathrm{W}/\Delta_1{}^\circ\mathrm{C}]$$

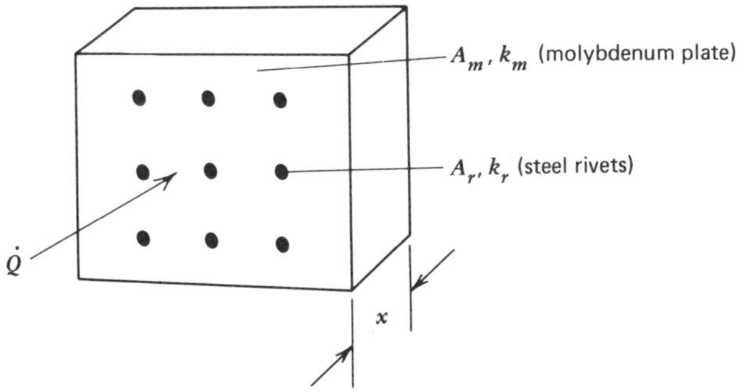

Figure 13-6 Sketch for Example Problem 13-3. Composite structure for conduction.

Answer

$$\overline{\mathcal{K}}_m = \mathbf{6.12\ W/\Delta_1{}^\circ C}$$

C-2. For the Steel Rivets:

Equation

$$\overline{\mathcal{K}}_r = \frac{k_r A_r}{x_r} \qquad [\mathrm{W}/\Delta_1{}^\circ\mathrm{C}] \qquad \text{(Eq. 13-6x)}$$

Parameters

$$k_r = 43\ \mathrm{W/m\,\Delta_1{}^\circ C} \qquad \text{(Appendix B-5)}$$

$$A_r = (1\ \mathrm{m})(0.005\ \mathrm{m}) = 0.005\ \mathrm{m}^2 \qquad \text{(given)}$$

$$x = 0.05\ \mathrm{m} \qquad \text{(given)}$$

Substitution

$$\overline{\mathcal{K}}_r = \frac{(43)(0.005)}{0.05} \qquad \frac{[\mathrm{W/m\,\Delta_1{}^\circ C}]}{[\mathrm{m}]} = [\mathrm{W}/\Delta_1{}^\circ\mathrm{C}]$$

Answer

$$\overline{\mathcal{K}}_r = \mathbf{4.30\ W/\Delta_1{}^\circ C}$$

C-3. For the Combination:

Equation

$$\overline{\mathcal{K}}_t = \overline{\mathcal{K}}_m + \overline{\mathcal{K}}_r \qquad [\mathrm{W}/\Delta_1{}^\circ\mathrm{C}] \qquad \text{(Eq. 13-10)}$$

Parameters

$$\overline{\mathcal{K}}_m = 6.12\ \mathrm{W}/\Delta_1{}^\circ\mathrm{C} \qquad \text{(from C-1)}$$

$$\overline{\mathcal{K}}_r = 4.30\ \mathrm{W}/\Delta_1{}^\circ\mathrm{C} \qquad \text{(from C-2)}$$

Substitution

$$\overline{\mathcal{K}}_t = 6.12 + 4.30 \qquad [\mathrm{W}/\Delta_1{}^\circ\mathrm{C}] + [\mathrm{W}/\Delta_1{}^\circ\mathrm{C}] = [\mathrm{W}/\Delta_1{}^\circ\mathrm{C}]$$

Answer

$$\overline{\mathcal{K}}_t = \mathbf{10.42\ W/\Delta_1{}^\circ C}$$

The Series Conduction Path

When different conduction path elements are connected so that the same quantity of heat flows through each conduction path element sequentially, the configuration is called a series conduction path and is illustrated in Figure 13-7. The heat transfer path element resistances are usually used with series heat transfer path systems instead of the conductances of the path elements.

Figure 13-7 shows a number of conduction path elements, each with its characteristic thermal resistance ($\overline{\mathcal{R}}_i$) connected sequentially with the same quantity of heat flow $\dot{Q}$ passing through each conduction path element. The heat flow through the series conduction path system is

$$\dot{Q} = \frac{\Delta T_t}{\overline{\mathcal{R}}_t} \qquad [\mathrm{W}] \qquad (13\text{-}11)$$

where

$$\overline{\mathcal{R}}_t = \overline{\mathcal{R}}_1 + \overline{\mathcal{R}}_2 + \overline{\mathcal{R}}_3 + \overline{\mathcal{R}}_n \qquad [\Delta°\mathrm{C/W}]$$

$$\Delta T_t = \Delta T_1 + \Delta T_2 + \Delta T_3 + \cdots \Delta T_n$$

Equation 13-11 is derived as follows: The temperature drop across each element is

$$\Delta T_i = \frac{\dot{Q}_i}{\mathcal{K}_{t_i}} = \dot{Q}_i \overline{\mathcal{R}}_t \qquad [\Delta°\mathrm{C}] \qquad (\text{Eq. 13-6x})$$

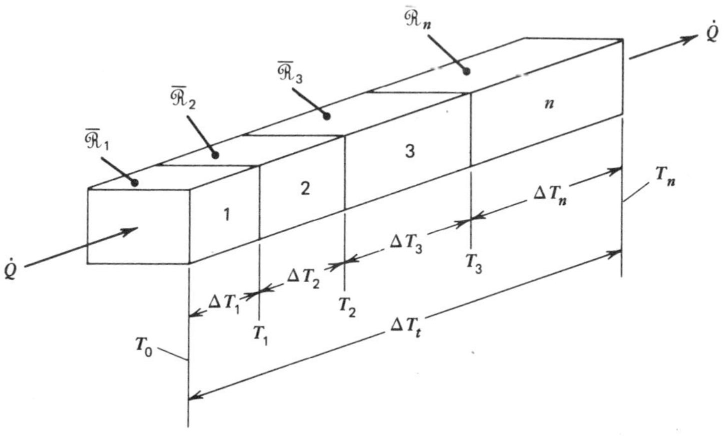

Figure 13-7 Series conduction path.

Then

$$\Delta T_t = \dot{Q}_1\overline{\mathcal{R}}_1 + \dot{Q}_2\overline{\mathcal{R}}_2 + \dot{Q}_3\overline{\mathcal{R}}_3 + \cdots \dot{Q}_n\overline{\mathcal{R}}_n \qquad [\Delta°\mathrm{C}]$$

Since the same heat flow passes through each series path element

$$\dot{Q} = \dot{Q}_1 = \dot{Q}_2 = \dot{Q}_3 = \dot{Q}_n$$

Then

$$\Delta T_t = \dot{Q}\left(\overline{\mathcal{R}}_1 + \overline{\mathcal{R}}_2 + \overline{\mathcal{R}}_3 + \cdots \overline{\mathcal{R}}_n\right) \qquad [\Delta°\mathrm{C}]$$

Since

$$\Delta T_t = \dot{Q}\overline{\mathcal{R}}_t \qquad [\Delta°\mathrm{C}] \qquad \text{(Eq. 13-4x)}$$

$$\overline{\mathcal{R}}_t = \overline{\mathcal{R}}_1 + \overline{\mathcal{R}}_2 + \overline{\mathcal{R}}_3 + \cdots \overline{\mathcal{R}}_n \qquad [\Delta°\mathrm{C/W}] \qquad (13\text{-}12)$$

The temperature drop to the interface between any two sections can be determined by using the sum of the thermal resistances between the selected interfaces. For example, if the temperature T_2 (at the interface between path elements 2 and 3) of the series linear conduction path illustrated in Figure 13-7 were desired, Equation 13-6 could be used. By substituting the thermal resistance $\overline{R}$ for x/Ak, the equation becomes

$$T_2 = T_0 - \dot{Q}\left(\overline{\mathcal{R}}_1 + \overline{\mathcal{R}}_2\right) \qquad [°\mathrm{C}] \qquad (13\text{-}13)$$

Various arrangements of the basic equation for heat flow in a series conduction path are given in Table 13-3.

Table 13-3 Arrangement of the Basic Equation (13-6) for the Series Conduction Path

Rate of heat flow:	$\dot{Q} = \dfrac{\Delta T}{\overline{\mathcal{R}}_t}$	[W]	
	$\dfrac{\dot{Q}}{A} = \dfrac{\Delta T}{A\overline{\mathcal{R}}}$	[W/m²]	
Temperature drop across the total path:	$\Delta T = \dot{Q}\overline{\mathcal{R}}$	[Δ°C]	
Total thermal resistance:	$\overline{\mathcal{R}}_t = \overline{\mathcal{R}}_1 + \overline{\mathcal{R}}_2 + \overline{\mathcal{R}}_3 + \cdots \overline{\mathcal{R}}_n$	[Δ_1°C/W]	(Eq. 13-12)
Thermal resistance of a path element:	$\overline{\mathcal{R}}_i = \dfrac{x_i}{A_i k_i}$	[Δ_1°C/W]	
	$\overline{\mathcal{R}}_i = \dfrac{x_i}{k_i}$	[m² Δ_1°C/W]	
Temperature at the interface between sections n and $n+1$:	$T_n = T_0 - \dot{Q}(\overline{\mathcal{R}}_1 + \overline{\mathcal{R}}_2 + \cdots \overline{\mathcal{R}}_n)$	[°C]	(Eq. 13-13)

In a series heat transfer path, a single high resistance section exercises a dominant control of the amount of heat flow. The consideration of the controlling thermal resistance section is a useful tool in a cursory analysis of a series heat transfer path both as to an evaluation of the amount of heat flow and the identification of modifications to increase or decrease the rate of heat flow. By way of an illustration, if a series heat flow path had one section with 80 percent of the total thermal resistance, relatively large changes in the thermal resistances of another section, such as doubling or halving its thermal resistance, would have relatively little effect upon the total series system thermal resistance.

In many applications, including space vehicles, thermal control is achieved by the use of heat transfer conduction paths. Such conduction paths often include mechanical joints, where adjoining pieces are clamped together. Such joints have a gap between the two mating surfaces of a greater or lesser width, which introduces a thermal resistance into the series conduction path. This thermal resistance, often called contact thermal resistance, can have large magnitudes even though the mating surfaces are smooth and flat by usual mechanical standards if there are no thermally conductive media in the gap such as air or thermal grease. The contact resistance of a dry joint in a vacuum becomes high and can well become the controlling resistance of the path.

Example Problem 13-4

The heat flow path in an automobile radiator can be considered as a series conduction path consisting of three thermal resistance elements.

- From the cooling water to the water surface of the radiator.
- Through the radiator wall (between the water surface and air surface.
- From the radiator air surface to the air.

(a) What is the heat transfer path thermal resistance per unit area ($\mathcal{R}$) for the following conditions?

1. The thermal conductance per unit area $\mathcal{K}$ for the heat transfer between the cooling water and the internal (water) surface of the radiator is 500 $[\mathrm{W/m^2\,\Delta_1{}^\circ C}]$
2. The thermal conductance per unit area $\mathcal{K}$ for the heat transfer between the radiator external surface and the air is 50 $[\mathrm{W/m^2\,\Delta_1{}^\circ C}]$.
3. The radiator is made of brass (70 percent Cu, 30 percent Zn) and the wall is 0.4 mm thick.

(b) What is the controlling thermal resistance? (See Figure 13-8.)

Equation

$$\mathcal{R}=\mathcal{R}_1+\mathcal{R}_2+\mathcal{R}_3 \qquad [\mathrm{m^2\,\Delta^\circ C/W}] \qquad \text{(Eq. 13-10x)}$$

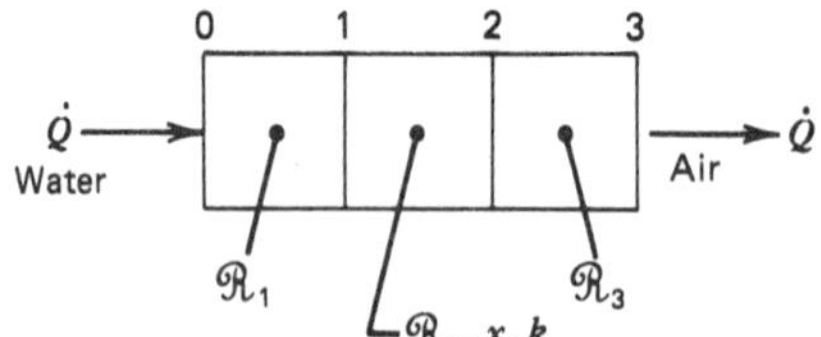

Figure 13-8 Sketch for Example Problem 13-4.

Parameters

$$\mathcal{R}_1 = \frac{1}{\mathcal{K}_1} \qquad [\text{m}^2\,\Delta°\text{C/W}] \qquad \text{(definition)}$$

$$\mathcal{K}_1 = 500 \qquad [\text{W/m}^2\,\Delta_1°\text{C}] \qquad \text{(given)}$$

$$\mathcal{R}_1 = \frac{1}{500} \rightarrow 0.002 \qquad [\text{m}^2\,\Delta°\text{C/W}]$$

$$\mathcal{R}_2 = \frac{x}{k} \qquad [\text{m}^2\,\Delta°\text{C/W}] \qquad \text{(Eq. 13-4x)}$$

$$x = 0.4\ \text{mm} \rightarrow 0.0004 \qquad [\text{m}] \qquad \text{(given)}$$

$$k = 128 \qquad [\text{W/m}\,\Delta_1°\text{C}] \qquad \text{(Appendix B-5 at 100°C)}$$

$$\mathcal{R}_2 = \frac{0.0004}{128} \qquad \frac{[\text{m}]}{[\text{W/m}\,\Delta_1°\text{C}]} = \frac{\text{m}^2\,\Delta°\text{C}}{\text{W}}$$

$$= 3.125 \times 10^{-6} \qquad [\text{m}^2\,\Delta°\text{C/W}]$$

$$\mathcal{R}_3 = \frac{1}{\mathcal{K}_3} \qquad [\text{m}^2\,\Delta°\text{C/W}]$$

$$\mathcal{K}_3 = 50 \qquad [\text{W/m}^2\,\Delta_1°\text{C}] \qquad \text{(given)}$$

$$\mathcal{R}_3 = \frac{1}{50} \rightarrow 0.02 \qquad [\text{m}^2\,\Delta°\text{C/W}]$$

Substitution

$$\mathcal{R} = 0.002 + 3.125 \times 10^{-6} + 0.02 \qquad [\text{m}^2\,\Delta°\text{C/W}] + [\text{m}^2\,\Delta°\text{C/W}] + [\text{m}^2\,\Delta°\text{C/W}]$$

(a) Answer

$$\boldsymbol{\mathcal{R} = 0.022 \qquad [\text{m}^2\,\Delta°\text{C/W}]}$$

(b) Answer

$\overline{\mathcal{R}}_3$ **between the radiator and the air**

Example Problem 13-5

If the radiator in Example Problem 13-4 had a heat transfer area of 0.9 m^2, were transferring 3.0 kW of heat, and the air temperature were 25°C, what would be the water temperature? (See Figure 13-9.)

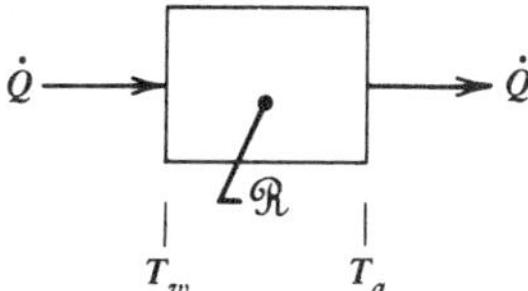

Figure 13-9 Sketch for Example Problem 13-5.

Equation

$$T_w = T_a + \frac{\dot{Q}\mathfrak{R}}{A} \qquad [°\text{C}] \qquad (\text{Eq. 13-13x})$$

Parameters

$$T_a = 25°\text{C} \qquad (\text{given})$$

$$\dot{Q} = 3.0\text{ kW} \rightarrow 3000 \qquad [\text{W}] \qquad (\text{given})$$

$$\mathfrak{R} = 0.022 \qquad [\text{m}^2\,\Delta°\text{C/W}] \qquad (\text{Problem 13-4})$$

$$A = 0.9 \qquad [\text{m}^2] \qquad (\text{given})$$

Substitution

$$T_w = 25 + \frac{(3000)(0.022)}{(0.9)} \qquad [°\text{C}] + \frac{[\text{W}][\text{m}^2\,\Delta°\text{C/W}]}{[\text{m}^2]}$$

$$= 25 + 73.3 \qquad [°\text{C}] + [\Delta°\text{C}] = [°\text{C}]$$

Answer

$$\boldsymbol{T_w = 98.3°\text{C}}$$

13-3 CONVECTION WITH NO PHASE CHANGE

Convection heat transfer occurs between a fluid and a solid surface when relative movement between the fluid and the solid surface occurs. Such fluid movement may be driven by buoyant forces generated by thermal gradients in the fluid (free convection) or by forced movement of the fluid (forced convection). The movement of the fluid has a dominant effect upon the convective heat transfer. Consequently, many factors (in addition to the temperature gradient, heat flow path area, and the material conductivity) become important. For example, in free convection the following fluid properties and geometrical parameters are involved:

L = significant length [m]

k = fluid thermal conductivity $[\mathrm{W/m \cdot \Delta_1 °C}]$

μ = fluid viscosity (absolute) $[\mathrm{Pa \cdot s}]$

c_p = fluid specific heat at constant pressure $[\mathrm{J/kg \cdot \Delta_1 °C}]$

ρ = fluid density $[\mathrm{kg/m^3}]$

β = coefficient of thermal expansion of the fluid [—]

g = acceleration of gravity $[\mathrm{m/s^2}]$

ΔT = temperature difference [Δ°C] or [ΔK]

$\mathbf{V}$ = fluid flow velocity [m/s]

In forced convection, the following properties are included:

- The fluid flow characteristics.
- The fluid properties.
- The fluid flow passage geometry.
- The solid surface condition.

The key parameter used for calculating the convective heat transfer is the *film coefficient* h_a. The basic equation for calculating the heat flow is

$$\dot{Q} = h_a A(T_f - T_s) = h_a A \Delta T \quad [\mathrm{W}] \tag{13-14}$$

where

$\dot{Q}$ = heat flow rate [W]

h_a = average convective surface (or film) coefficient $[\mathrm{W/m^2 \Delta_1 °C}]$

A = surface area $[\mathrm{m^2}]$

T_f = fluid bulk (main mass) temperature [°C]

T_s = solid surface temperature [°C]

To evaluate h_a for any set of fluid and flow conditions, we use experimental test data. To apply experimental test data to other fluid and flow conditions, the use of *dimensionless ratios* has been developed. It was found that by plotting these data on the basis of various dimensionless ratios, useful relationships could be established for applying test data from one set of conditions to another.

The primary dimensionless ratios used in convective heat transfer are

Nusselt number: $$\mathrm{Nu} = \frac{h_c L}{k} \quad [-] \tag{13-15}$$

Prandtl number: $$\mathrm{Pr}=\frac{\mu c_p}{k} \quad [-] \tag{13-16}$$

Grashof number: $$\mathrm{Gr}=\frac{L^3\rho^2\beta g\,\Delta T}{\mu^2} \quad [-] \tag{13-17}$$

Reynolds number: $$\mathrm{Re}=\frac{VL\rho}{\mu} \quad [-] \tag{13-18}$$

where

h_c = heat transfer film coefficient $[\mathrm{W/m^2}\,\Delta_1{}^\circ\mathrm{C}]$

L = significant length $[\mathrm{m}]$

k = fluid thermal conductivity $[\mathrm{W/m}\,\Delta_1{}^\circ\mathrm{C}]$

μ = fluid viscosity (absolute) $[\mathrm{Pa\cdot s}]$

c_p = fluid specific heat (at constant pressure) $[\mathrm{J/kg}\cdot\Delta_1{}^\circ\mathrm{C}]$

ρ = fluid density $[\mathrm{kg/m^3}]$

β = coefficient of thermal expansion (volumetric) of the fluid $[-]$

g = acceleration of gravity $[\mathrm{m/s^2}]$

ΔT = temperature difference $[\Delta^\circ\mathrm{C}]$

V = fluid flow velocity $[\mathrm{m/s}]$

These dimensionless numbers can be given the following general physical interpretations.

The *Nusselt number* can be interpreted physically as the ratio of the temperature gradient in the fluid immediately in contact with the surface to a reference temperature gradient. For a given value of the Nusselt number, the convective surface coefficient h_c is directly proportional to the thermal conductivity of the fluid and inversely proportional to the significant length L.

The *Prandtl number* is a ratio of the physical properties of the fluid.

The *Grashof number* can be interpreted physically as the ratio of buoyant to viscous forces in the convective system.

The *Reynolds number* can be interpreted physically as the ratio of the fluid flow inertia forces to viscous forces.

13-3.1 STEADY STATE FREE CONVECTION

Convection heat transfer occurs between a solid body and a fluid (a liquid or gas). The heat transfer is effected by a combination of the molecular conduction within the fluid in combination with energy transport resulting from the motion (circulation) of fluid particles. The convective process is termed *free convection* when the circulation (or flow) of the fluid is caused by changes in the fluid density resulting from temperature gradients between the surface of the solid body and the main mass of the fluid.

Free convection can only occur in gravitational fields because the fluid circulation results from density gradient forces. In space vehicles with a zero gravity flight trajectory (such as orbiting satellites), free convection is nonexistent because there are no density forces.

In free convection heat transfer application problems, using Equation 13-14, two of the three equation parameters ($\dot{Q}, A, \Delta T$) are known. In order to solve for the other parameter, the value of h_a must be determined. An equation form making use of dimensionless numbers was developed in the following manner.

In the usual free convection circumstances with no phase change, the fluid flow velocities are sufficiently small so that inertia forces in the flow are negligible in comparison with the forces of friction and buoyancy. For the usual free convection circumstances the following dimensionless numbers apply:

1. Nusselt.
2. Prandtl.
3. Grashof.

The basic equation developed from dimensional analysis for use in determining the value of h is

$$\mathrm{Nu} = C(\mathrm{Gr})^a(\mathrm{Pr})^b \tag{13-19}$$

To evaluate the exponents a and b in Equation 13-19, the results of tests of free convection with various fluids (both liquids and gases) flowing over single horizontal cylinders and vertical flat plates were used. The exponents a and b were found to be the same, which permits Equation 13-19 to be written as:

$$\mathrm{Nu} = C[\mathrm{Gr} \times \mathrm{Pr}]^m \tag{13-20}$$

By the use of these dimensionless numbers, an excellent correlation between a wide variety of fluids and conditions can be obtained as illustrated in the plot of Nu versus (Gr $\times$ Pr) presented in Figure 13-10 for free convection heat transfer to cylinders. The curve of a similar plot of data for another geometric configuration would be different. From these plots, values for the coefficient C and the exponent m can be derived. The values of these parameters are affected by:

- The boundary layer configuration.
- The fluid properties.
- The temperature difference.

The student should consult a suitable heat transfer book to obtain values of C and m for the application.[17]

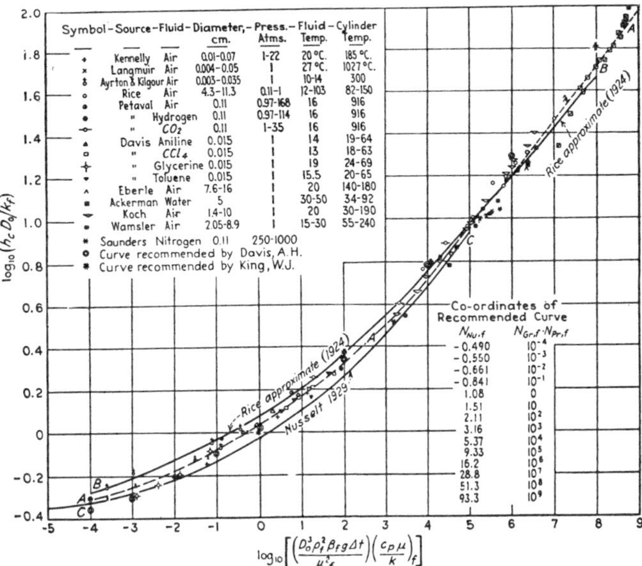

Figure 13-10 Correlation of data for free convection heat transfer from horizontal cylinders in gases and liquids. (By permission from W. H. McAdams, *Heat Transmission*, 3rd ed., McGraw-Hill Book Company, New York, 1954.

Typical ranges in the values of h_a (the average convective film coefficient) are

- Air 3 to 7 $[\text{W/m}^2\,\Delta_1{}^\circ\text{C}]$.
- Gases 2 to 20 $[\text{W/m}^2\,\Delta_1{}^\circ\text{C}]$.
- Liquids 30 to 300 $[\text{W/m}^2\,\Delta_1{}^\circ\text{C}]$.

13-3.2 STEADY STATE FORCED CONVECTION

The convective heat transfer occurring between a solid body and a fluid is termed *forced convection* when the circulation of the fluid (liquid or gas) is caused and controlled by some mechanical means, commonly a pump or a blower. A large portion of heat transfer applications make use of forced convection.

The heat transfer equation for forced convection is the same form as the free convection Equation 13-14 presented in Section 13-3.1. That is,

$$\dot{Q}=h_aA(T_f-T_s)=h_aA\,\Delta T \qquad [\text{W}] \qquad (\text{Eq. 13-14})$$

However, the forced convection average film coefficient is evaluated in a significantly different manner and has significantly different ranges in values. The mechanical forcing of the fluid flow supplants the thermal buoyancy of the free convective system to generate the fluid flow characteristics.

In typical forced convective heat transfer application problems, two of the three equation parameters ($\dot{Q}$, A, ΔT) in Equation 13-14 are known. In order to solve for the other parameters, the value of h_a must be determined.

The value of h_a for a forced convection heat transfer configuration is a function of:

- The flow pattern characteristics and boundary layer.
- The fluid properties.
- The flow passage geometry.
- The surface condition.

Fluid Flow Pattern Characteristics

The fluid flow pattern is characterized by being one of the following:

- Laminar.
- Critical.
- Transitional.
- Fully turbulent.

Flow patterns are significantly different for various geometrical configurations. Because flow through round tubes is a common configuration, it is used as the basis for the following discussion. What is true for flow inside tubes cannot be directly applied per se to other configurations.

Laminar Flow

In laminar, or streamline flow, the fluid particles flow in paths that are parallel to each other and to the solid surface. A laminar boundary extends from the solid surface out into the fluid flow. The flow velocity on the solid surface is zero, and the flow velocity increases across the boundary layer until it reaches the undisturbed fluid flow velocity. This point is the edge of the boundary layer and may extend to the center of the pipe or duct in which the fluid is flowing.

Critical Flow

Critical flow is the term applied to laminar flow in a pipe or duct when the boundary layer extends across the entire flow duct and the flow velocity is the maximum for the laminar flow mode. Higher flow velocities would cause turbulence to appear in the flow. In the critical flow region, small variations in the flow may cause significant changes in the heat transfer rates. Consequently, the critical flow region should be avoided in heat transfer equipment, as the heat transfer performance is subject to significant fluctuations.

Transitional Flow

In transitional flow, a varying amount of turbulence exists. In the transitional flow region, large changes in the fluid flow and heat transfer occur as the degree of turbulence increases from the laminar critical flow pattern to fully developed turbulent flow pattern.

Fully Turbulent Flow

In fully developed turbulent flow, strong velocity components normal to the nominal direction of flow exist. The path of any individual particle is zigzag and irregular. On a statistical basis the overall motion of the aggregate of fluid particles is regular and predictable.

Flow Similarities

Osborne Reynolds investigated the problem of quantizing the *similar* flow of fluids of different properties and flow geometries and velocities. In this study he considered the circumstances that would make the flow of different fluids with different velocities similar. He showed that the flow of fluids is similar when a specific ratio of variables is the same. This ratio is dimensionless, and is called the Reynolds number (Re).

$$\mathrm{Re} = \frac{\mathbf{V} D \rho}{\mu} \qquad [-] \qquad \text{(Eq. 13-9)}$$

The Reynolds number includes terms of a characteristic length, a velocity, fluid density, and fluid viscosity. These can be arranged in numerous forms such as substituting the fluid mass flow rate per unit cross-sectional area of the fluid flow for the product of the flow velocity and the density. All arrangements for the same set of fluid flow conditions give the same numerical value.

The values of the Reynolds numbers for the different flow regimes, that is, laminar, critical, transitional, and full turbulent flows, vary for different geometrical configurations of the fluid flow. For flow inside circular pipes, the ranges of the Reynolds number associated with the four flow regimes are listed below. In this case the Reynolds number is based on the pipe diameter for the length dimension D and the average flow velocity for the velocity dimension $\mathbf{V}$.

Laminar flow <2100 [—]
Critical flow 2100–3000 [—]
Transitional flow 3000–4000 [—]
Turbulent flow >4000 [—]

The corresponding values of the Reynolds number for other flow configurations are different.

In forced convection with no phase change the fluid flow velocities are sufficiently high so that the inertia forces in the flow are important. The buoyancy forces are negligible in comparison with the forces of inertia and friction. For the usual forced convection circumstances the following dimensionless numbers apply:

1. Nusselt number.
2. Prandtl number.
3. Reynolds number.

The basic equation developed from dimensional analysis for use in determining the value of h is

$$\mathrm{Nu} = C(\mathrm{Re})^a(\mathrm{Pr})^b \tag{13-21}$$

The numerical values for the constant C and the exponents a and b are determined by obtaining the best fit to experimental data. These values are different for various flow configurations such as flow inside pipes, flow along a flat plate, flow across the outside of a cylinder, and flow through a bank of tubes passing across the outside of the tubes. An example of a plot of test data for turbulent water flow inside a circular pipe is shown in Figure 13-11. The ordinate is $(\mathrm{Nu})(\mathrm{Pr})^{0.40}$ versus Re as the abscissa. The solid line AA represents the values obtained from the McAdams equation (13-22).

Test data for turbulent water flow inside pipes were correlated by McAdams by the following equation:

$$h = 0.023\frac{k}{D}\left(\frac{D\mathbf{V}_\rho}{\mu}\right)^{0.8}\left(\frac{C_p\mu}{k}\right)^{0.4} \qquad [\mathrm{W/m^2\,\Delta_1{}^\circ C}] \tag{13-22}$$

Equation 13-22 is generally known as the McAdams equation. It applies to the following conditions:

1. $N_{\mathrm{Re}} = 10{,}000$ to 150,000.
2. $N_{\mathrm{Pr}} = 0.7$ to 120.
3. pipe length > 60 L/D.
4. The temperature difference across the film is not large (less than 10°).

The properties of the fluid should be evaluated at the bulk temperature, which is defined as the arithmetic mean of the cup temperatures (mixed average) at the beginning and at the end of the heat transfer section.

As the forced convection characteristics vary greatly between types of fluids, that is, gases, waterlike liquids, viscous liquids, and liquid metals, as well as the flow passage configuration and the fluid flow regime, the student should consult a suitable heat transfer textbook to obtain the equation for his application.

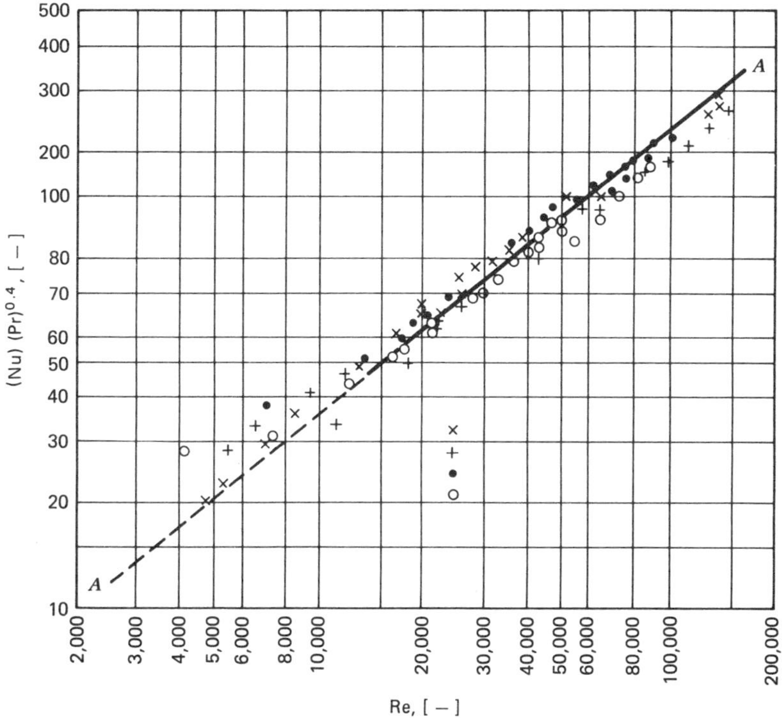

Figure 13-11 Heating of water in tubes ranging in length from 59 to 224 diameters (Equation 13-22). Lawrence and Sherwood data: $x = 3.4$ m, $+ = 9.1$ m, $\cdot = 18.3$ m, and $0 = 0.9$ m.

Typical ranges in the nominal values of h_a are

1. Air and superheated steam 30–300 $[\mathrm{W/m^2\,\Delta_1{}^\circ C}]$.
2. Oil 60–3000 $[\mathrm{W/m^2\,\Delta_1{}^\circ C}]$.
3. Water 300–10,000 $[\mathrm{W/m^2\,\Delta_1{}^\circ C}]$.

13-4 RADIATION HEAT TRANSFER

In the modes of heat transfer discussed thus far (i.e., conduction and convection), the heat has been transmitting the increased energy of molecular motion progressively along a temperature gradient extending through the solid, liquid, or gaseous media involved. Heat may also be transmitted by thermal radiations between two bodies without an intermediate material to act as a carrier of energy. This mode of heat transfer is called radiation heat transfer. The thermal radiant energy is transmitted in a wavelength bandwidth of 10^{-4} to 10^{-7} meters in the electromagnetic radiation spectrum as indicated in Figure 13-12. The wavelength of thermal radiations is a function of the temperature of the emitting body.

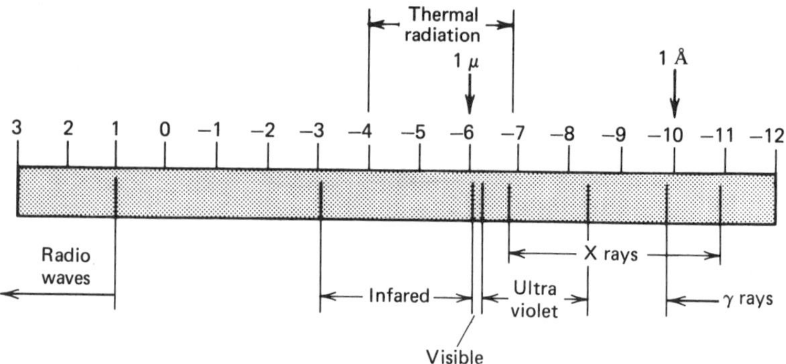

Figure 13-12 Electromagnetic spectrum.

The long wavelength edge of the thermal radiation bandwidth is limited by low temperatures (absolute zero), and the short wavelength edge shown in Figure 13-12 is established by the temperature of the sun. The wavelength bandwidth of thermal radiations associated with temperatures from 20 to 2500°C, for example, lies largely between 10^{-5} to 10^{-6} meters, extending into visible light at the high temperature end. Thermal radiations exhibit characteristics similar to those of visible light and follow optical laws. They travel in a straight line at the speed of light; they can be reflected and refracted, and are subject to scattering and absorption when they pass through media.

Key concepts used in radiant heat transfer are

- Blackbody.
- Emissivity of a surface. [ε]
- Absorptivity of a surface and/or material. [α]
- Reflectivity of a surface. [ρ]
- Transmissivity of a material. [τ]

The blackbody concept and emissivity are concerned with radiant energy *emission* by a body. Absorptivity, reflectivity, and transmissivity are concerned with radiant energy, coming from another source, *incident* upon a body. The key concepts listed above are discussed in the following paragraphs.

A *blackbody* has ideal radiation heat transfer characteristics, namely

- A blackbody radiates the *maximum energy possible* at the temperature of the body.
- A blackbody absorbs *all* of the incident radiant energy regardless of the wavelength.

The *emissivity* of a surface ε is the ratio of the total radiant emissive power per unit area W of the surface of interest to the total radiant emissive power per unit area W_b of a blackbody at the same temperature. Values for emissivities range from 0.0 to 1.0.

$$\varepsilon = \frac{W}{W_b} \tag{13-23}$$

The *absorptivity* of a surface and/or material α is the fraction of the total incident radiant energy absorbed by the body.

The *reflectivity* of a surface ρ is the fraction of the total incident radiant energy reflected from the surface of a body.

The *transmissivity* of a material τ is the fraction of the total incident radiant energy transmitted through the body.

The basic mechanics of radiant heat transfer is simple. A body radiates energy proportional to the fourth power of its absolute temperature K and its surface emissivity ε. A body absorbs incident radiation in accordance with its absorptivity α. As the incident radiant energy is absorbed, reflected, and transmitted to various degrees in accordance with the absorptivity α, reflectivity ρ, and transmissivity τ characteristics of the body; the sum of the absorptivity, reflectivity, and transmissivity of a surface always equals 1.0.

$$\alpha + \rho + \tau = 1 \tag{13-24}$$

The net amount of radiant energy gain to, or loss from, a body is the difference between the radiant energy emissions by the body (energy loss) and the absorbed incident radiations (energy gain). When a body radiates more energy than it absorbs, the net heat transfer is away from the body with a net energy loss by the body. When the body absorbs more energy than it radiates, the net heat transfer is to the body with a net energy gain by the body. For steady state conditions (no temperature change of the body with respect to time) the radiant energy emissions (energy loss) must equal the incident radiant energy absorption (energy gain) if radiation is the only mode of heat transfer involved.

13-4.1 ENERGY EMISSIONS FROM A BODY

A body radiates energy in accordance with:

- The fourth power of its absolute temperature.
- The surface emissivity ε.

The total energy radiated per unit area of a body surface (radiant energy flux density, W) is

$$W = \varepsilon \sigma T^4 \qquad [\mathrm{W/m^2}] \tag{13-25}$$

where

ε = the surface emissivity [—]

σ = the Stefan–Boltzmann constant, 5.669×10^{-8} [W/m^2·K^4]

T = the surface temperature of the blackbody [K]

Figure 13-13 is a plot of the blackbody radiant energy flux density versus the absolute temperature.

The total energy radiated E_R by a blackbody ($\varepsilon = 1.0$) is

$$E_R = AW_b = A\sigma T^4 \quad [\text{W}] \tag{13-26}$$

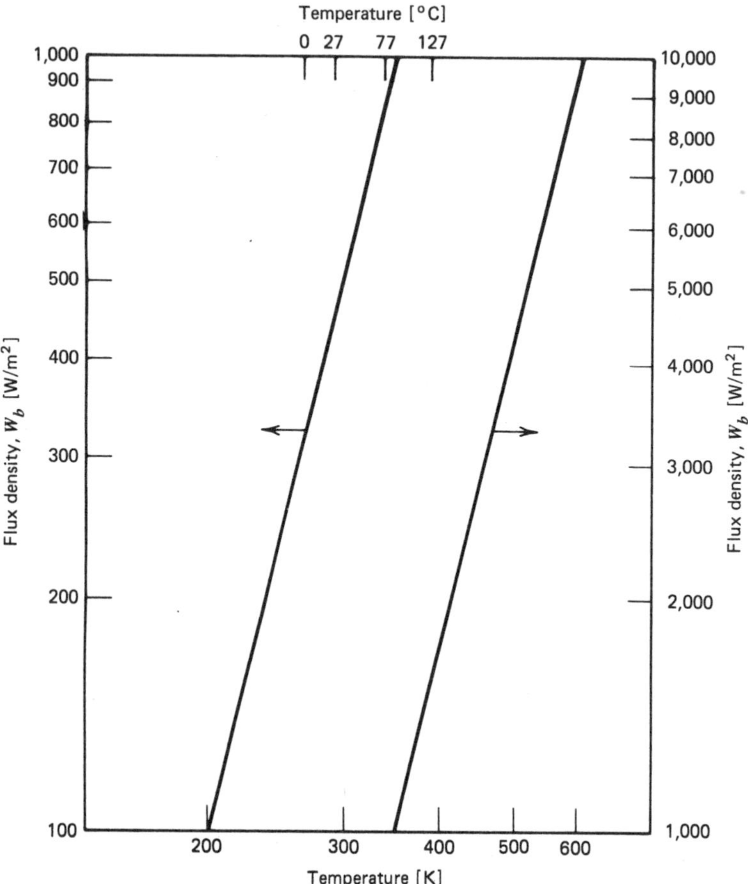

Figure 13-13 Blackbody radiant energy flux density versus temperature (absolute).

where

A = the area of the radiating surface $[m^2]$

W_b = the blackbody radiant energy flux density $[W/m^2]$

σ = the Stefan–Boltzmann constant 5.67×10^{-8} $[W/m^2 \cdot K^4]$

Example Problem 13-6

What is the total radiant energy emitted by a blackbody in the form of a cube 1 m long on the side when the temperature of the blackbody is 1000 K? (See Figure 13-14.)

Equation

$$E_R = A W_b \quad [W] \quad \text{(Eq. 13-26)}$$

Parameters

$$A = 6A' \quad [m^2]$$

$$A' = 1 \times 1 \quad [m][m] = [m^2]$$

$$A' = 1\ m^2$$

$$A = (6)(1) \quad [m^2]$$

$$A = 6\ m^2$$

$$W_b = \sigma T^4 \quad [W/m^2] \quad \text{(Eq. 13-25)}$$

$$\sigma = 5.670 \times 10^{-8} \quad [W/m^2 \cdot K^4]$$

$$T = 1000\ K \quad \text{(given)}$$

$$W_b = (5.670 \times 10^{-8})(1000^4) \quad [W/m^2 \cdot K^4][K^4] = [W/m^2]$$

$$W_b = 5.670 \times 10^4 \quad [W/m^2]$$

Substitution

$$E_R = (6)(5.670 \times 10^4) \quad [m^2][W/m^2] = [W]$$

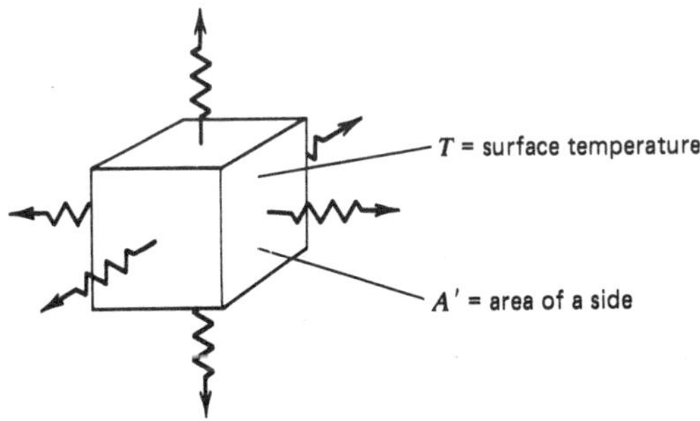

Figure 13-14 Sketch for Example Problem 13-6.

Answer

$$E_R = 3.402 \times 10^5 \text{ W} \qquad \text{or} \qquad 340.2 \text{ kW}$$

Energy Emissions from a Real Solid Surface

A real surface may not necessarily have a constant emissivity value for all wavelengths of emissions (monochromatic emissivity). If it does, it is called a graybody and has a characteristic emissivity value that does not vary with temperature. The value of the graybody emissivity is a characteristic of the surface and must be less than 1.0.

The graybody radiant energy flux density W is

$$W = \varepsilon W_b \qquad [\text{W/m}^2] \tag{13-27}$$

where

W_b = blackbody radiant energy flux density $[\text{W/m}^2]$

ε = graybody (or effective) emissivity $[—]$

Most real surfaces, however, do not have a constant value of monochromatic emissivity ε_λ with respect to the wavelength as illustrated in Figure 13-15.

As the temperature of a radiating body changes, the magnitude of the blackbody monochromatic radiant flux density shifts. Consequently, as the temperature of a radiating body changes, the total (integrated) radiant flux density will change relative to the blackbody radiant flux density. This means that the effective emissivity changes. Consequently, in tables of values of emissivities for various surfaces, the associated surface temperature is also given.

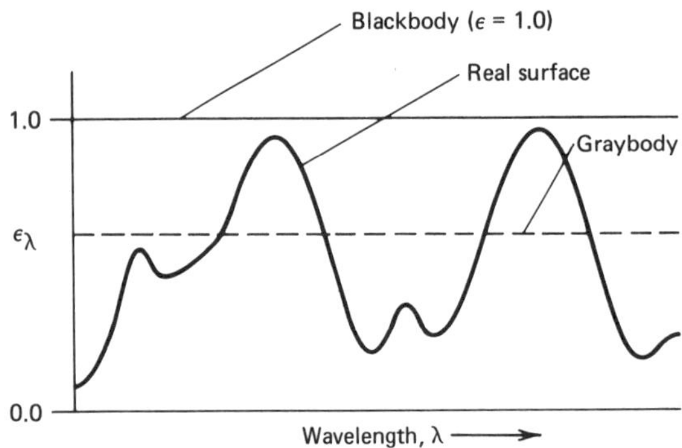

Figure 13-15 Comparison between a real surface and blackbody monochromatic emissivities.

13-4.2 INCIDENT RADIATION

When radiant energy falls on the surface of a body, part may be transmitted, part reflected, and the remainder absorbed as illustrated by the sketch in Figure 13-16. The incident radiations are completely independent of, and uneffected by, the emissions from the body, which are occurring simultaneously in accordance with the equations we have just studied. The mathematical sum of the transmissivity, reflectivity, and absorptivity must equal 1 as indicated in Equation 13-24.

$$\alpha+\rho+\tau=1 \qquad [-] \qquad \text{(Eq. 13-24)}$$

where

α = absorptivity or the fraction of total energy absorbed [—]

ρ = reflectivity or the fraction of the total energy reflected [—]

τ = transmissivity or the fraction of the total energy transmitted through the body [—]

For an incident radiation flux $\mathcal{G}_i$ upon a body, the radiation flux energy absorbed is

$$\mathcal{G}_{ia}=\alpha\mathcal{G}_i \qquad [\text{W/m}^2] \tag{13-28}$$

The radiation flux energy reflected is

$$\mathcal{G}_{i\rho}=\rho\mathcal{G}_i \qquad [\text{W/m}^2] \tag{13-29}$$

The radiation flux energy transmitted through is

$$\mathcal{G}_{i\tau}=\tau\mathcal{G}_i \qquad [\text{W/m}^2] \tag{13-30}$$

The absorptivity, reflectivity, and transmissivity of the incident radiation flux are discussed in the following sections.

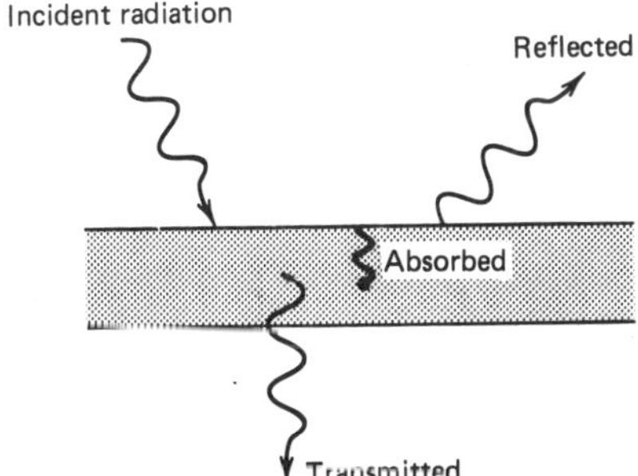

Figure 13-16 Sketch showing the distribution of incident radiant energy.

Absorptivity

In opaque materials the absorptivity is a surface phenomenon. In transparent materials the absorption occurs throughout the material. The absorptivity of a real material is usually a function of the wavelength of the incident radiations. Consequently, the absorptivity of a surface to incident solar energy may be significantly different from the absorptivity to incident radiations from a 300 K surface. This is a result of the wide difference in the spectral energy distribution in the radiations from the two different sources.

The absorptivity of a surface was established to be equal to the emissivity of the surface, at the same temperature, by Kirchhoff in 1859. The Kirchhoff law states:

If a surface at any given temperature absorbs *n* times more of an arriving radiation than another surface of equal area at the same temperature, then at the same temperature the first surface emits *n* times more of the same kind of radiation than does the second surface. Simply stated:

$$\varepsilon=\alpha \qquad [-] \tag{13-31}$$

where

$\varepsilon=$ emissivity of the material surface $[-]$

$\alpha=$ absorptivity of the material surface $[-]$

Values for the absorptivities of materials at various temperatures are the same as the emissivities listed for the materials and temperatures in Appendix B, Table B-8.

The amount of heat $\dot{Q}_a$ absorbed by a body is a function of the radiant energy incident upon the surface E_i and the surface absorptivity α.

$$\dot{Q}_\alpha=\alpha E_i \qquad [\mathrm{W}] \tag{13-32}$$

where

$\dot{Q}_\alpha=$ the heat absorbed by the body $[\mathrm{W}]$

$\alpha=$ the absorptivity of the body $[-]$

$E_i=$ the energy incident upon the surface $[\mathrm{W}]$

Reflectivity

The reflectivity characteristics of the surface of a material follow conventional optics. A smooth surface will give a specular reflection, with the angle of reflection equal to the angle of incidence of the radiations. If the surface is rough, a diffuse reflection may occur. Depending upon the nature of the surface of the solid material, the reflectivity can vary from 0 to 1. If the material is transparent, a reflection will also occur. The magnitude of this reflection at an air/material interface (for transparent surfaces without an antireflection coating) is a function of the index of refraction (n) of the transparent material. The equation for the reflectivity for normal incident radiation is

$$\rho=\left(\frac{n-1}{n+1}\right)^2 \quad [—] \qquad (13\text{-}33)$$

As the angle of incidence (measured from the normal to the surface) becomes larger, an increased reflectance will generally be experienced. At grazing incidence angles, the reflectivity usually approaches (100 percent).

Transmissivity

The majority of solid materials encountered in engineering, except for glasses, are opaque. Incident radiant energy is not transmitted through opaque materials. However, many liquids and gases have relatively high transmissivities.

The monochromatic transmittance characteristics of materials is a function of the wavelength of the incident radiation. For example, common glass is highly transparent to visible light but essentially opaque to long wavelength infrared radiations.

For opaque bodies (wherein the transmissivity is zero), Equation 13-24 becomes

$$\alpha+\rho=1 \quad [—] \qquad (13\text{-}34)$$

Example Problem 13-7

A plate-type solar energy collector has an absorbing surface covered by a glass plate to minimize convective heat transfer losses. The glass plate has a reflectivity of 0.15 and a transmissivity of 0.75. The absorbing surface has an absorptivity of 0.95, and the area of the collector is 3 m^2.

When the incident solar flux density on the glass plate surface is 750 W/m^2:

(a) What solar energy flux density is incident on the collector absorbing surface?

(b) How much solar energy is absorbed by the collector? (See Figure 13-17.)

Equation

$$\mathcal{G}_{i\tau}=\tau\mathcal{G}_i \quad [W/m^2] \quad \text{(Eq. 13-30)}$$

Parameters

$$\tau=0.75 \quad [—] \quad \text{(given)}$$

$$\mathcal{G}_i=750 \quad [W/m^2] \quad \text{(given)}$$

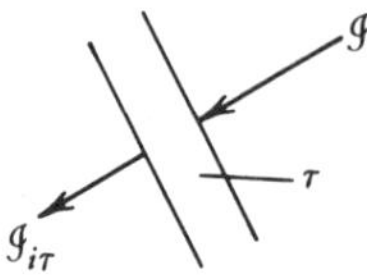

Figure 13-17 Sketch for Example Problem 13-7a.

Substitution

$$\mathcal{G}_{i_\tau}=(0.75)(750) \qquad [—][\mathrm{W/m^2}]=[\mathrm{W/m^2}]$$

Answer

$$\mathbf{\mathcal{G}_{i_\tau}=562.5} \qquad [\mathrm{W/m^2}]$$

(See Figure 13-18.)
Equation

$$\dot{Q}_\alpha=\alpha A\mathcal{G}_{i_\tau} \qquad [\mathrm{W}] \qquad \text{(Eq. 13-32x)}$$

Parameters

$$\alpha=0.95 \qquad [—] \qquad \text{(given)}$$

$$A=3 \qquad [\mathrm{m^2}] \qquad \text{(given)}$$

$$\mathcal{G}_{i_\tau}=562.5 \qquad [\mathrm{W/m^2}] \qquad \text{(from Prob. 13-7a)}$$

Substitution

$$\dot{Q}_\alpha=(0.95)(3)(562.5) \qquad [—][\mathrm{m^2}][\mathrm{W/m^2}]=[\mathrm{W}]$$

Answer

$$\mathbf{\dot{Q}_\alpha=1603} \qquad [\mathrm{W}]$$

13-5 PROBLEMS

13-1. What are the three basic mechanisms of heat transfer? What are their key features? How do they differ?

13-2. What is the relationship between the thermal conductivity k of a material and the heat transfer path thermal conductance $\overline{\mathcal{K}}$?

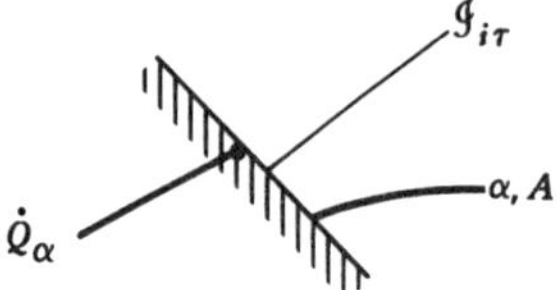

Figure 13-18 Sketch for Example Problem 13-7b.

13-3. What is the relationship between a thermal heat transfer path conductance and resistance?
13-4. What is the general difference in the relative magnitude of thermal conductivities for gases, liquids, and solids?
13-5. Give an equation for the heat flow through a conduction heat transfer path.
13-6. What is the thermal conductance of a copper bar conduction path with the following configuration and properties?

length of the copper bar = 30 cm.

cross-sectional area of the bar = 0.0006 m^2.

thermal conductivity of the copper = 250 W/m Δ_1°C.

13-7. How much heat will flow along the copper bar of Problem 13-6 when the temperature difference between the two ends is 20 Δ°C?
13-8. If the copper bar of Problem 13-6 were to be replaced with an aluminum bar of the same length, what cross-sectional area would be required of the aluminum bar to have the same heat transfer path conductance $\overline{\mathcal{K}}$?
13-9. Describe a parallel conduction path system and give an example, showing how your example meets the criteria for a parallel path system.
13-10. A liquid nitrogen tank is suspended inside a vacuum shell on three stainless steel (18 Cr, 8 Ni) rods 0.5 cm in diameter and 3 meters long. If the liquid nitrogen temperature is 80 K and the ambient air outside the vacuum tank shell is 25°C, what is the magnitude of the conductive heat flow along the support rods?
13-11. Describe a series conduction path system and give an example, showing how your example meets the criteria for a series path system.
13-12. If one mechanical joint with a conductive thermal resistance in a vacuum of 2500 [Δ°C/W] were put in each rod, what would be the magnitude of the conduction heat flow into the liquid nitrogen tank described in Problem 13-10?
13-13. What are dimensionless numbers? How are they used with respect to convection heat transfer? List four (4) dimensionless numbers commonly used in heat transfer, giving its name, symbol, and component parameters.
13-14. Describe convective heat transfer. What is the boundary layer and how does it affect the convective heat transfer?
13-15. In free convection what factors have a primary influence and why? What dimensionless numbers are used to correlate test data and in what usual relationship?
13-16. In forced convection what factors have a primary influence and why? What dimensionless numbers are used to correlate test data and in what usual relationship?
13-17. For forced convection heat transfer to a fluid flowing inside a round tube, what length dimension is used to calculate the Reynolds number?
13-18. Describe laminar, transitional, and fully developed turbulent flow inside a circular tube. What Reynolds number range is associated with each?
13-19. Why is the critical and transitional fluid flow region usually avoided in heat transfer design?
13-20. Using the McAdams equation, what would be the indicated value of h_a for

the following conditions:

Fluid liquid water.
Bulk temperature, 80°C.
Tube inside diameter, 2.54 cm.
Flow velocity, 3 m/s.

13-21. Calculate the Re, Pr, and Nu numbers for the conditions of Problem 13-20. What fluid flow regime is indicated?

13-22. Describe the basic mechanism of radiant heat transfer.

13-23. What is a blackbody? How do real surfaces differ from a blackbody? What is a graybody?

13-24. What factors affect the amount of radiant energy emitted from a surface? Give the formula for emitted radiant energy flux.

13-25. The stainless steel (Cr-Ni alloy) tail pipe of an aircraft engine has a surface temperature of 225°C. What is the emitted radiant energy flux density from this surface?

13-26. The exterior surface of an automobile radiator is at a temperature of 80°C. What is the emitted radiant energy flux density from this surface?

13-27. If the external surface area of the tailpipe in Problem 13-25 were in the form of a cylinder 1.5 m in diameter and 1 m long, how much energy would be radiated from the external surface of the tail pipe?

13-28. If an automobile radiator had an area 50 by 60 cm and the surface temperature were 95°C, how much energy would be radiated from the front and back surfaces?

13-29. Radiations incident upon a surface may be subject to absorption, reflection, and or transmission. Describe or define each of these. What is the relationship between emissivity and absorptivity?

13-30. A solar energy collector uses a glass plate to reduce convective heat transfer losses. If the glass plate absorptivity is 0.08 and its index of refraction n is 1.3, what fraction of the incident solar energy is transmitted.

13-31. If the transmitted solar energy in Problem 13-30 is incident on an absorbing surface with an α of 0.98, how much of the solar energy incident upon the glass plate surface is absorbed by the collector?

13-32. If the solar collector in Problem 13-30 is a concentrating collector with a 5-m diameter entrance aperture focusing on a 0.15-m diameter glass covered collector surface (absorber) with an α of 0.98, what cooling heat flux $\dot{Q}/A$ would be required to equal the incident solar radiations absorbed by the collector when the solar radiation flux density incident upon the entrance aperture is 750 [W/m^2]? The reflectivity of the concentrating collector surface is 0.95.

13-33. If the equilibrium temperature of the absorber is 550°C and at this temperature it has an ε of 0.90, how much radiant energy is emitted from the absorber surface?

13-10 - $q = \frac{kA\Delta T}{L}$ for 3 bars = 3(q)

13-20 - $h = 0.023\left(\frac{k}{D}\right)\left(\frac{DV\rho}{\mu}\right)^{0.8}\left(\frac{C_p\mu}{k}\right)^{0.4}$

CHAPTER 14

APPLICATIONS OF HEAT TRANSFER

Heat transfer is basic to the implementation of all thermodynamic systems wherein heat Q enters or leaves the thermodynamic system, or where cooling of the physical parts of the thermodynamic machine is necessary. The following discussion of heat transfer is organized from an applications approach, starting with the required heat exchange function, identifying the needed heat transfer parameters, followed by a discussion of the characteristics of each required parameter. The scope of the discussions include:

- Heat exchangers (fluid to fluid with no phase change).
- Boilers/evaporators (phase change in one fluid).
- Condensers (phase change in one fluid).
- Heat pipes.

The primary objective of the following sections on applied heat transfer is to give the student an overview of several key applications. If the reader is interested in making detailed heat transfer analyses or designs, it is recommended that he or she make use of a supplementary text treating the heat transfer aspects of interest in appropriate detail.

14-1 HEAT EXCHANGERS

14-1.1 GENERAL

Heat exchangers comprise the bulk of heat transfer applications. The purpose of a heat exchanger is to either (1) add heat to a fluid, or (2) remove heat from a fluid. This fluid is called the primary (or dominant) fluid flow, and the other fluid supplying or removing the heat from the primary fluid is called the secondary fluid.

The heat exchange is accomplished by transferring the heat from the hotter fluid, through the wall of the heat exchanger structure, into the

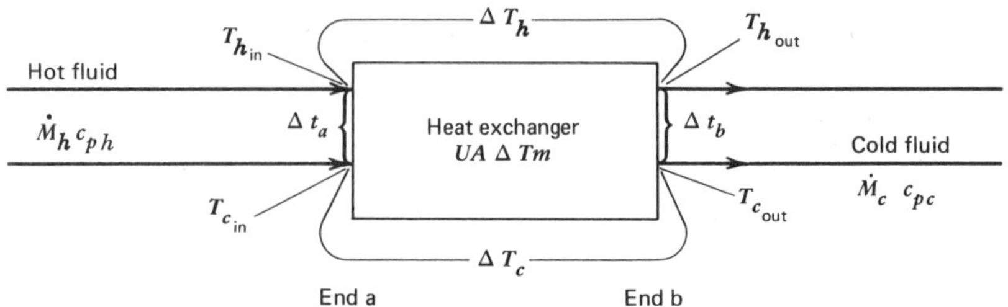

Figure 14-1 Block diagram of a heat exhanger.

colder fluid. The mechanism of this heat transfer is usually conduction/convection.

A heat exchanger can be represented by the block diagram shown in Figure 14-1.

The indicated parameters are

Hot Fluid

$\dot{M}_h$ = mass flow [kg/s]

c_{ph} = specific heat [J/kgΔ_1°C]

$T_{h_{in}}$ = fluid temperature entering the heat exchanger [°C]

$T_{h_{out}}$ = fluid temperature leaving the heat exchanger [°C]

ΔT_h = temperature drop of the fluid across the heat exchanger [Δ°C]

Coolant

$\dot{M}_c$ = mass flow [kg/s]

c_{pc} = specific heat [J/kgΔ_1°C]

$T_{c_{in}}$ = coolant temperature entering the heat exchanger [°C]

$T_{c_{out}}$ = coolant temperature leaving the heat exchanger [°C]

ΔT_c = temperature rise of the coolant across the heat exchanger [Δ°C]

Heat Exchanger Structure.

Δt_a = temperature difference between the hot fluid and coolant fluid at the end of the heat exchanger where the hot fluid enters (end a) [Δ°C]

$\Delta t_b =$ temperature difference between the hot fluid and coolant fluid at the end of the heat exchanger where the hot fluid exits (end b) $[\Delta °C]$

$U =$ overall heat transfer coefficient between the two fluids (through the respective films and the heat transfer wall of the heat exchanger) $[W/m^2\,\Delta_1 °C]$

$A =$ effective direct heat transfer surface area $[m^2]$

$\Delta T_m =$ effective mean temperature differences across the heat exchanger structure for transferring heat through the heat exchanger structure $[\Delta °C]$

In the design and analysis of heat exchangers, the following aspects must be considered individually:

- The hot fluid (being cooled).
- The coolant (being heated).
- The heat exchanger structure.

The *hot fluid* gives up heat.

$$\dot{Q}_h = \dot{M}_h c_{ph}(T_{h_{in}} - T_{h_{out}}) \qquad [W] \tag{14-1}$$

The *coolant* picks up heat.

$$\dot{Q}_c = \dot{M}_c c_{pc}(T_{c_{out}} - T_{c_{in}}) \qquad [W] \tag{14-2}$$

The *heat exchanger structure* transfers the heat from the hot fluid to the coolant.

$$\dot{Q}_e = UA\,\Delta T_m \qquad [W] \tag{14-3}$$

The amount of heat given up by the hot fluid goes into the coolant by being transferred through the heat exchanger structure

$$\dot{Q}_h = \dot{Q}_c = \dot{Q}_e \qquad [W] \tag{14-4}$$

The ΔT_m, which is determined from the initial and final temperature differences between the two fluid stream temperatures, must be sufficient to effect the required heat transfer through the heat exchanger structure.

14-1.2 BASIC FLOW CONFIGURATION TYPES

For purposes of heat transfer performance design, heat exchangers can be grouped according to the characteristic geometric arrangement between the direction of the flow paths of the two fluids passing through the heat exchanger. The usual four groupings are

- Parallel flow.
- Counterflow.
- Cross-flow.
- Multipass (tube and shell).

In a performance and design analysis based on Equation 12-3, each of these four groups of heat exchangers has substantially different performance characteristics. The values of U and A are determined in essentially the same manner for all of the four heat exchanger groupings. The differences in the performance characteristics of each of the four groups lie in the values of the effective mean temperature difference ΔT_m. This is discussed in detail in Section 14-1.5.

Fundamentally, the counterflow design is preferred for the following reasons:

1. The exchange of heat may raise the temperature of the cold fluid to more nearly the initial temperature of the hot fluid or cool the hot fluid to more nearly the initial temperature of the cold fluid depending upon the purpose of the heat exchanger.
2. To transfer the same amount of heat, a smaller surface area is required due to the higher ΔT_m.

However, geometric or other requirements for specific designs may result in the selection of one of the other heat exchanger configurations.

The primary characteristics of the four heat exchanger groupings are discussed in the following sections.

Parallel Flow Configuration

A parallel flow heat exchanger is one in which the two fluids flow in the same direction. In this configuration the final temperature ($T_{c_{out}}$) of the cold fluid does not reach the final temperature $T_{h_{out}}$ of the hot fluid, as indicated in Figure 14-2.

Counterflow Configuration

A counterflow heat exchanger is one in which the two fluids flow in opposite directions. In this configuration the final temperature $T_{c_{out}}$ of the cold fluid may be higher than the final temperature $T_{h_{out}}$ of the hot fluid, as indicated in Figure 14-3.

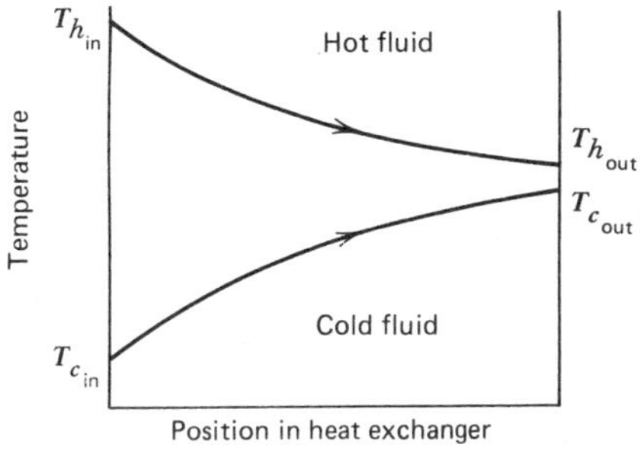

Figure 14-2 Fluid temperatures in a parallel flow heat exchanger.

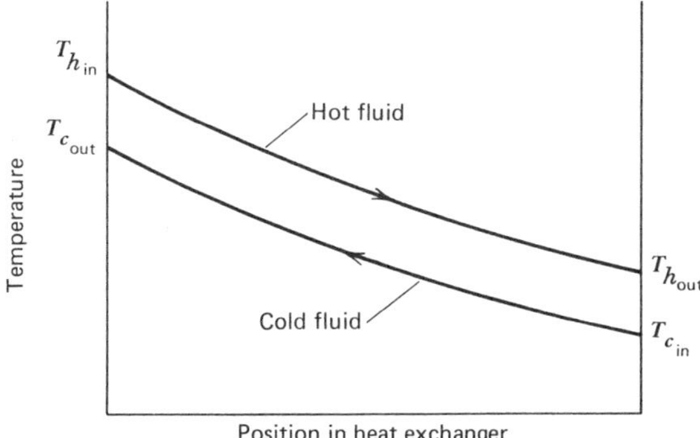

Figure 14-3 Fluid temperatures in a counterflow heat exchanger.

Cross-Flow Configuration

A cross-flow heat exchanger is one in which the two fluids flow in paths essentially at right angles to each other as illustrated in Figure 14-4.

In a cross-flow heat exchanger, the fluids leaving the heat exchanger are not of a uniform temperature. The cold fluid filament adjacent to the hot fluid entrance will be heated to a higher temperature than the cold fluid filament adjacent to the hot fluid exit ($T_{c_a} > T_{c_b}$). Similarly, the hot fluid filament adjacent to the cold fluid entrance will be cooled to a lower temperature than the hot fluid filament adjacent to the cold fluid exit

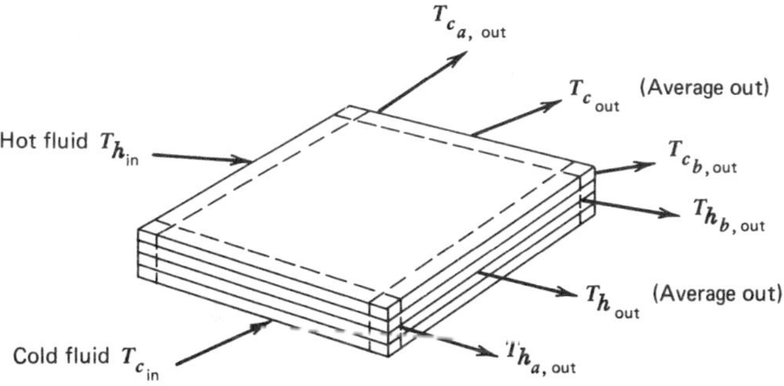

Figure 14-4 Cross-flow exchanger configuration.

($T_{h_a} < T_{h_b}$). In some cases this temperature difference in the exiting fluids can be significant, and subsequent mixing of the exiting fluid stream to obtain a uniform temperature $T_{h_{(avg)}}$ or $T_{c_{(avg)}}$ may be required.

Multipass (Tube and Shell) Heat Exchanger

Tube and shell heat exchangers use a tube bundle inside a shell with one fluid flowing inside the tubes and the other fluid flowing around the outside of the tubes, being confined within the shell. Baffles within the shell are often provided to cause the fluid surrounding the tubes to flow in a path considered to be the most advantageous for the required heat transfer. There are many arrangements of the tube bundle, flow baffles, and shell design, resulting in the general term *multipass*.

The tube and shell configuration is often selected for heat exchanger applications for reasons including the following:

- Lower cost of manufacture.
- Ease of disassembly for cleaning or repair.
- Reduced thermal stress due to expansion.

14-1.3 THE OVERALL HEAT TRANSFER COEFFICIENT

The heat flow path in a heat exchanger is a simple linear series conduction path, as illustrated in Figure 14-5, which shows heat flowing from a fluid inside a tube into a fluid around the outside of the tube.

For most heat transfer surface geometries of thin-walled tubes or flat plates, a linear series conduction heat transfer path with a constant cross-sectional area normal to the heat flow can be assumed. The thermal resistances comprising the heat transfer path are

- Convection resistance on the tube inside the wall ($1/h_i$).
- Wall conduction resistance (x/k).
- Convection resistance on the tube outside the wall ($1/h_o$).

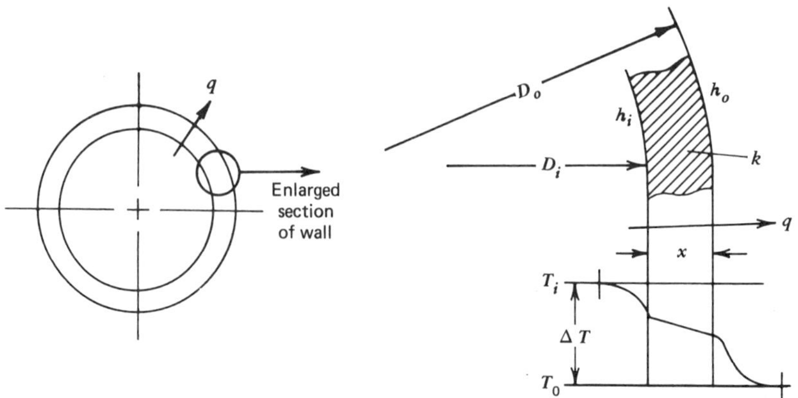

Figure 14-5 Typical heat transfer path through a heat exchanger tube.

The total thermal resistance ($\mathcal{R}_t$) of the heat transfer path is the sum of the thermal resistances of the heat transfer path components, as discussed in Section 13-2 and is

$$\mathcal{R}_t = \frac{1}{h_i} + \frac{x}{k} + \frac{1}{h_o} \qquad \left[\frac{m^2\,\Delta_1{}^\circ C}{W}\right] \tag{14-5}$$

where

$h_i =$ the film heat transfer coefficient on the inside surface $[W/m^2\,\Delta_1{}^\circ C]$

$h_o =$ the film heat transfer coefficient on the outside surface $[W/m^2\,\Delta_1{}^\circ C]$

$x =$ the length of the heat transfer path through the heat exchanger wall (direct surface) $[m]$

$k =$ the thermal conductivity of the heat exchanger wall material $[W/m\,\Delta_1{}^\circ C]$

As a thermal path conductance $[W/m^2\,\Delta_1{}^\circ C]$ is the inverse of the thermal path resistance $\mathcal{R}_t$, the overall heat transfer coefficient U (which is a conductance) is

$$U = 1/\mathcal{R}_t = \left[\frac{1}{1/h_i + x/k + 1/h_o}\right] \qquad [W/m^2\,\Delta_1{}^\circ C] \tag{14-6}$$

In cases where one of the film coefficients h is very low, this film becomes the "controlling resistance" and often fins or other configurations of extended surface are added to reduce the effective thermal resistance of that film. This condition is encountered in most heat transfer between liquids and air, such as in automobile radiators, direct engine air cooling, and air conditioning "coils." The evaluation of the effect of extended surface is relatively complex. The basic principles of extended surfaces are not discussed in this book. The reader is referred to other texts treating this aspect in sufficient detail for his or her purposes.

Typical overall coefficients of heat transfer for metal tubes, as applied in a variety of forms of heat exchangers, are listed in Table 14-1.

14-1.4 HEAT TRANSFER PATH AREA

The heat transfer path area A for use in Equation 14-3 is the area (normal to the flow of the heat) of the surface that is between the two fluids and is in direct contact with both fluids passing through the heat exchanger. This heat transfer area (called the "direct" surface area) is the wetted surface in contact with each fluid. In flat plate walls the wetted surface is the same

Table 14-1 Nominal Ranges of Typical Overall Coefficients for Metal Tubes in Powerplant and Refrigeration Apparatus

	U [$W/m^2\,\Delta_1{}^{\circ}C$]
Steam boiler tubes	20–80
Economizers	10–60
Steam superheaters, close to furnace	45–85
Surface condenser (steam)	1100–5700
Feed-water heaters	1100–8500
Air preheaters	5–20
Ammonia condensers:	
Double pipe	850–1400
Shell and tube	850–1700
Atmospheric	350–1100
Brine coolers:	
Double pipe	850–1700
Shell and tube	500–600
Water cooler, shell and coil	85–140
Cooling coils:	
Water to air in unit coolers	30–50
Boiling refrigerant to air in unit coolers	25–45
Brine to unagitated air	10–15

for each of the two fluids. In a tubular heat exchanger the tube walls are usually relatively thin (the ratio of the inside to the outside diameters of the tube is essentially 1.0); the effective direct surface area A is based on the average tube diameter (between the tube inside and outside diameters).

In cases where extended surfaces are involved, the effect on the heat transfer is usually indicated in terms of an increased effective film coefficient applied to the basic direct surface area.

14-1.5 EFFECTIVE MEAN TEMPERATURE DIFFERENCE

The effective mean temperature difference ΔT_m as used in Equation 14-3 can readily be calculated (using calculus) for pure parallel flow and counterflow, using the principle that the rate of heat flow (transfer) at any point along the heat exchanger is directly proportional to the temperature difference between the hot and cold fluids at that point. The resulting equation (known as the log mean temperature difference) is

$$\Delta T_{\mathrm{lm}} = \frac{\Delta t_a - \Delta t_b}{\ln(\Delta t_a / \Delta t_b)} \qquad [\Delta^{\circ}C] \tag{14-7}$$

where

ΔT_{lm} = log mean temperature difference [Δ°C]

Δt_a = temperature difference at one end of the heat exchange [Δ°C]

$\Delta t_b =$ temperature difference at the other end of the heat exchange [Δ°C]

ln = natural logarithm [—]

The Δt_a and Δt_b for parallel flow and counterflow are illustrated in Figure 14-6.

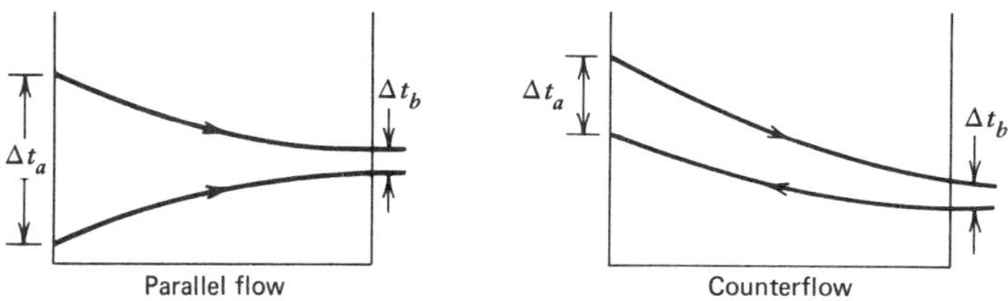

Figure 14-6 Fluid temperature difference at heat exchanger ends.

Example Problem 14-1

Calculate the mean temperature difference in a double-pipe *parallel flow* heat exchanger, where 500 kg/hr of water enters at 80°C, flows through the outer annulus and exits at 50°C. Inside the inner pipe, water at 1000 kg/hr enters at 20°C and exits at 35°C.

Equation

$$\Delta T_{\text{lm}} = \frac{\Delta t_a - \Delta t_b}{\ln(\Delta t_a / \Delta t_b)} \qquad [\Delta°\text{C}] \qquad (\text{Eq. 14-7})$$

Parameters

The temperature difference at each end will be

$$\Delta t_a = (80 - 20) = 60 \qquad [\Delta°\text{C}]$$

$$\Delta t_b = (50 - 35) = 15 \qquad [\Delta°\text{C}]$$

Substitution

$$\Delta T_{\text{lm}} = \frac{(60 - 15)}{\ln(60/15)} \qquad \frac{[\Delta°\text{C} - \Delta°\text{C}]}{[-]} = [\Delta°\text{C}]$$

$$= \frac{45}{\ln 4.00} = \frac{45}{1.3862}$$

Answer

$$\mathbf{\Delta T_{lm} = 32.5\ \Delta°C}$$

Example Problem 14-2

Calculate the mean temperature difference in a double-pipe *counterflow* heat exchanger, where 500 kg/hr of water enters at 80°C, flows through the outer annulus and exits at 50°C. Inside the inner pipe 1000 kg/hr of water enters at 20°C and exits at 35°C.

Equation

$$\Delta T_{\text{lm}} = \frac{\Delta t_a - \Delta t_b}{\ln(\Delta t_a / \Delta t_b)} \qquad [\Delta °\text{C}] \qquad (\text{Eq. 14-7})$$

Parameters

The temperature difference at each end will be

$$\Delta t_a = (80 - 35) = 45\ \Delta °\text{C}$$

$$\Delta t_b = (50 - 20) = 30\ \Delta °\text{C}$$

Substitution

$$\Delta T_{\text{lm}} = \frac{45 - 30}{\ln(45/30)} \qquad \frac{[\Delta °\text{C}] - [\Delta °\text{C}]}{[-]} = [\Delta °\text{C}]$$

$$= \frac{15}{\ln 1.500} = \frac{15}{0.4055}$$

Answer

$$\mathbf{\Delta T_{lm} = 37.0\ \Delta °C}$$

Cross-Flow Heat Exchangers

In the case of the cross-flow heat exchangers, the relative heat capacity of the two fluid streams has a significant effect upon the temperature spread in the fluid leaving the heat exchanger, and consequently upon the effective mean temperature used in Equation 14-3. The effective temperature difference ΔT_m is obtained by applying a factor F to the log mean temperature difference for a *counterflow* heat exchanger with the fluid stream inlet and average outlet temperatures of the cross-flow heat exchanger.

$$\Delta T_m = F \Delta T_{\text{lm}} \qquad [\Delta °\text{C}] \qquad (14\text{-}8)$$

The ΔT_{lm} is calculated as for the counterflow configuration. Referring to Figure 14-4, we find that

$$\Delta T_{\text{lm}} = \frac{(T_{h_{\text{in}}} - T_{c_{\text{out}}}) - (T_{h_{\text{out}}} - T_{c_{\text{in}}})}{\ln \dfrac{T_{h_{\text{in}}} - T_{c_{\text{out}}}}{T_{h_{\text{out}}} - T_{c_{\text{in}}}}} \qquad [\Delta °\text{C}] \qquad (14\text{-}9)$$

Authors, such as Bowman, Mueller, and Nagle,[19] have determined the value of F for the relative heat capacities of the two fluid streams by using the inlet and average outlet temperatures of the two streams in two ratios,

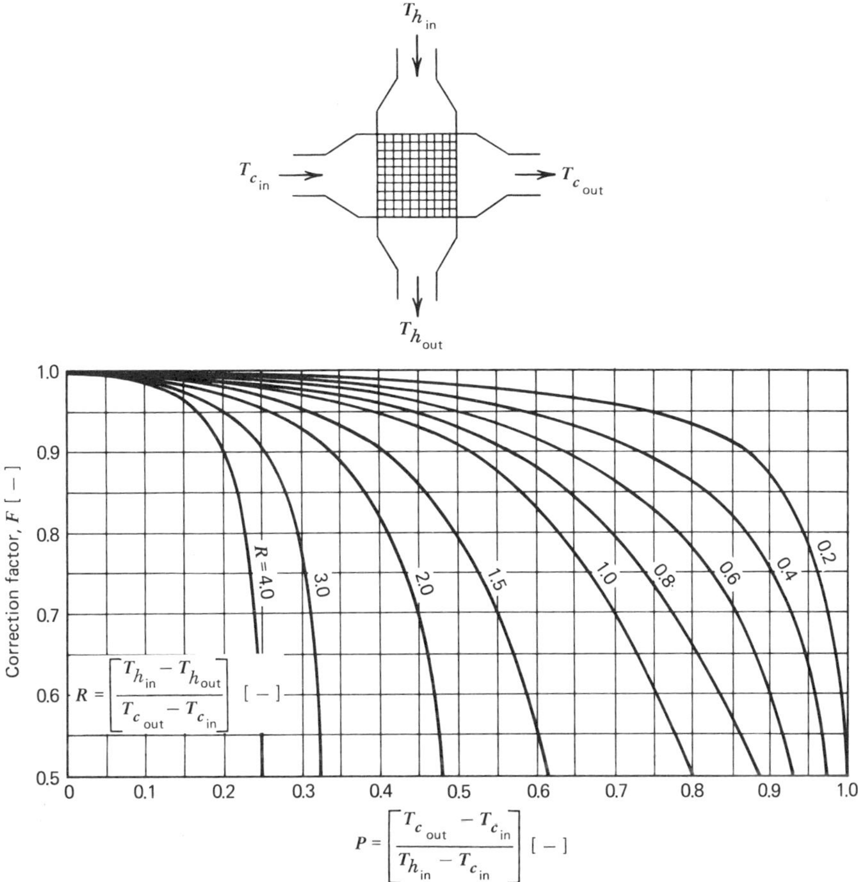

Figure 14-7 Correction factor plot for a single-pass crossflow exchanger, both fluids unmixed.

which have been designated P and R. These ratios are defined in Figure 14-7, which is a plot of the correction factor F versus the ratios P and R for the configuration of a single-pass cross-flow heat exchanger with the two fluids flowing in unmixed filaments through the heat exchanger, with a subsequent averaging of the outlet temperature. If one or both of the fluids are continuously mixed as they flow through a cross-flow heat exchanger, the correction factor F will be different from the case shown in Figure 14-7.

Example Problem 14-3

Calculate the effective temperature difference ΔT_m for a cross-flow heat exchanger when 500 kg·s of oil with a c_p of 2.1 kJ/kg·Δ_1°C enters at 110°C and is cooled to 70°C with a flow of 350 kg·s of water entering at 30°C. (See Figure 14-8.)

Equation

$$\Delta T_m = F \Delta T_{lm} \qquad [\Delta°C] \qquad \text{(Eq. 14-8)}$$

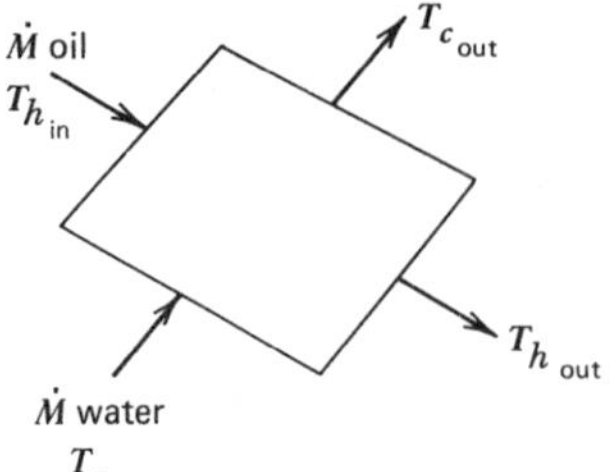

Figure 14-8 Sketch for Example Problem 14-3.

Parameters

Determine F by using Figure 14-7. To evaluate P and R, we need $T_{c_{out}}$.

(C-1). $T_{c_{out}}$

As $\dot{Q}_h = \dot{Q}_c$ [W]

$$\dot{M}_h c_{p_h}(T_{h_{in}} - T_{h_{out}}) = \dot{M}_c c_{p_c}(T_{c_{out}} - T_{c_{in}}) \qquad \text{(Eqs. 14-1 and 14-2)}$$

Solving for $T_{c_{out}}$, we obtain

$$T_{c_{out}} = T_{c_{in}} + \frac{\dot{M}_h c_{p_h}}{\dot{M}_c c_{p_c}}(T_{h_{in}} - T_{h_{out}}) \qquad [°C]$$

$T_{c_{in}} = 30°C$ (given)

$\dot{M}_h = 500$ kg/s (given)

$c_{p_h} = 2.1$ kJ/kg·Δ_1°C (given)

$\dot{M}_c = 350$ kg/s (given)

$c_{p_c} = 4.2$ kJ/kg·Δ_1°C (Table B-3 at 4°C)

$T_{h_{in}} = 110°C$ (given)

$T_{h_{out}} = 70°C$ (given)

$$T_{c_{out}} = 30 + \frac{(500)(2.1)}{(350)(4.2)}(110 - 70) \qquad [°C] + \frac{[kg/s]}{[kg/s]}kJ/kg \cdot \Delta_1°C$$

$$[°C] + \frac{[kg/s][kJ/kg \cdot \Delta_1°C]}{[kg/s][kJ/kg \cdot \Delta_1°C]}[°C - °C] = [°C] + [\Delta°C] = [°C]$$

$$= 30 + 28.6$$

$$T_{c_{out}} = 58.6°C$$

(C-2). Calculation

$$R = \frac{T_{h_{in}} - T_{h_{out}}}{T_{c_{out}} - T_{c_{in}}} \qquad [—]$$

Using the parameters listed in (C-1).

$$R = \frac{110-70}{58.6-30} = \frac{40}{28.6} \qquad \frac{[°C]-[°C]}{[°C]-[°C]} = \frac{[\Delta°C]}{[\Delta°C]} = [—]$$

$$R = 1.4 \qquad [—]$$

(C-3). Calculate P.

$$P = \frac{T_{c_{out}} - T_{c_{in}}}{T_{h_{in}} - T_{c_{in}}} \qquad [—]$$

Using the parameters listed in (C-1).

$$P = \frac{58.6-30}{110-30} = \frac{28.6}{80} \qquad \frac{[°C]-[°C]}{[°C]-[°C]} = \frac{[\Delta°C]}{[\Delta°C]} = [—]$$

$$P = 0.358$$

(C-4). Determine F from Figure 14-7. For $R = 1.4$ and $P = 0.358$,

$$F = 0.94 \qquad [—]$$

(C-5). Calculate ΔT_{lm}.

$$\Delta T_{lm} = \frac{(T_{h_{in}} - T_{c_{out}}) - (T_{h_{out}} - T_{c_{in}})}{\ln(T_{h_{in}} - T_{c_{out}})/(T_{h_{out}} - T_{c_{in}})} \qquad [\Delta°C] \qquad \text{(Eq. 14-9)}$$

Use the parameters from (C-1).

$$\Delta T_{lm} = \frac{(110-58.6)-(70-30)}{\ln(110-58.6)/(70-30)}$$

$$\frac{([°C]-[°C])-([°C]-[°C])}{\ln([°C]-[°C])/([°C]-[°C])} = \frac{[\Delta°C]}{[—]} = [\Delta°C]$$

$$= \frac{51.4-40}{\ln 51.4/40} = \frac{11.4}{\ln 1.285} = \frac{11.4}{0.25}$$

$$\Delta T_{lm} = 45.6\ \Delta°C$$

Substitution

$$\Delta T_m = (0.94)(45.6) \qquad [—][\Delta°C] = [\Delta°C]$$

Answer

$$\mathbf{\Delta T_m = 42.9 \qquad [\Delta°C]}$$

Multipass (Tube and Shell) Heat Exchanger

For the multipass designs, a correction factor F applied to the counterflow, log mean temperature difference (Equation 14-9) is used in the same manner as for the cross-flow heat exchanger. Plots of the correction factor F are given in Figures 14-9 and 14-10 for two basic tube and shell heat exchanger configurations. Tube and shell configurations are described in terms of:

- Shell passes.
- Tube passes.

A shell pass is the flow of a fluid down the length of a shell passing over the outside of the tubes. The fluid enters one end of the shell and exits at the other end. Baffles force the fluid to cross through the tube bundle a number of times. If the fluid enters one end of the shell and leaves the other end, as illustrated in Figure 14-9, the heat exchanger is characterized as a "one-shell pass." If the fluid flows down a shell and returns in the same shell or in another shell, as illustrated in Figure 14-10, the heat exchanger is characterized as a "two-shell pass."

A tube pass is the flow of the fluid inside the tubes down the length of the shell. A common tube and shell design, to minimize thermal stresses,

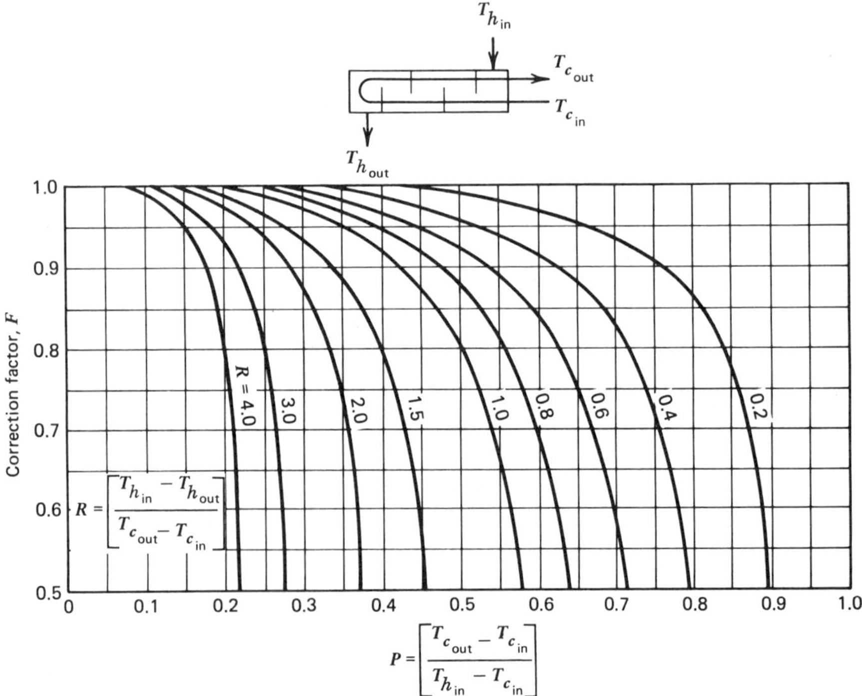

Figure 14-9 Correction factor plot for an exchanger with a one-shell pass and two, four, or any multiple of two-tube passes.

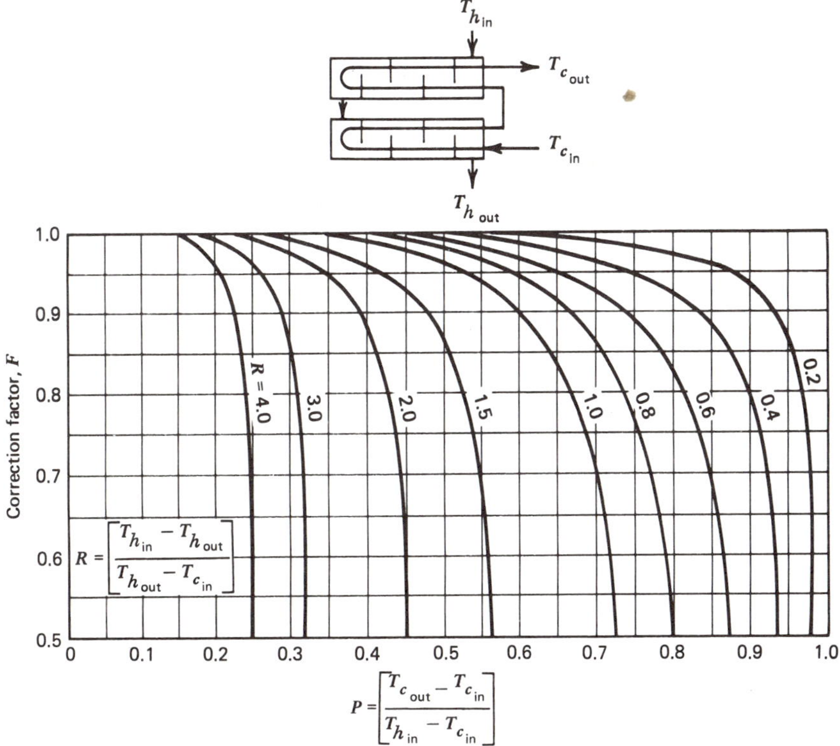

Figure 14-10 Correction factor plot for an exchanger with two-shell passes and four, eight, or any multiple of four-tube passes.

arranges the tube bundle as a U so that the fluid flowing in the tubes enters and leaves at the same end of the heat exchanger. This configuration is known as a *two-tube pass*.

The one-shell, two-tube pass configuration is illustrated in Figure 14-9. The two-shell, four-tube pass configuration is illustrated in Figure 14-10.

The plots of the correction factor F for one- and two-shell passes are given in Figures 14-9 and 14-10.

- A one-shell pass and any multiple of two-tube passes is given in Figure 14-9.
- A two-shell pass and any multiple of four-tube passes is given in Figure 14-10.

14-1.6 PERFORMANCE ANALYSIS

General Discussion

In general, there are two approaches to the design and analysis of heat exchangers. The first is to establish a design for a particular set of requirements. The other is to determine what an existing heat exchanger with a certain performance for one set of conditions will do under another set of conditions. The mean effective temperature method, based upon Equation 14-3, was initially the general method used for the design and analysis of heat exchangers. However, it was found to be a relatively cumbersome method to use for the analysis of a heat exchanger design such as determining what would be the change in the amount of heat transferred by a heat exchanger if, for example, the mass flow rate of the cooling fluid were doubled or if an inlet temperature were changed. To make this type of analysis of a heat exchanger easier, Kays and London[20] introduced what is known as the Effectiveness-NTU method. Both of these methods are discussed in the following sections, beginning with the mean effective temperature method.

The Mean Effective Temperature Method

The mean effective temperature method is based upon the use of Equation 14-3.

$$Q_e = UA\,\Delta T_m \qquad [\mathrm{W}] \qquad \text{(Eq. 14-3)}$$

where

Q_e = the heat flow transferred through the heat exchanger structure [W]

U = the overall heat transfer coefficient for the heat transfer path from one fluid, through the heat exchanger structure, and into the other fluid $[\mathrm{W/m^2}\,\Delta_1{}^\circ\mathrm{C}]$

A = the direct heat transfer surface of the heat exchanger $[\mathrm{m^2}]$

ΔT_m = the mean effective temperature difference between the two fluids in the heat exchanger $[\Delta{}^\circ\mathrm{C}]$
(a) for parallel and counterflow—the log mean temperature difference ΔT_{lm} (Eq. 14-7)
(b) for cross-flow and multipass—the log mean temperature difference for counter flow modified by a factor F.

For determining design and performance parameters of a heat exchanger, Equation 14-3 can be used; with the supplemental use of Equations 14-1, 14-2, 14-4, and 14-7, all of the heat exchanger parameters can be established.

Example Problem 14-4

If a mass flow of hot oil of 1 kg/s were cooled from 150°C to 80°C by water entering at 25°C and leaving at 90°C in a counterflow heat exchanger, what would

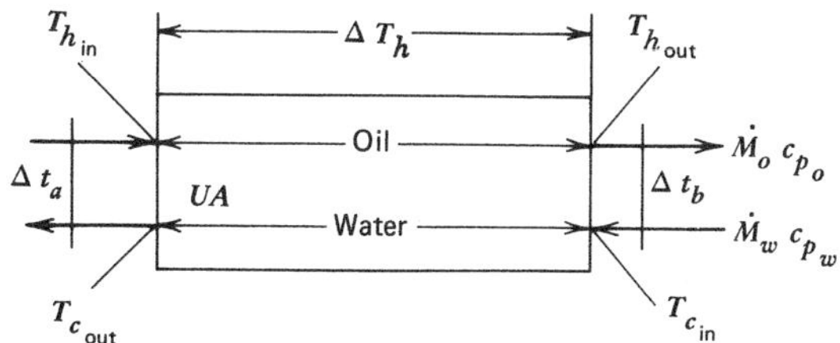

Figure 14-11 Sketch for Example Problem 14-4.

be the direct surface area A required if the value of U were 500 [W/m^2 Δ_1°C]? The specific heat of the oil is 2 kJ/kg Δ_1°C. (See Figure 14-11.)

Equation

$$A = \frac{\dot{Q}_e}{U \Delta T_m} \qquad [\text{m}^2] \qquad \text{(Eq. 14-3x)}$$

Parameters

$$\dot{Q}_e = \dot{Q}_o = \dot{M}_o c_{p_o} \Delta T_o \qquad [\text{W}] \qquad \text{(Eqs. 14-1 and 14-4)}$$

$\dot{M}_o = 1$ kg/s (given)

$c_{p_o} = 2$ kJ/kg Δ_1°C (given)

$\Delta T_h = (T_{h_{in}} - T_{h_{out}})$ [Δ°C]

$T_{h_{in}} = 150$°C (given)

$T_{h_{out}} = 80$°C (given)

$\Delta T_h = 150 - 80$ [°C]−[°C]=[Δ°C]

$\Delta T_h = 70$ [Δ°C]

$\dot{Q}_e = (1)(2)(70)$ [kg/s] [kJ/kg·Δ_1°C] [Δ°C]=[kJ/s]=kW

$\dot{Q}_e = 140$ kW = 140,000 W

$U = 500$ W/m^2 Δ_1°C (given)

$$\Delta T_{lm} = \frac{\Delta t_a - \Delta t_b}{\ln \Delta t_a / \Delta t_b} \qquad [\Delta°\text{C}] \qquad \text{(Eq. 14-7)}$$

$\Delta t_a = (T_{h_{in}} - T_{c_{out}})$ [Δ°C]

$T_{h_{in}} = 150$°C (given)

$T_{c_{out}} = 90$°C (given)

$\Delta t_a = 150 - 90$ [°C]−[°C]=[Δ°C]

$\Delta t_a = 60$ [Δ°C]

$\Delta t_b = (T_{h_{out}} - T_{c_{in}})$ [Δ°C]

$T_{h_{out}} = 80$ [°C] (given)

$T_{c_{in}} = 25$ [°C] (given)

$$\Delta t_b = 80 - 25 \qquad [°C] - [°C] = [\Delta °C]$$

$$\cdot \Delta t_b = 55 \qquad [\Delta °C]$$

$$\Delta T_{lm} = \frac{60 - 55}{\ln 60/55} = \frac{[\Delta °C][\Delta °C]}{[—]} = [\Delta °C]$$

$$= \frac{5}{\ln 1.0909} = \frac{5}{0.0870}$$

$$\cdot \Delta T_{lm} = 57.47 \qquad [\Delta °C]$$

Substitution

$$A = \frac{140{,}000}{(500)(57.47)} \qquad \frac{[W]}{[W/m^2 \cdot \Delta_1 °C][\Delta °C]} = [m^2]$$

Answer

$$\mathbf{A = 4.87\ m^2}$$

Example Problem 14-5

If we use the heat exchanger in Problem 14-4 and keeping the inlet oil flow, inlet oil temperature, and the U value the same, to what temperature would the oil be cooled by doubling the cooling water mass flow rate with an inlet temperature of 25°C?

Solution

The temperatures T_2, and t_2 can be related by the use of Equations 14-1, 14-2, 14-3, 14-4, and 14-7.

$$\dot{Q}_h = \dot{Q}_c = \dot{Q}_e \qquad [W] \qquad \text{(Eq. 14-4)}$$

$$\dot{Q}_h = \dot{M}_h c_{p_h}(T_{h_{in}} - T_{h_{out}}) \qquad [W]$$

$$\dot{Q}_c = \dot{M}_c c_{p_c}(T_{c_{in}} - T_{c_{out}}) \qquad [W]$$

Equating $\dot{Q}_h = \dot{Q}_c$

$$\dot{M}_h c_{p_h}(T_{h_{in}} - T_{h_{out}}) = \dot{M}_c c_{p_c}(T_{c_{in}} - T_{c_{out}})$$

The two equations can be obtained:

1. $\dot{Q}_h = \dot{Q}_c$
2. $\dot{Q}_h = \dot{Q}_e$

Using the first equation ($\dot{Q}_h = \dot{Q}_c$) the temperatures T_2 and t_2 are related by a "basic equation":

$$\dot{Q}_h = \dot{M}_h c_{p_h}(T_{h_{in}} - T_{h_{out}}) \qquad [W] \qquad \text{(Eq. 14-1)}$$

$$\dot{Q}_c = \dot{M}_c c_{p_c}(T_{c_{out}} - T_{c_{in}})$$

Basic Equation

$$\dot{M}_h c_{p_h}(T_{h_{\text{in}}} - T_{h_{\text{out}}}) = \dot{M}_c c_{p_c}(T_{c_{\text{out}}} - T_{c_{\text{in}}}) \qquad [\text{W}]$$

Parameter Evaluation

$$\dot{M}_h = 1\ \text{kg/s} \qquad \text{(given)}$$

$$c_{p_h} = 2000\ \text{J/kg}\cdot\Delta_1{}^\circ\text{C} \qquad \text{(given)}$$

$$T_{h_{\text{in}}} = 150^\circ\text{C} \qquad \text{(given)}$$

Using Equation 12-4, we obtain

$$\dot{Q}_e = \dot{Q}_c \qquad [\text{W}]$$

$$\dot{Q}_e = \dot{M}_c c_{p_c}(T_{c_{\text{out}}} - T_{c_{\text{in}}}) \qquad [\text{W}]$$

Therefore,

$$\dot{M}_c = c_{p_c} \frac{\dot{Q}_e}{(T_{c_{\text{out}}} - T_{c_{\text{in}}})}\ \text{kg/s}$$

$$\dot{M}_c = 2 \text{ times } \dot{M}_c \text{ for Example Problem 14-4}$$

$$\dot{M}_c = 2\left[\frac{\dot{Q}_e}{c_{p_c}(T_{c_{\text{out}}} - T_{c_{\text{in}}})}\right] \qquad [\text{kg/s}]$$

$$\dot{Q}_e = 140{,}000 \qquad [\text{W}] \qquad \text{(Example Problem 14-4)}$$

$$c_{p_c} = 4200\ \text{J/kg}\cdot\Delta_1{}^\circ\text{C}$$

$$T_{c_{\text{in}}} = 25^\circ\text{C} \qquad \text{(given)}$$

$$T_{c_{\text{out}}} = 90^\circ\text{C} \qquad \text{(given)}$$

$$\dot{M}_c = 2\left[\frac{140{,}000}{4200(25-90)}\right] \qquad \frac{[\text{W}]}{[\text{J/kg}\cdot\Delta_1{}^\circ\text{C}][\Delta^\circ\text{C}]} = [\text{kg/s}]$$

$$= 1.026\ \text{kg/s}$$

$$c_{p_c} = 4200\ \text{J/kg}\cdot\Delta_1{}^\circ\text{C} \qquad \text{(Appendix B, Table B-3)}$$

$$T_{c_{\text{in}}} = 25^\circ\text{C} \qquad \text{(given)}$$

Substituting in the basic equation yields

$$(1)(2000)(150 - T_{h_{\text{out}}}) = (1.026)(4200)(T_{c_{\text{out}}} - 25)$$

Equation A. $T_{h_{\text{out}}} + 2.155\,T_{c_{\text{out}}} = 203.9$

Using the second equation ($\dot{Q}_h = \dot{Q}_e$)

$$\dot{M}_h c_{p_h}(T_{h_{\text{in}}} - T_{h_{\text{out}}}) = UA\cdot\Delta T_{\text{m}} \qquad [\text{W}]$$

$$\dot{M}_h c_{p_h}(T_{h_{\text{in}}} - T_{h_{\text{out}}}) UA\left[\frac{(T_{h_{\text{in}}} - T_{c_{\text{out}}}) - (T_{h_{\text{out}}} - T_{h_{\text{in}}})}{\ln 150 - T_{c_{\text{out}}}/T_{h_{\text{out}}} - 25}\right] \qquad [\text{W}]$$

Substituting the values listed above yields

$$(1)(2000)(150-T_{h_{out}})=(500)(4.87)\left[\frac{(150-T_{c_{out}})-(T_{h_{out}}-25)}{\ln\left(\dfrac{150-T_{c_{out}}}{T_{h_{out}}-25}\right)}\right]$$

Equation B.

$$T_{h_{out}}+1.218\left[\frac{(150-T_{c_{out}})-(T_{h_{out}}-25)}{\ln 150-T_{c_{out}}/T_{h_{out}}-25}\right]=150$$

Equations A and B are two equations with the two unknowns $T_{h_{out}}$ and $T_{c_{out}}$, which satisfies the mathematical requirements for solution. However, Equations A and B do not lend themselves readily to a direct mathematical solution. Consequently, the practical method for solution is to solve by iteration, as illustrated below:

First Iteration. Assume that $T_{h_{out}}=70°C$

Solve Equation A for $T_{c_{out}}$

Substitute $T_{h_{out}}$ and $T_{c_{out}}$ in Equation B to see if it is valid.

$$\begin{aligned}T_{c_{out}}&=94.62-0.464T_{h_{out}}\\&=94.62-(0.464)(70)=94.62-32.48\\&=62.14°\text{C}\end{aligned}$$

Substitute $T_{h_{out}}$ and $T_{c_{out}}$ in Equation B.

$$70+1.218\left[\frac{(150-62.14)-(70-25)}{\ln 150-62.14/70-25}\right]$$

$$70+1.218\left[\frac{87.86-45}{\ln 1.952}\right]$$

$$70+1.218\left[\frac{42.86}{0.669}\right]=70+78.03$$

$$=148.03\neq 150$$

Second Iteration. Assume that $T_{h_{out}}=72°C$

Equation A: $\quad T_{c_{out}}=94.62-(0.464)(72)=94.62-33.41$

$$T_{c_{out}}=61.21°\text{C}$$

Equation B: $\quad 72+1.218\left[\dfrac{(150-62.21)-(72-25)}{\ln(150-61.21)/(72-25)}\right]$

$$72+1.218\frac{88.79-47}{\ln 1.889}=72+1.218\left(\frac{41.79}{0.636}\right)$$

$$72+80.03=152.03\neq 150.$$

Third Iteration. Assume $T_{h_{out}}$ halfway between 70 and 72°C

Answer

$$T_{h_{out}} = 71°\text{C}$$

The Effectiveness–NTU Method

The Effectiveness–NTU method (NTU method) offers many advantages for the evaluation of different types of heat exchangers. Although the log mean temperature difference (LMTD) method of heat exchanger analysis is useful when all of the inlet and outlet fluid temperatures are known, the use of the LMTD method to predict the performance of a given heat exchanger design for other conditions involves an iterative procedure (due to the logarithmic function) as illustrated by Example Problem 14-5.

The Effectiveness–NTU method involves the following parameters:

- The heat exchanger effectiveness $[\varepsilon]$.
- The fluid flow capacity rates $[C_{min}, C_{max}]$.
- The parameter $[AU/C_{min}]$ called NTU, meaning *N*umber of *T*ransfer *U*nits. This parameter is related to the size of the heat exchanger.

The effectiveness of a heat exchanger is defined as:

$$\text{Effectiveness}[\varepsilon] = \frac{\text{actual heat transfer}}{\text{maximum possible heat transfer}}$$

The *actual heat transfer* $\dot{Q}_a$ for the parallel flow configuration shown in Figure 14-12 is

$$\dot{Q}_a = \dot{M}_h c_{p_h}(T_{h_{in}} - T_{h_{out}}) \qquad \text{(Eqs. 14-1 and 14-2)}$$
$$= \dot{M}_c c_{p_c}(T_{c_{out}} - T_{c_{in}})$$

The *maximum possible heat transfer* $\dot{Q}_{mp}$ is the concept of the heat transferred if the fluid were heated or cooled (as the case may be) from its inlet temperature to the inlet temperature of the other fluid. Hence

$$\dot{Q}_{mp_h} = \dot{M}_h c_{p_h}(T_{h_{in}} - T_{c_{in}}) \qquad (14\text{-}10)$$

$$\dot{Q}_{mp_c} = \dot{M}_c c_{p_c}(T_{c_{in}} - T_{h_{in}}) \qquad (14\text{-}11)$$

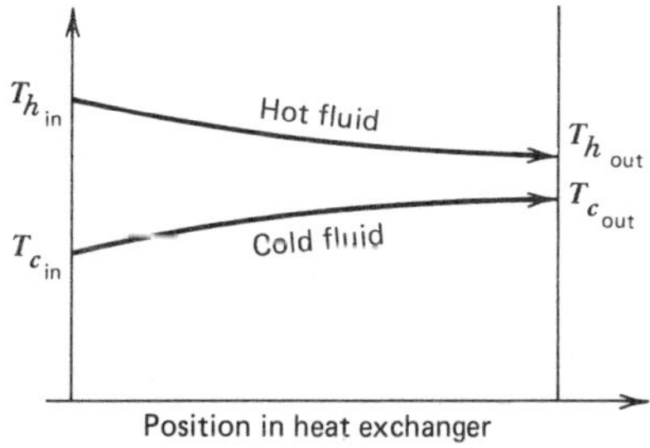

Figure 14-12 Parallel flow configuration.

The maximum possible heat transfer ($\dot{Q}_{mp}$) is different for the two fluid flows in the heat exchanger. Consequently, the effectiveness for one fluid flow is different from the effectiveness of the other fluid flow.

In the Effectiveness–NTU method, the *effectiveness of the fluid stream having the smaller heat capacity* ($\dot{M}cp$) *is used*. The effectiveness is readily calculated by one of the following equations:

$$\varepsilon_h = \frac{(T_{h_{\text{in}}} - T_{h_{\text{out}}})}{(T_{h_{\text{in}}} - T_{c_{\text{in}}})} \qquad [-] \tag{14-12}$$

or

$$\varepsilon_c = \frac{(T_{c_{\text{out}}} - T_{c_{\text{in}}})}{(T_{h_{\text{in}}} - T_{c_{\text{in}}})} \qquad [-] \tag{14-13}$$

where the subscript h refers to the fluid flow with the smaller heat capacity rate and the subscript c refers to the fluid flow with the larger heat capacity rate.

The effectiveness is then simply the ratio of the temperature change of the fluid flow with the smaller heat capacity to the maximum temperature difference in the heat exchanger.

The heat capacity rate C of the fluid flow is defined as:

$$C = \dot{M}c_p \qquad [\text{W}/\Delta_1{}^\circ\text{C}] \tag{14-14}$$

where

$$\dot{M} = \text{mass flow rate} \qquad [\text{kg/s}]$$
$$c_p = \text{specific heat} \qquad [\text{J/kg}\cdot\Delta_1{}^\circ\text{C}]$$

Each of the two fluids has its own heat capacity rate:

$$C_h = \dot{M}_h c_{p_h}, \qquad C_c = \dot{M}_c c_{p_c}$$

where the subscripts c and h refer to the cold and hot fluid streams, respectively.

The higher value of C, regardless of whether it is for the hot or cold fluid stream becomes C_{max}, and the lower value becomes C_{min}. The effectiveness–NTU method makes use of both:

- The ratio $C_{\text{min}}/C_{\text{max}}$.
- C_{min}.

Kays and London[20] have published graphs of effectiveness versus the NTU parameter (AU/C_{min}) for various heat exchanger configurations.

Graphs for some of the more common heat exchanger configurations are shown in Figures 14-13 through 14-17.

These curves give a relationship between:

- Effectiveness.
- C_{min}/C_{max}.
- AU/C_{min}.

With any two of these three parameters known, the third can be determined from these graphs. The three parameters serve to completely define the heat exchanger and its heat transfer performance.

Example Problem 14-6

If a mass flow of hot oil of 1 kg/s were cooled from 150°C to 80°C by water entering at 25°C and leaving at 90°C in a counterflow heat exchanger, what would be the direct surface area A required if the value of U were 500 [W/m² Δ_1°C]? The specific heat of the oil is 2 kJ/kg Δ_1°C.

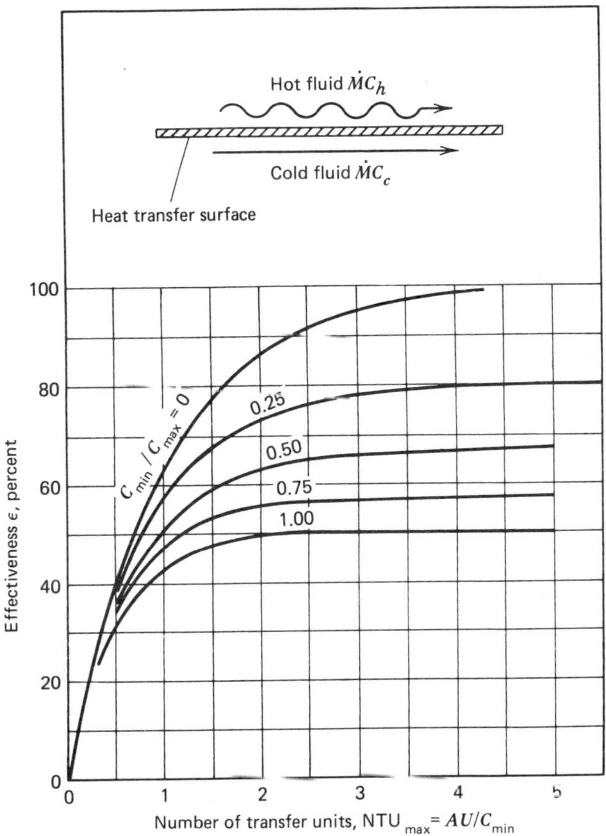

Figure 14-13 Effectiveness for parallel flow exchanger performance.

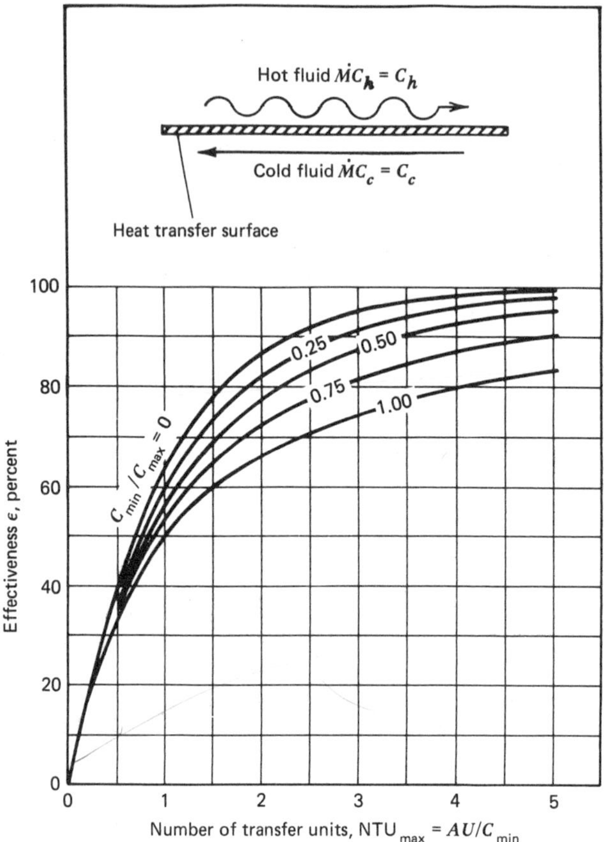

Figure 14-14 Effectiveness for counterflow exchanger performance.

Solution

Use Figure 14-14, which requires:

$$\text{Effectiveness } (\varepsilon) \qquad [\text{—}]$$

$$\frac{C_{\text{min}}}{C_{\text{max}}} \qquad [\text{—}]$$

$$C_{\text{min}} \qquad [\text{W}/\Delta_1{}^\circ\text{C}]$$

Equation

$$\varepsilon = \frac{(T_{h_{\text{in}}} - T_{h_{\text{out}}})}{(T_{h_{\text{in}}} - T_{c_{\text{in}}})} \qquad [\text{—}] \qquad \text{(Eq. 14-12)}$$

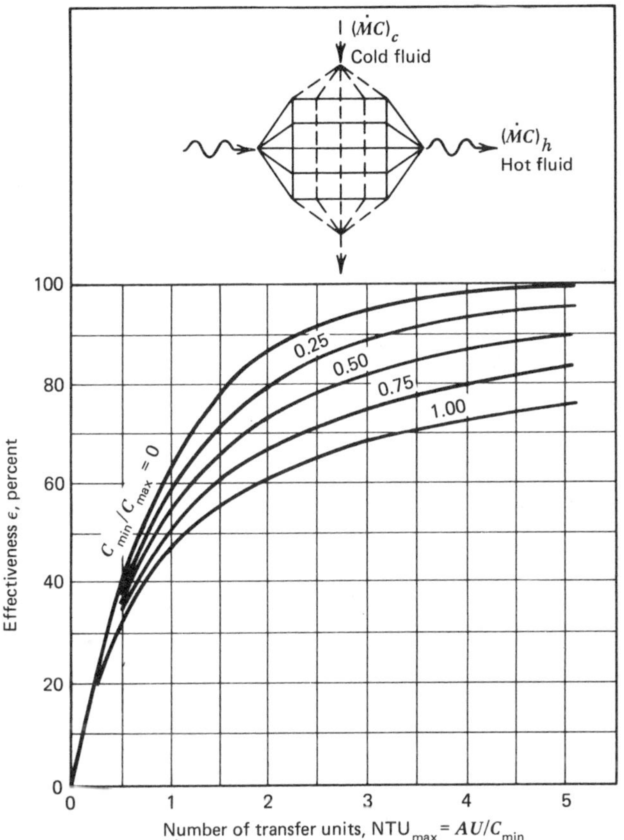

Figure 14-15 Effectiveness for cross-flow exchanger with fluids unmixed.

Parameters

To establish which temperatures must be used in Equation 14-11, determine which fluid has $C_{\min}$.

$$C_{\text{oil}} = \dot{M}_o c_{p_o} \qquad [\text{W}/\Delta_1{}^\circ\text{C}]$$

$$\dot{M}_o = 1 \qquad [\text{kg/s}] \qquad \text{(given)}$$

$$c_{p_o} = 2000 \qquad [\text{J/kg}\,\Delta_1{}^\circ\text{C}] \qquad \text{(given)}$$

$$C_{\text{oil}} = (1)(2000) \qquad [\text{kg/s}] \qquad [\text{J/kg}\,\Delta_1{}^\circ\text{C}] = [\text{J/s}\,\Delta_1{}^\circ\text{C}] = [\text{W}/\Delta_1{}^\circ\text{C}]$$

$$= 2000 \qquad \text{W}/\Delta_1{}^\circ\text{C}]$$

To determine C_w, use:

$$Q_h = Q_c \qquad [\text{W}]$$

$$Q_h = C_{\text{oil}}\,\Delta T_{\text{oil}} \qquad [\text{W}]$$

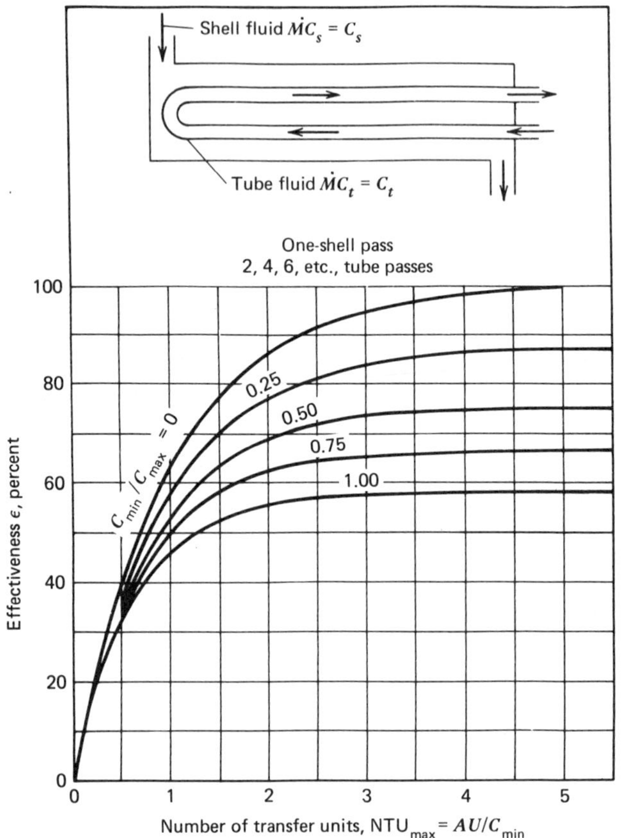

Figure 14-16 Effectiveness for 1–2 parallel counterflow exchanger performance.

$$Q_c = C_w \Delta T_w \qquad [\text{W}]$$

$$C_w = C_{oil} \frac{\Delta T_{oil}}{\Delta T_w} \qquad [\text{W/°C}]$$

$$\Delta T_{oil} = T_{h_{in}} - T_{h_{out}} \qquad [\Delta°\text{C}]$$

$$T_{h_{in}} = 150°\text{C} \qquad (\text{given})$$

$$T_{h_{out}} = 80°\text{C} \qquad (\text{given})$$

$$\Delta T_{oil} = 150 - 80 \qquad [°\text{C}] - [°\text{C}] = [\Delta°\text{C}]$$

$$\Delta T_{oil} = 70 \qquad [\Delta°\text{C}]$$

$$\Delta T_w = T_{c_{out}} - T_{c_{in}} \qquad [\Delta°\text{C}]$$

$$= T_{c_{out}} - 90 \qquad [°\text{C}] \qquad (\text{given})$$

$$T_{c_{in}} = 25 \qquad [°\text{C}] \qquad (\text{given})$$

$$\Delta T_w = 90 - 25 \qquad [°\text{C}] - [°\text{C}] = [\Delta°\text{C}]$$

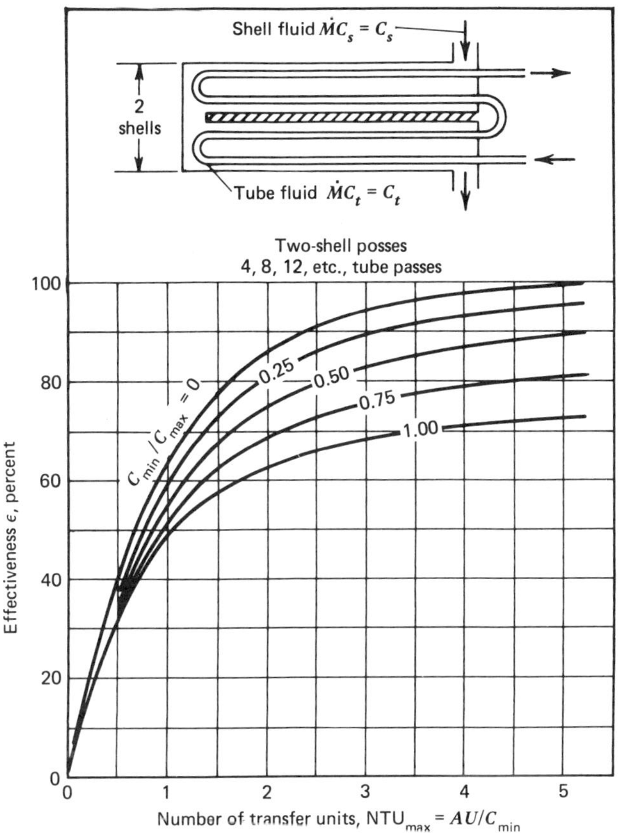

Figure 14-17 Effectiveness for 2–4 multipass counterflow exchanger performance.

$$\Delta T_w = 65 \qquad [\Delta°C]$$

$$C_w = 2000\frac{70}{65} \qquad [W/\Delta_1°C]\frac{[\Delta°C]}{[\Delta°C]} = [W/\Delta_1°C]$$

$$C_w = 2154 \qquad [W/\Delta_1°C]$$

Therefore,

$$C_{oil} = C_{min}$$

The oil temperatures are T_h; water temperatures are T_c.

Substitution

$$\varepsilon = \frac{150-80}{150\ \ 25} \qquad \frac{[°C]-[°C]}{[°C]-[°C]} = \frac{[\Delta°C]}{[\Delta°C]} = [—]$$

$$\varepsilon = \frac{70}{125}$$

Answer

$$\cdot \varepsilon = \mathbf{0.56} \qquad [\text{—}]$$

Using Figure 14-14 to determine AU/C_{min} at $\varepsilon = 0.56$ [—]

$$\frac{C_{min}}{C_{max}} = C_{oil}/C_w = \frac{2000}{2154} \qquad \frac{[\text{W}/\Delta_1{}^\circ\text{C}]}{[\text{W}/\Delta_1{}^\circ\text{C}]} = [\text{—}]$$

$$\cdot \frac{C_{min}}{C_{max}} = 0.929 \qquad [\text{—}]$$

$$\cdot C_{min} = 2000 \qquad [\text{W}/\Delta_1{}^\circ\text{C}]$$

$$\frac{AU}{C_{min}} = 1.22 \qquad [\text{—}]$$

Solving for area gives

$$A = \frac{1.22 C_{min}}{U} \qquad [\text{m}^2]$$

$$U = 500 \qquad \left[\text{W}/\text{m}^2\,\Delta_1{}^\circ\text{C}\right] \qquad \text{(given)}$$

$$A = \frac{(1.22)(2000)}{(500)} \qquad [\text{—}] \qquad \frac{[\text{W}/\Delta_1{}^\circ\text{C}]}{\left[\text{W}/\text{m}^2\,\Delta_1{}^\circ\text{C}\right]} = [\text{m}^2]$$

Final Answer

$$\boldsymbol{A = 4.88} \qquad [\mathbf{m^2}]$$

Example Problem 14-7

Using the heat exchanger in the Problem 14-4 and keeping the inlet oil flow, inlet oil temperature, and the U value the same, to what temperature would the oil be cooled by doubling the cooling water mass flow rate with an inlet temperature of 25°C?

Solution

Use Figure 14-14.

- The heat exchanger characterization $AU/C_{min} = 1.22$ [—] remains the same as in Problem 14-6.
- C_{min}/C_{max} changes as the coolant water flow is increased.
- Determine ε from Figure 14-14.
- Then calculate the oil temperature out.

The cooling water mass flow is doubled. Therefore, C_w is doubled (C_{max}).

$$C_w = (2154)(2) \qquad [\text{W}/\Delta_1{}^\circ\text{C}]$$

$$C_w = 4308 \qquad [\text{W}/\Delta_1{}^\circ\text{C}]$$

There is no change in the oil mass flow rate.

$$C_{min} = 2000 \quad [W/\Delta_1 °C] \quad \text{(from Problem 14-6)}$$

$$\frac{C_{min}}{C_{max}} = \frac{2000}{4308} \quad \frac{[W/\Delta_1 °C]}{[W/\Delta_1 °C]} = [—]$$

$$\frac{C_{min}}{C_{max}} = 0.464 \quad [—]$$

Using Figure 14-14 to determine ε

$$\frac{AU}{C_{min}} = 1.22 \quad [—]$$

$$\frac{C_{min}}{C_{max}} = 0.464 \quad [—]$$

$$\varepsilon = 0.63 \quad [—]$$

Calculating $T_{1_{out}}$.

$$\varepsilon = \frac{T_{h_{in}} - T_{h_{out}}}{T_{h_{in}} - T_{c_{in}}} \quad [—]$$

$$T_{h_{out}} = T_{h_{in}} - \varepsilon(T_{h_{in}} - T_{c_{in}}) \quad [°C]$$

Parameters

$$T_{h_{in}} = 150°C \quad \text{(given)}$$

$$T_{c_{in}} = 25°C \quad \text{(given)}$$

$$\varepsilon = 0.63 \quad [—]$$

Substitution

$$T_{h_{out}} = 150 - 0.63(150 - 25) \quad [°C] - [—][°C - °C] = [°C] - [\Delta °C] = [°C]$$

$$= 150 - 78.75$$

Answer

$$T_{h_{out}} = 71.25°C$$

14-2 BOILERS

Boilers play an important part in power generation, chemical plants, oil refineries, and processing activities. The primary function of a boiler is to generate steam. A boiler (or steam generator) is a thermodynamic device that converts water into steam by the transfer or energy (heat) from combustion gases. Typical boilers are illustrated in Figures 14-18 and 14-19.

Boiling is not a simple process because it involves a phase change from liquid to vapor. Because of the relatively large differences in the properties of the liquid and vapor phases, there are many factors that affect the boiling heat transfer. These include:

- Specific heat of the saturate liquid.
- Temperature difference between the wall and saturation.
- Prandtl number of the saturated liquid.
- Liquid viscosity.
- Surface tension of the liquid–vapor interface.
- Surface wetability factor.

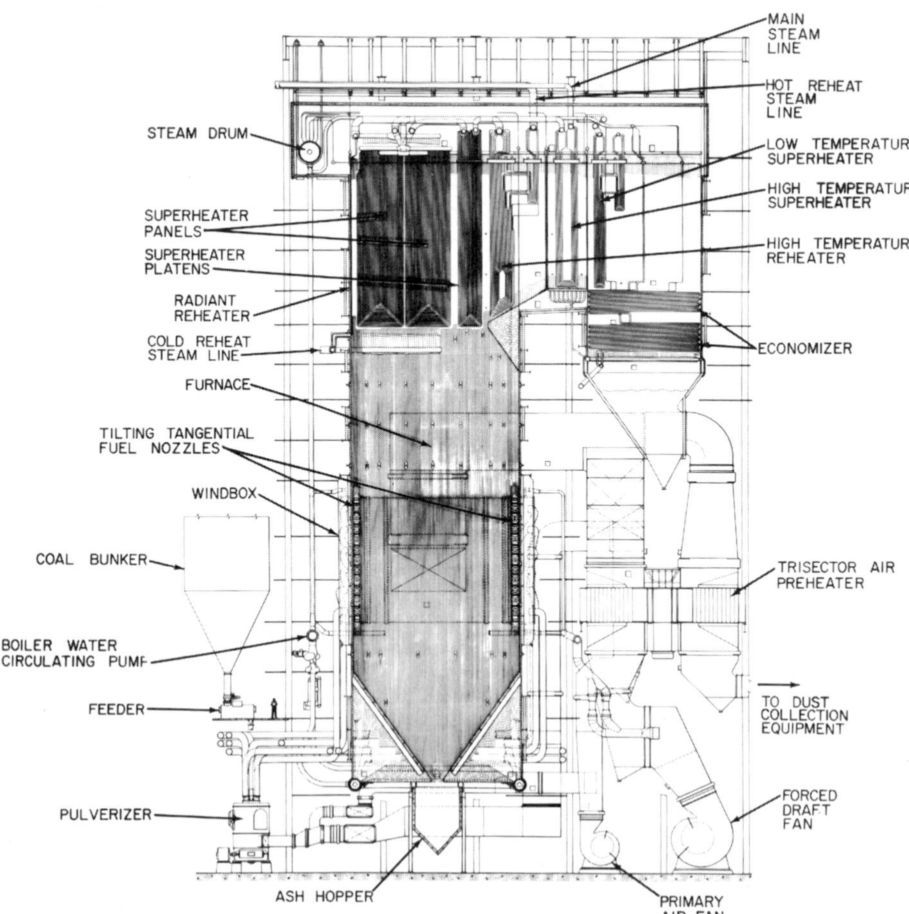

Figure 14-18 Field-erected, coal-fired, C-E "Controlled Circulation steam generator. Nominal 650 Megawatts. (Photo courtesy of Combustion Engineering Inc.)

Figure 14-19 Shop assembled boiler. C-E "VP-14."(Photo courtesy of Combustion Engineering Inc.)

- Density of saturated liquid.
- Density of saturated vapor.
- Gravitational acceleration.
- Latent heat of vaporization.

The basic boiling heat transfer equation has the general form of:

$$\dot{Q}=h_c A\,\Delta T \qquad [\mathrm{W}] \tag{14-15}$$

where

$\dot{Q}$ = heat flow rate [W]

h_c = boiling film coefficient $[\mathrm{W/m^2 \cdot \Delta_1 {}^\circ C}]$

A = surface area $[\mathrm{m^2}]$

ΔT = temperature excess $(T_{\mathrm{wall}} - T_{\mathrm{sat.}})$ [°C]

Frequently, the heat flux ($\dot{Q}/A$ [W/m²]) is used for plotting boiling data. From Equation 14-15

$$\frac{\dot{Q}}{A}=h_c\,\Delta T \qquad [\mathrm{W/m^2}] \tag{14-16}$$

The two basic types of boiling heat transfer are called *pool* and *forced convection*.

Pool boiling occurs when a body of liquid is in a container with no mechanical stirring. Any circulation of the fluid is caused by changes in the density of the heated fluid and by the generation of vapor bubbles.

Forced convection boiling occurs when the fluid is pumped and forced to flow across heat transfer surfaces in a controlled manner.

Whether boiling occurs in pool boiling or in forced convection boiling, there are six definite regimes of boiling associated with progressively increasing heat transfer fluxes as illustrated in Figure 14-20. In this figure the heat transfer flux is plotted against the temperature difference between the wall surface and the saturation temperature of the liquid. In the first regime (I) no vapor bubbles in the liquid are involved. Evaporative cooling occurs at the surface of the liquid. As the heat transfer flux to the boiling surface is increased, bubbles are generated. In the boiling regime II the bubbles condense before they reach the liquid surface. At the end of the boiling regime III, the increased amount of vapor at the boiling surface prevents an increased amount of heat flux until the temperature difference between the wall and the fluid saturation temperature becomes very high as indicated by the dashed line in Figure 14-20. The heat flux level at which this jump occurs is called the *critical heat flux*.

Pressure has a significant effect upon nucleate boiling. Extension of the performance at one pressure to another usually is relatively inaccurate.

14-2.1 CRITICAL HEAT FLUX

As the temperature difference between the wall and the boiling fluid is increased in the nucleate boiling regime, the heat flux is increased, as indicated by the curve in Figure 14-20. The heat flux reaches a maximum at the end of the nucleate boiling regime as indicated by point *a*. The heat flux corresponding to the heat flux at point *a* is called the critical heat flux. Any further increase in the heat flux results in a shift in the boiling regime to point *b*. At point *b* there is a significant increase in the temperature difference between the wall temperature and the fluid temperature, which results in a marked increase in the wall temperature. If this temperature exceeds the temperature limit of the wall material, "burnout" of the wall (i.e., structural failure) will occur. If this temperature does not exceed the temperature limit of the wall material, boiling will continue under the conditions of the higher wall temperature.

If the boiling is occurring when a constant heat flux is being applied (such as by an electrical resistance, rocket combustion gases, an atomic reactor, etc.), burnout usually results when the critical heat flux is reached. However, in other configurations, such as with a constant temperature heat source, the increased wall temperature will result in a reduced input heat flux. The boiling will then occur at a reduced heat flux and a resultant wall

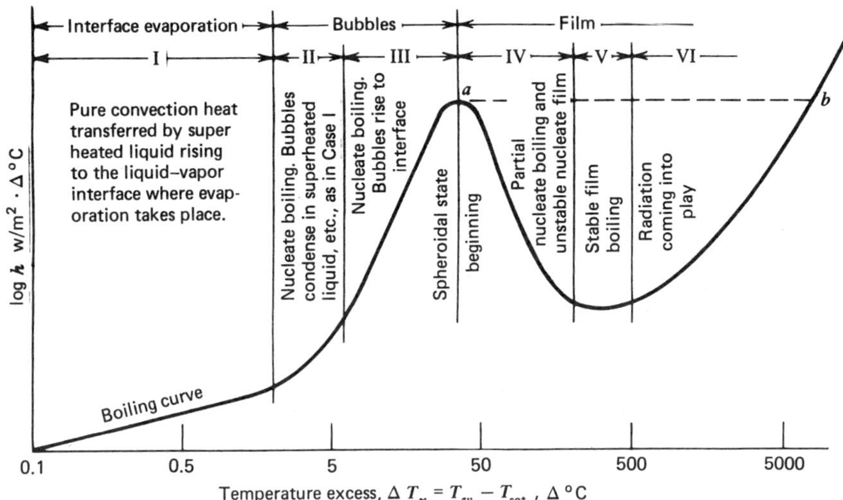

Figure 14-20 Boiling regimes. Heat flux data from an electrically heated platinum wire. (From Farber and Scorah.[17])

temperature that may be below the burnout limits of the wall material. In this case burnout may not occur.

14-3 CONDENSERS

A condenser is a heat exchanger in which the hot fluid enters as a gaseous vapor and leaves as a liquid. Typical applications include heat removal in the Rankine power generation cycle and in chemical processing and oil refining. Figures 14-21 and 14-22 show examples of typical condensers.

Condensing heat transfer has been characterized by two types of condensation:

- Film condensation.
- Dropwise condensation.

In the film condensation, which is the normal stable type of condensation, the condensation wets the solid cooling surface and forms a film of saturated liquid covering the solid surface.

The heat transfer path along which the heat $\dot{Q}$ flows from the condensing vapor to the cooling fluid as illustrated in Figure 14-23.

The heat transfer path is described by the following parameters:

- Surface area of the tubes A [m^2].
- Overall heat transfer coefficient U [W/m$^2 \cdot \Delta_1$°C].
- Temperature difference ΔT—the difference between the steam temperature T_s and the cooling water temperature T_w [Δ°C].

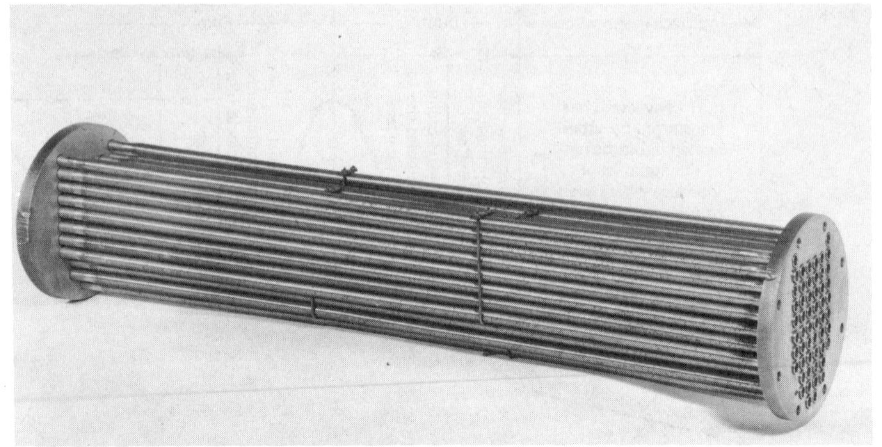

Figure 14-21 Condenser used in large power plant. (Photo courtesy of Marley Heat Transfer Co.)

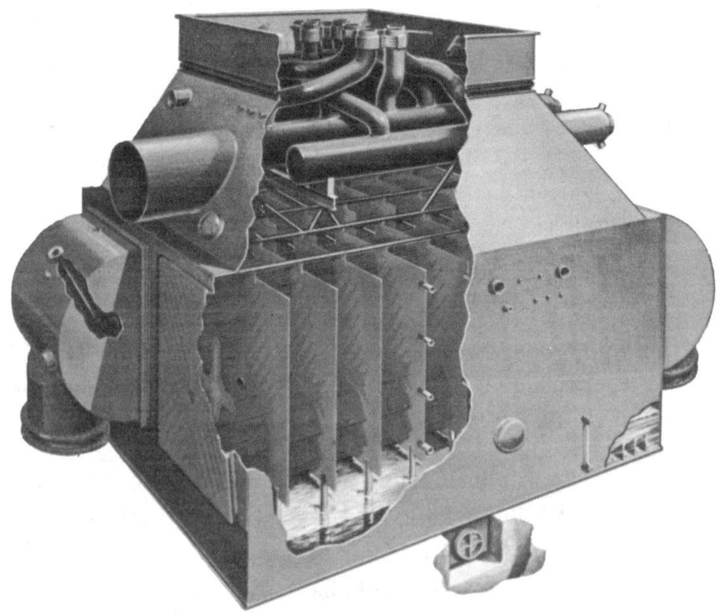

Figure 14-22 Cutaway of an air conditioner condenser. (Photo courtesy of ITT Fluid Handling Division.)

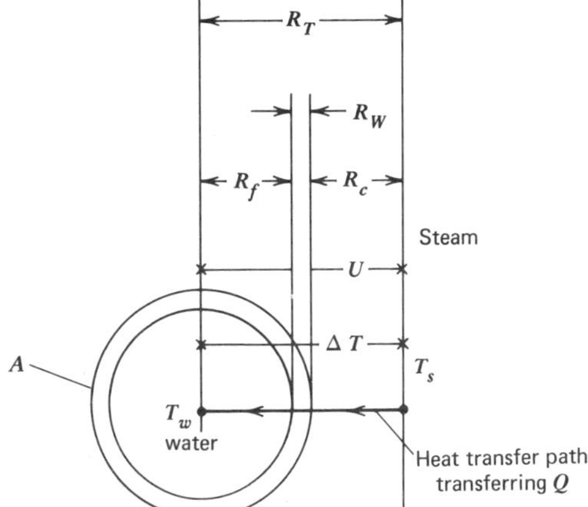

Figure 14-23 Typical heat transfer path in a tubular steam condenser.

The basic equation for the heat transferred $\dot{Q}$ is

$$\dot{Q} = AU\Delta T_{\text{m}} \qquad [\text{W}] \tag{14-17}$$

where

A = the surface area of the tubes $[\text{m}^2]$

U = the overall heat transfer coefficient $[\text{W/m}^2 \cdot \Delta_1 {}^\circ\text{C}]$

ΔT_m = the effective temperature difference between the saturation temperature of the steam and the cooling water $[\Delta {}^\circ\text{C}]$

Effective Mean Temperature Difference ΔT_{lm}

In a condenser the usual heat transfer model assumed that the condensing vapors have unrestricted access to the condensation surfaces resulting in a uniform vapor saturation temperature throughout the condenser. The cooling water, as it picks up heat, increases in temperature as it passes through the condenser. This is illustrated in Figure 14-24, which is a plot of the temperatures along the length of the condenser.

The effective temperature difference ΔT_{lm} can be determined mathematically on the basis that at any point along the tube the rate of the heat transfer is proportional to the temperature difference between the condenser steam inlet saturation temperature and the cooling water at that

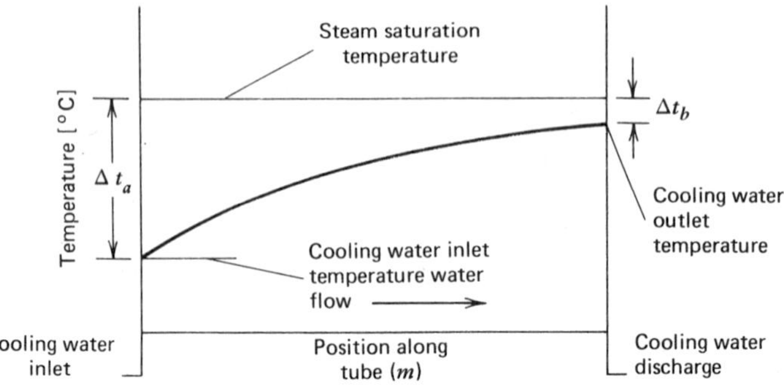

Figure 14-24 Temperatures along the length of a condenser tube.

point. The resulting temperature difference ΔT_{lm} is the log mean temperature difference (LMTD), developed in Section 14-1.4. The equation is

$$\Delta T_{\text{lm}} = \frac{\Delta t_a - \Delta t_b}{\ln(\Delta t_a / \Delta t_b)} \qquad [°\text{C}] \qquad (\text{Eq. 14-7})$$

where

ΔT_{lm} = the log mean temperature difference (LMTD) [Δ°C]

Δt_a = the inlet temperature difference (steam to inlet water) [Δ°C]

Δt_b = the outlet temperature difference (steam to outlet water) [Δ°C]

ln = the natural logarithm [—]

However, if the condenser tube bundle geometry significantly restricts the vapor flow, some regions in the condenser will be at a lower pressure than the nominal condenser inlet pressure and then the associated vapor saturation temperature will be lower. This condition will reduce the overall condenser effective mean temperature difference, which results in an increase in the condensing surface area to transfer the required amount of heat, or an increase in the cooling water flow.

The Overall Condensing Heat Transfer Coefficient *U*

The overall condensing heat transfer coefficient *U*, as discussed in Section 14-1.3, is a series conduction path comprising the following factors:

- Condensing film thermal resistance $\left(\frac{1}{h_c}\right)$.
- Condensing wall conduction resistance $\left(\frac{x}{k}\right)$.
- Cooling fluid, film thermal resistance $\left(\frac{1}{h}\right)$.

The cooling fluid, film heat transfer coefficient h was discussed in Section 13-3; the condensing wall conduction resistance was discussed in Section 13-2. The condensing film coefficient h_c for the condensation of a pure saturated vapor on the outside of a single horizontal tube has been described by an equation derived from the generalized Nusselt equation as follows:

$$h_c = 0.725\left[\frac{g\rho^2 k^3 r}{D\mu\,\Delta T}\right]^{0.25} \qquad [\mathrm{W/m^2 \cdot \Delta_1 °C}] \qquad (14\text{-}18)$$

where

g = gravitational constant, 9.807 $[\mathrm{m/s^2}]$

ρ = density of the liquid condensate film $[\mathrm{Kg/m^3}]$

k = thermal conductivity of the liquid condensate film $[\mathrm{W/m \cdot \Delta_1 °C}]$

r = vapor latent heat of condensation $[\mathrm{J/Kg}]$

D = diameter of tube $[\mathrm{m}]$

μ = absolute viscosity $[\mathrm{Pa \cdot s}]$

ΔT = temperature difference between vapor saturation and the tube surface $[°\mathrm{C}]$

The numerical values for k, ρ, and μ are determined at a temperature that is the average of the condensing surface temperature and saturated steam temperature.

Representative approximate values of condensing film coefficients for several fluids are given in Table 14-2.

Table 14-2 Typical Values of Heat Transfer Coefficients for the Film Condensation of Water

Fluid (vapor)	Film Coefficient [W/m² °C]
Steam condensing inside tubes	6,800
Steam condensing on horizontal tubes	11,400

Even small amounts of noncondensible gases decrease the rate of condensation. This is attributed to the noncondensible gases forming a gaseous insulating layer over the cooling surfaces.

Example Problem 14-7

Using Equation 14-18, find the condensing film coefficient for steam at a pressure of 20 kPa on the surface of a 2-cm-outside-diameter horizontal tube with a temperature of 38°C. (See Figure 14-25, page 380.)

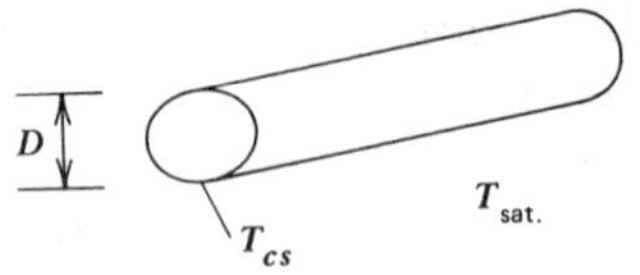

Figure 14-25 Sketch for Example Problem 14-7.

Equation

$$h_c = 0.725\left[\frac{g\rho^2 k^3 r}{D\mu\,\Delta T}\right]^{0.25} \qquad [\mathrm{W/m^2 \cdot \Delta_1 {}^\circ C}]$$

Parameters

Values for k, ρ, and μ are at the average temperature T_a between the steam saturation and the tube wall surface; so, to determine T_a, use the following procedure.

$$T_a = \frac{T_{sat} + T_{cs}}{2} \qquad [{}^\circ\mathrm{C}].$$

$T_{sat} = 60°C$ (Appendix A-3.2, saturated, at 20 kPa)

$T_{cs} = 38°C$ (given)

$$T_a = \frac{60 + 38}{2} \qquad [{}^\circ\mathrm{C}] - [{}^\circ\mathrm{C}] = [{}^\circ\mathrm{C}]$$

$T_a = 49°C$

$k = 0.644\ \mathrm{W/m \cdot \Delta_1 {}^\circ C}$ (Appendix B-3, liquid, at 49°C)

$\rho = 988.8\ \mathrm{kg/m^3}$ (Appendix B-3, liquid, at 49°C)

$\mu = 5.62 \times 10^{-4}\ \mathrm{kg/m \cdot s}$ (Appendix B-3, liquid, at 49°C)

$r = h_{fg} = 2358.3\ \mathrm{kJ/kg}$ (Appendix A-3.2, h_{fg} at 20 kPa)

$\rightarrow 2358300\ \mathrm{J/kg}$

$D = 2\ \mathrm{cm} \rightarrow 0.02\ \mathrm{m}$ (given)

$\Delta T = T_{sat} - T_{cs} \qquad [\Delta {}^\circ\mathrm{C}]$

$T_{sat} = 60°C$ (C-1)

$T_{cs} = 38°C$ (given)

$\Delta T = 60 - 38 \qquad [{}^\circ\mathrm{C}] - [{}^\circ\mathrm{C}] = [\Delta {}^\circ\mathrm{C}]$

$\Delta T = 22\ \Delta°C$

Substitution

$$h_c=0.725\left[\frac{(9.807)(988.8)^2(0.644)^3(2{,}358{,}300)}{(0.02)(5.62\times10^{-4})(22)}\right]^{0.25}$$

$$=[-]\left[\frac{[\mathrm{m/s^2}][\mathrm{kg/m}]^2[\mathrm{W/m\cdot\Delta_1{}^\circ C}]^3[\mathrm{J/kg}]}{[\mathrm{m}][\mathrm{kg/m\cdot s}][\Delta{}^\circ\mathrm{C}]}\right]^{0.25}$$

$$=[-]\left[\frac{(\mathrm{W}^4)}{(\mathrm{m}^8)({}^\circ\mathrm{C}^4)}\right]^{0.25}=\left[\mathrm{W/m^2\cdot\Delta_1{}^\circ C}\right]$$

$$h_c=0.725\left[\frac{6.0396\times10^{12}}{2.4728\times10^{-4}}\right]^{0.25}$$

$$=0.725[2.442\times10^{16}]^{0.25}$$

$$=(0.725)(12{,}501)$$

Answer

$$\mathbf{h_c=9063\ W/m^2\cdot\Delta_1{}^\circ C}$$

14-4 PROBLEMS

14-1. In a heat exchanger there are three separate heat transfer considerations that must be met. What are they? What relationship must they have with each other?

14-2. Name the basic flow types of heat exchangers. Give the salient heat transfer characteristics of each. What flow type is considered to be the most efficient, and why?

14-3. In a heat exchanger there are several key parameters affecting the heat transfer. Name and describe three.

14-4. Give the formula for the log mean temperature difference (ΔT_{lm}).

14-5. Heat is transferred from hot water to an oil in a double-pipe counterflow heat exchanger. The water enters the outer pipe at 90°C and exits at 40°C, while the oil with a c_p of 2090 J/kg·Δ_1 K enters the inner pipe at 25°C and exits at 45°C. Calculate the log mean temperature difference.

14-6. Heat is transferred from hot oil to water in a concentric pipe parallel flow heat exchanger. The water enters the outer pipe at 20°C and exits at 40°C, while the oil with a c_p of 2090 J/kg·Δ_1 K enters the inner pipe at 100°C and exits at 60°C. Calculate the log mean temperature difference.

14-7. If the water flow rate in Problem 14-5 were doubled with the water entering at 90°C and the oil flow remained the same, entering at 25°C, what would be the water and oil outlet temperatures (assuming the value of U remained constant)? Calculate using the log mean temperature method.

14-8. If the water flow rate in Problem 14-6 were doubled with the water entering at 20°C and the oil at 100°C, what would be the water and oil outlet temperatures (assuming that the value of U remained constant)? Calculate using the log mean temperature method.

14-9. Using the effectiveness–NTU method, calculate the AU value for the heat exchanger of Problem 14-5 when the oil flow is 0.1 kg/s.

14-10. Using the effectiveness–NTU method, calculate the AU value for the heat exchanger of Problem 14-6 when the oil flow is 0.1 kg/s.

14-11. Using the AU value obtained for the heat exchanger in Problem 14-9, determine the oil outlet temperature for the conditions of Problem 14-7.

14-12. Using the AU value obtained for the heat exchanger in Problem 14-10, determine the oil outlet temperature for the conditions of Problem 14-8.

14-13. List 10 factors that affect the boiling heat transfer and describe how you think each may affect the boiling process.

14-14. Describe the six regimes of boiling.

14-15. What is the critical heat flux? How does the critical heat flux relate to burn-out?

14-16. In nucleate boiling how does the pressure level generally affect the heat transfer flux for a given temperature difference between the surface and the fluid saturation, and the critical heat flux?

14-17. In the condensing process, what is the controlling resistance in the heat transfer path?

14-18. In the equations for the condensing film coefficient, are the fluid parameters evaluated for the liquid or vapor phase of the condensing substance? Why?

14-19. What is the difference between film and dropwise condensation? Which has higher heat transfer film coefficients? Which type of condensation is usually in condenser design? Why?

APPENDIX A THERMODYNAMIC DATA

Table A-1 Common Thermodynamic Parameters, Dimensions, Units and Symbols

Parameter			SI (Metric) System	
Name	Text Symbol	Dimensions	Unit Name	Unit Symbol
Acceleration	a, g	L/t^2	meter/sec^2	m/s^2
Area	A	L^2	meter^2	m^2
Atmospheres (pressure)	atm	—	atmospheres	atm
Compressibility factor	**Z**	—	dimensionless	—
Density	ρ	M/L^3	kilogram/meter^3	kg/m^3
Elevation (Head)	Z	L	meter	m
Energy	E	FL	joule=[N·m]	J
Enthalpy (total)	H	FL or Q	joule=[N·m]	J
Entropy (total)	S	Q/T	joule/kelvin	J/K
Flow work	pV	FL	joule=[N·m]	J
Force	F	F	newton=[kg·m/s^2]	N
Gas constant	R	Q/MT	joule/kilogram·kelvin	J/kg·Δ_1K
Heat (total)	Q	Q	joule	J
Heat (unit)	q	Q/M	joule/kilogram	J/kg
Heat transfer rate (total)	$\dot{Q}$	Q/t	watt=[J/s]	W
Heat transfer rate (unit)	$\dot{q}$	Q/Mt	watt/kilogram	W/kg
Internal energy (total)	U	Q	joule=[N·m]	J
Isentropic exponent	γ	—	dimensionless	—
Kinetic energy	E_k	FL	joule=[N·m]	J
Latent heat	r	Q/M	joule/kilogram	J/kg
Length	x, L	L	meter	m
Mass	M	M	kilogram	kg
Mass flow rate	$\dot{M}$	M/t	kilogram/sec	kg/s
Molecular weight	$\mathfrak{M}$	mol^{-1}	per mole	mol^{-1}
Polytropic Exponent	n	—	dimensionless	—
Potential Energy	E_p	FL	joule=[N·m]	J
Power	$P, \dot{W}_k$	FL/t	watt=[J/s]	W
Pressure	p	F/L^2	pascal=[N/m^2]	Pa
Quality	x	—	dimensionless	—
Specific Enthalpy	h	Q/M	joule/kilogram	J/kg
Specific Entropy	s	Q/MT	joule/kilogram·kelvin	J/kg·Δ_1K
Specific Gravity	s.g.	—	dimensionless	—

Table A-1 (Continued)

Parameter			SI (Metric) System	
Name	Text Symbol	Dimensions	Unit Name	Unit Symbol
Specific Heat				
(constant pressure)	c_p	Q/MT	joule/kilogram·kelvin	$J/kg \cdot \Delta_1 K$
(constant volume)	c_v	Q/MT	joule/kilogram·kelvin	$J/kg \cdot \Delta_1 K$
Specific Volume	v	L^3/M	meter^3/kilogram	m^3/kg
Specific Weight	w	F/L^3	newton/meter^3	N/m^3
Temperature-Difference	ΔT	T	kelvin or degree Celsius	ΔK or Δ°C
Temperature-Level	T	T	kelvin or degree Celsius	K or °C
Temperature "per Degree"	—	$1/T$	kelvin^{-1} or degree Celsius^{-1}	$1/\Delta_1$ K or $1/\Delta_1$°C
Thrust	F	F	newton	N
Time	t	t	seconds, hours	s, h
Universal gas constant	$\overline{R}$	Q/M mol T	joule/kilogram·mol ·kelvin	J/kg·mol·K
Velocity	**V**	L/t	meter/sec	m/s
Viscosity (absolute)	μ	M/L	pascal·second	Pa·s
Viscosity (kinematic)	ν	L^2/t	meter^2/sec	m^2/s
Volume	V	L^3	meter^3	m^3
Weight	W	F	newton	N
Work	Wk	FL	joule=[N·m]	J

Table A-2 Gravitational Acceleration near the Earth

Altitude above	Location on the Earth's Surface, Degrees Latitude (m/s^2)					
Sea Level (m)	0°	20°	40°	60°	80°	90°
0	9.780	9.787	9.802	9.819	9.831	9.832
250	9.780	9.786	9.801	9.818	9.830	9.831
500	9.779	9.785	9.800	9.817	9.829	9.830
1,000	9.777	9.784	9.799	9.816	9.828	9.829
2,000	9.774	9.780	9.796	9.814	9.826	9.827
3,000	9.771	9.777	9.793	9.810	9.822	9.823
30,000	9.686	9.695	9.710	9.727	9.739	9.740

Source: U.S. Coast and Geodetic Survey, 1912.
Note: Gravitational constant, $g_c = 1$ kg·m/N·s^2

Table A-3.1 (Continued)

TEMP °C ~~Press. kPa~~ P	PRESS MPa ~~Temp. °C~~ T	Specific Volume		Internal Energy			Enthalpy			Entropy		
		Sat. Liquid v_f	Sat. Vapor v_g	Sat. Liquid u_f	Evap. u_{fg}	Sat. Vapor u_g	Sat. Liquid h_f	Evap. h_{fg}	Sat. Vapor h_g	Sat. Liquid s_f	Evap. s_{fg}	Sat. Vapor s_g
215	2.104	0.001 181	0.094 79	918.14	1682.9	2601.1	920.62	1879.9	2800.5	2.4714	3.8507	6.3221
220	2.318	0.001 190	0.086 19	940.87	1661.5	2602.4	943.62	1858.5	2802.1	2.5178	3.7683	6.2861
225	2.548	0.001 199	0.078 49	963.73	1639.6	2603.3	966.78	1836.5	2803.3	2.5639	3.6863	6.2503
230	2.795	0.001 209	0.071 58	986.74	1617.2	2603.9	990.12	1813.8	2804.0	2.6099	3.6047	6.2146
235	3.060	0.001 219	0.065 37	1009.89	1594.2	2604.1	1013.62	1790.5	2804.2	2.6558	3.5233	6.1791
240	3.344	0.001 229	0.059 76	1033.21	1570.8	2604.0	1037.32	1766.5	2803.8	2.7015	3.4422	6.1437
245	3.648	0.001 240	0.054 71	1056.71	1546.7	2603.4	1061.23	1741.7	2803.0	2.7472	3.3612	6.1083
250	3.973	0.001 251	0.050 13	1080.39	1522.0	2602.4	1085.36	1716.2	2801.5	2.7927	3.2802	6.0730
255	4.319	0.001 263	0.045 98	1104.28	1496.7	2600.9	1109.73	1689.8	2799.5	2.8383	3.1992	6.0375
260	4.688	0.001 276	0.042 21	1128.39	1470.6	2599.0	1134.37	1662.5	2796.9	2.8838	3.1181	6.0019
265	5.081	0.001 289	0.038 77	1152.74	1443.9	2596.6	1159.28	1634.4	2793.6	2.9294	3.0368	5.9662
270	5.499	0.001 302	0.035 64	1177.36	1416.3	2593.7	1184.51	1605.2	2789.7	2.9751	2.9551	5.9301
275	5.942	0.001 317	0.032 79	1202.25	1387.9	2590.2	1210.07	1574.9	2785.0	3.0208	2.8730	5.8938
280	6.412	0.001 332	0.030 17	1227.46	1358.7	2586.1	1235.99	1543.6	2779.6	3.0668	2.7903	5.8571
285	6.909	0.001 348	0.027 77	1253.00	1328.4	2581.4	1262.31	1511.0	2773.3	3.1130	2.7070	5.8199
290	7.436	0.001 366	0.025 57	1278.92	1297.1	2576.0	1289.07	1477.1	2766.2	3.1594	2.6227	5.7821
295	7.993	0.001 384	0.023 54	1305.2	1264.7	2569.9	1316.3	1441.8	2758.1	3.2062	2.5375	5.7437
300	8.581	0.001 404	0.021 67	1332.0	1231.0	2563.0	1344.0	1404.9	2749.0	3.2534	2.4511	5.7045
305	9.202	0.001 425	0.019 948	1359.3	1195.9	2555.2	1372.4	1366.4	2738.7	3.3010	2.3633	5.6643
310	9.856	0.001 447	0.018 350	1387.1	1159.4	2546.4	1401.3	1326.0	2727.3	3.3493	2.2737	5.6230
315	10.547	0.001 472	0.016 867	1415.5	1121.1	2536.6	1431.0	1283.5	2714.5	3.3982	2.1821	5.5804
320	11.274	0.001 499	0.015 488	1444.6	1080.9	2525.5	1461.5	1238.6	2700.1	3.4480	2.0882	5.5362
330	12.845	0.001 561	0.012 996	1505.3	993.7	2498.9	1525.3	1140.6	2665.9	3.5507	1.8909	5.4417
340	14.586	0.001 638	0.010 797	1570.3	894.3	2464.6	1594.2	1027.9	2622.0	3.6594	1.6763	5.3357
350	16.513	0.001 740	0.008 813	1641.9	776.6	2418.4	1670.6	893.4	2563.9	3.7777	1.4335	5.2112
360	18.651	0.001 893	0.006 945	1725.2	626.3	2351.5	1760.5	720.5	2481.0	3.9147	1.1379	5.0526
370	21.03	0.002 213	0.004 925	1844.0	384.5	2228.5	1890.5	441.6	2332.1	4.1106	.6865	4.7971
374.14	22.09	0.003 155	0.003 155	2029.6	0	2029.6	2099.3	0	2099.3	4.4298	0	4.4298

Table A-3.2 Saturated Steam: Pressure Table

0.6113	0.01	0.001 000	206.14	.00	2375.3	2375.3	.01	2501.3	2501.4	.0000	9.1562	9.1562
1.0	6.98	0.001 000	129.21	29.30	2355.7	2385.0	29.30	2484.9	2514.2	.1059	8.8697	8.9756
1.5	13.03	0.001 001	87.98	54.71	2338.6	2393.3	54.71	2470.6	2525.3	.1957	8.6322	8.8279
2.0	17.50	0.001 001	67.00	73.48	2326.0	2399.5	73.48	2460.0	2533.5	.2607	8.4629	8.7237
2.5	21.08	0.001 002	54.25	88.48	2315.9	2404.4	88.49	2451.6	2540.0	.3120	8.3311	8.6432
3.0	24.08	0.001 003	45.67	101.04	2307.5	2408.5	101.05	2444.5	2545.5	.3545	8.2231	8.5776
4.0	28.96	0.001 004	34.80	121.45	2293.7	2415.2	121.46	2432.9	2554.4	.4226	8.0520	8.4746
5.0	32.88	0.001 005	28.19	137.81	2282.7	2420.5	137.82	2423.7	2561.5	.4764	7.9187	8.3951
7.5	40.29	0.001 008	19.24	168.78	2261.7	2430.5	168.79	2406.0	2574.8	.5764	7.6750	8.2515
10	45.81	0.001 010	14.67	191.82	2246.1	2437.9	191.83	2392.8	2584.7	.6493	7.5009	8.1502
15	53.97	0.001 014	10.02	225.92	2222.8	2448.7	225.94	2373.1	2599.1	.7549	7.2536	8.0085
20	60.06	0.001 017	7.649	251.38	2205.4	2456.7	251.40	2358.3	2609.7	.8320	7.0766	7.9085
25	64.97	0.001 020	6.204	271.90	2191.2	2463.1	271.93	2346.3	2618.2	.8931	6.9383	7.8314
30	69.10	0.001 022	5.229	289.20	2179.2	2468.4	289.23	2336.1	2625.3	.9439	6.8247	7.7686
40	75.87	0.001 027	3.993	317.53	2159.5	2477.0	317.58	2319.2	2636.8	1.0259	6.6441	7.6700
50	81.33	0.001 030	3.240	340.44	2143.4	2483.9	340.49	2305.4	2645.9	1.0910	6.5029	7.5939
75	91.78	0.001 037	2.217	384.31	2112.4	2496.7	384.39	2278.6	2663.0	1.2130	6.2434	7.4564
MPa												
0.100	99.63	0.001 043	1.6940	417.36	2088.7	2506.1	417.46	2258.0	2675.5	1.3026	6.0568	7.3594
0.125	105.99	0.001 048	1.3749	444.19	2069.3	2513.5	444.32	2241.0	2685.4	1.3740	5.9104	7.2844
0.150	111.37	0.001 053	1.1593	466.94	2052.7	2519.7	467.11	2226.5	2693.6	1.4336	5.7897	7.2233
0.175	116.06	0.001 057	1.0036	486.80	2038.1	2524.9	486.99	2213.6	2700.6	1.4849	5.6868	7.1717
0.200	120.23	0.001 061	0.8857	504.49	2025.0	2529.5	504.70	2201.9	2706.7	1.5301	5.5970	7.1271
0.225	124.00	0.001 064	0.7933	520.47	2013.1	2533.6	520.72	2191.3	2712.1	1.5706	5.5173	7.0878

Table A-3.2 (Continued)

MPa	T °C											
0.250	127.44	0.001 067	0.7187	535.10	2002.1	2537.2	535.37	2181.5	2716.9	1.6072	5.4455	7.0527
0.275	130.60	0.001 070	0.6573	548.59	1991.9	2540.5	548.89	2172.4	2721.3	1.6408	5.3801	7.0209
0.300	133.55	0.001 073	0.6058	561.15	1982.4	2543.6	561.47	2163.8	2725.3	1.6718	5.3201	6.9919
0.325	136.30	0.001 076	0.5620	572.90	1973.5	2546.4	573.25	2155.8	2729.0	1.7006	5.2646	6.9652
0.350	138.88	0.001 079	0.5243	583.95	1965.0	2548.9	584.33	2148.1	2732.4	1.7275	5.2130	6.9405
0.375	141.32	0.001 081	0.4914	594.40	1956.9	2551.3	594.81	2140.8	2735.6	1.7528	5.1647	6.9175
0.40	143.63	0.001 084	0.4625	604.31	1949.3	2553.6	604.74	2133.8	2738.6	1.7766	5.1193	6.8959
0.45	147.93	0.001 088	0.4140	622.77	1934.9	2557.6	623.25	2120.7	2743.9	1.8207	5.0359	6.8565
0.50	151.86	0.001 093	0.3749	639.68	1921.6	2561.2	640.23	2108.5	2748.7	1.8607	4.9606	6.8213
0.55	155.48	0.001 097	0.3427	655.32	1909.2	2564.5	655.93	2097.0	2753.0	1.8973	4.8920	6.7893
0.60	158.85	0.001 101	0.3157	669.90	1897.5	2567.4	670.56	2086.3	2756.8	1.9312	4.8288	6.7600
0.65	162.01	0.001 104	0.2927	683.56	1886.5	2570.1	684.28	2076.0	2760.3	1.9627	4.7703	6.7331
0.70	164.97	0.001 108	0.2729	696.44	1876.1	2572.5	697.22	2066.3	2763.5	1.9922	4.7158	6.7080
0.75	167.78	0.001 112	0.2556	708.64	1866.1	2574.7	709.47	2057.0	2766.4	2.0200	4.6647	6.6847
0.80	170.43	0.001 115	0.2404	720.22	1856.6	2576.8	721.11	2048.0	2769.1	2.0462	4.6166	6.6628
0.85	172.96	0.001 118	0.2270	731.27	1847.4	2578.7	732.22	2039.4	2771.6	2.0710	4.5711	6.6421
0.90	175.38	0.001 121	0.2150	741.83	1838.6	2580.5	742.83	2031.1	2773.9	2.0946	4.5280	6.6226
0.95	177.69	0.001 124	0.2042	751.95	1830.2	2582.1	753.02	2023.1	2776.1	2.1172	4.4869	6.6041
1.00	179.91	0.001 127	0.194 44	761.68	1822.0	2583.6	762.81	2015.3	2778.1	2.1387	4.4478	6.5865
1.10	184.09	0.001 133	0.177 53	780.09	1806.3	2586.4	781.34	2000.4	2781.7	2.1792	4.3744	6.5536
1.20	187.99	0.001 139	0.163 33	797.29	1791.5	2588.8	798.65	1986.2	2784.8	2.2166	4.3067	6.5233
1.30	191.64	0.001 144	0.151 25	813.44	1777.5	2591.0	814.93	1972.7	2787.6	2.2515	4.2438	6.4953
1.40	195.07	0.001 149	0.140 84	828.70	1764.1	2592.8	830.30	1959.7	2790.0	2.2842	4.1850	6.4693
1.50	198.32	0.001 154	0.131 77	843.16	1751.3	2594.5	844.89	1947.3	2792.2	2.3150	4.1298	6.4448
1.75	205.76	0.001 166	0.113 49	876.46	1721.4	2597.8	878.50	1917.9	2796.4	2.3851	4.0044	6.3896
2.00	212.42	0.001 177	0.099 63	906.44	1693.8	2600.3	908.79	1890.7	2799.5	2.4474	3.8935	6.3409
2.25	218.45	0.001 187	0.088 75	933.83	1668.2	2602.0	936.49	1865.2	2801.7	2.5035	3.7937	6.2972
2.5	223.99	0.001 197	0.079 98	959.11	1644.0	2603.1	962.11	1841.0	2803.1	2.5547	3.7028	6.2575
3.0	233.90	0.001 217	0.066 68	1004.78	1599.3	2604.1	1008.42	1795.7	2804.2	2.6457	3.5412	6.1869

Table A-3.2 (Continued)

Press. MPa P	Temp. °C T	Specific Volume: Sat. Liquid v_f	Specific Volume: Sat. Vapor v_g	Internal Energy: Sat. Liquid u_f	Internal Energy: Evap. u_{fg}	Internal Energy: Sat. Vapor u_g	Enthalpy: Sat. Liquid h_f	Enthalpy: Evap. h_{fg}	Enthalpy: Sat. Vapor h_g	Entropy: Sat. Liquid s_f	Entropy: Evap. s_{fg}	Entropy: Sat. Vapor s_g
3.5	242.60	0.001 235	0.057 07	1045.43	1558.3	2603.7	1049.75	1753.7	2803.4	2.7253	3.4000	6.1253
4	250.40	0.001 252	0.049 78	1082.31	1520.0	2602.3	1087.31	1714.1	2801.4	2.7964	3.2737	6.0701
5	263.99	0.001 286	0.039 44	1147.81	1449.3	2597.1	1154.23	1640.1	2794.3	2.9202	3.0532	5.9734
6	275.64	0.001 319	0.032 44	1205.44	1384.3	2589.7	1213.35	1571.0	2784.3	3.0267	2.8625	5.8892
7	285.88	0.001 351	0.027 37	1257.55	1323.0	2580.5	1267.00	1505.1	2772.1	3.1211	2.6922	5.8133
8	295.06	0.001 384	0.023 52	1305.57	1264.2	2569.8	1316.64	1441.3	2758.0	3.2068	2.5364	5.7432
9	303.40	0.001 418	0.020 48	1350.51	1207.3	2557.8	1363.26	1378.9	2742.1	3.2858	2.3915	5.6772
10	311.06	0.001 452	0.018 026	1393.04	1151.4	2544.4	1407.56	1317.1	2724.7	3.3596	2.2544	5.6141
11	318.15	0.001 489	0.015 987	1433.7	1096.0	2529.8	1450.1	1255.5	2705.6	3.4295	2.1233	5.5527
12	324.75	0.001 527	0.014 263	1473.0	1040.7	2513.7	1491.3	1193.6	2684.9	3.4962	1.9962	5.4924
13	330.93	0.001 567	0.012 780	1511.1	985.0	2496.1	1531.5	1130.7	2662.2	3.5606	1.8718	5.4323
14	336.75	0.001 611	0.011 485	1548.6	928.2	2476.8	1571.1	1066.5	2637.6	3.6232	1.7485	5.3717
15	342.24	0.001 658	0.010 337	1585.6	869.8	2455.5	1610.5	1000.0	2610.5	3.6848	1.6249	5.3098
16	347.44	0.001 711	0.009 306	1622.7	809.0	2431.7	1650.1	930.6	2580.6	3,7461	1.4994	5.2455
17	352.37	0.001 770	0.008 364	1660.2	744.8	2405.0	1690.3	856.9	2547.2	3.8079	1.3698	5.1777
18	357.06	0.001 840	0.007 489	1698.9	675.4	2374.3	1732.0	777.1	2509.1	3.8715	1.2329	5.1044
19	361.54	0.001 924	0.006 657	1739.9	598.1	2338.1	1776.5	688.0	2464.5	3.9388	1.0839	5.0228
20	365.81	0.002 036	0.005 834	1785.6	507.5	2293.0	1826.3	583.4	2409.7	4.0139	.9130	4.9269
21	369.89	0.002 207	0.004 952	1842.1	388.5	2230.6	1888.4	446.2	2334.6	4.1075	.6938	4.8013
22	373.80	0.002 742	0.003 568	1961.9	125.2	2087.1	2022.2	143.4	2165.6	4.3110	.2216	4.5327
22.09	374.14	0.003 155	0.003 155	2029.6	0	2029.6	2099.3	0	2099.3	4.4298	0	4.4298

Table A-3.3 Superheated Vapor

	P = .010 MPa (45.81)				P = .050 MPa (81.33)				P = .10 MPa (99.63)			
T	*v*	*u*	*h*	*s*	*v*	*u*	*h*	*s*	*v*	*u*	*h*	*s*
Sat.	14.674	2437.9	2584.7	8.1502	3.240	2483.9	2645.9	7.5939	1.6940	2506.1	2675.5	7.3594
50	14.869	2443.9	2592.6	8.1749								
100	17.196	2515.5	2687.5	8.4479	3.418	2511.6	2682.5	7.6947	1.6958	2506.7	2676.2	7.3614
150	19.512	2587.9	2783.0	8.6882	3.889	2585.6	2780.1	7.9401	1.9364	2582.8	2776.4	7.6134
200	21.825	2661.3	2879.5	8.9038	4.356	2659.9	2877.7	8.1580	2.172	2658.1	2875.3	7.8343
250	24.136	2736.0	2977.3	9.1002	4.820	2735.0	2976.0	8.3556	2.406	2733.7	2974.3	8.0333
300	26.445	2812.1	3076.5	9.2813	5.284	2811.3	3075.5	8.5373	2.639	2810.4	3074.3	8.2158
400	31.063	2968.9	3279.6	9.6077	6.209	2968.5	3278.9	8.8642	3.103	2967.9	3278.2	8.5435
500	35.679	3132.3	3489.1	9.8978	7.134	3132.0	3488.7	9.1546	3.565	3131.6	3488.1	8.8342
600	40.295	3302.5	3705.4	10.1608	8.057	3302.2	3705.1	9.4178	4.028	3301.9	3704.7	9.0976
700	44.911	3479.6	3928.7	10.4028	8.981	3479.4	3928.5	9.6599	4.490	3479.2	3928.2	9.3398
800	49.526	3663.8	4159.0	10.6281	9.904	3663.6	4158.9	9.8852	4.952	3663.5	4158.6	9.5652
900	54.141	3855.0	4396.4	10.8396	10.828	3854.9	4396.3	10.0967	5.414	3854.8	4396.1	9.7767
1000	58.757	4053.0	4640.6	11.0393	11.751	4052.9	4640.5	10.2964	5.875	4052.8	4640.3	9.9764
1100	63.372	4257.5	4891.2	11.2287	12.674	4257.4	4891.1	10.4859	6.337	4257.3	4891.0	10.1659
1200	67.987	4467.9	5147.8	11.4091	13.597	4467.8	5147.7	10.6662	6.799	4467.7	5147.6	10.3463
1300	72.602	4683.7	5409.7	11.5811	14.521	4683.6	5409.6	10.8382	7.260	4683.5	5409.5	10.5183
	P = .20 MPa (120.23)				P = .30 MPa (133.55)				P = .40 MPa (143.63)			
Sat.	.8857	2529.5	2706.7	7.1272	.6058	2543.6	2725.3	6.9919	.4625	2553.6	2738.6	6.8959
150	.9596	2576.9	2768.8	7.2795	.6339	2570.8	2761.0	7.0778	.4708	2564.5	2752.8	6.9299
200	1.0803	2654.4	2870.5	7.5066	.7163	2650.7	2865.6	7.3115	.5342	2646.8	2860.5	7.1706
250	1.1988	2731.2	2971.0	7.7086	.7964	2728.7	2967.6	7.5166	.5951	2726.1	2964.2	7.3789
300	1.3162	2808.6	3071.8	7.8926	.8753	2806.7	3069.3	7.7022	.6548	2804.8	3066.8	7.5662
400	1.5493	2966.7	3276.6	8.2218	1.0315	2965.6	3275.0	8.0330	.7726	2964.4	3273.4	7.8985

Table A-3.3 (Continued)

T	v	u	h	s	v	u	h	s	v	u	h	s
	P = .20 MPa (120.23)				P = .30 MPa (133.55)				P = .40 MPa (143.63)			
500	1.7814	3130.8	3487.1	8.5133	1.1867	3130.0	3486.0	8.3251	.8893	3129.2	3484.9	8.1913
600	2.013	3301.4	3704.0	8.7770	1.3414	3300.8	3703.2	8.5892	1.0055	3300.2	3702.4	8.4558
700	2.244	3478.8	3927.6	9.0194	1.4957	3478.4	3927.1	8.8319	1.1215	3477.9	3926.5	8.6987
800	2.475	3663.1	4158.2	9.2449	1.6499	3662.9	4157.8	9.0576	1.2372	3662.4	4157.3	8.9244
900	2.706	3854.5	4395.8	9.4566	1.8041	3854.2	4395.4	9.2692	1.3529	3853.9	4395.1	9.1362
1000	2.937	4052.5	4640.0	9.6563	1.9581	4052.3	4639.7	9.4690	1.4685	4052.0	4639.4	9.3360
1100	3.168	4257.0	4890.7	9.8458	2.1121	4256.8	4890.4	9.6585	1.5840	4256.5	4890.2	9.5256
1200	3.399	4467.5	5147.3	10.0262	2.2661	4467.2	5147.1	9.8389	1.6996	4467.0	5146.8	9.7060
1300	3.630	4683.2	5409.3	10.1982	2.4201	4683.0	5409.0	10.0110	1.8151	4682.8	5408.8	9.8780
	P = .50 MPa (151.86)				P = .60 MPa (158.85)				P = .80 MPa (170.43)			
Sat.	.3749	2561.2	2748.7	6.8213	.3157	2567.4	2756.8	6.7600	.2404	2576.8	2769.1	6.6628
200	.4249	2642.9	2855.4	7.0592	.3520	2638.9	2850.1	6.9665	.2608	2630.6	2839.3	6.8158
250	.4744	2723.5	2960.7	7.2709	.3938	2720.9	2957.2	7.1816	.2931	2715.5	2950.0	7.0384
300	.5226	2802.9	3064.2	7.4599	.4344	2801.0	3061.6	7.3724	.3241	2797.2	3056.5	7.2328
350	.5701	2882.6	3167.7	7.6329	.4742	2881.2	3165.7	7.5464	.3544	2878.2	3161.7	7.4089
400	.6173	2963.2	3271.9	7.7938	.5137	2962.1	3270.3	7.7079	.3843	2959.7	3267.1	7.5716
500	.7109	3128.4	3483.9	8.0873	.5920	3127.6	3482.8	8.0021	.4433	3126.0	3480.6	7.8673
600	.8041	3299.6	3701.7	8.3522	.6697	3299.1	3700.9	8.2674	.5018	3297.9	3699.4	8.1333
700	.8969	3477.5	3925.9	8.5952	.7472	3477.0	3925.3	8.5107	.5601	3476.2	3924.2	8.3770
800	.9896	3662.1	4156.9	8.8211	.8245	3661.8	4156.5	8.7367	.6181	3661.1	4155.6	8.6033
900	1.0822	3853.6	4394.7	9.0329	.9017	3853.4	4394.4	8.9486	.6761	3852.8	4393.7	8.8153
1000	1.1747	4051.8	4639.1	9.2328	.9788	4051.5	4638.8	9.1485	.7340	4051.0	4638.2	9.0153
1100	1.2672	4256.3	4889.9	9.4224	1.0559	4256.1	4889.6	9.3381	.7919	4255.6	4889.1	9.2050

Table A-3.3 (Continued)

1200	1.3596	4466.8	5146.6	9.6029	1.1330	4466.5	5146.3	9.5185	.8497	4466.1	5145.9	9.3855
1300	1.4521	4682.5	5408.6	9.7749	1.2101	4682.3	5408.3	9.6906	.9076	4681.8	5407.9	9.5575
	$P = 1.00$ MPa (179.91)				$P = 1.20$ MPa (187.99)				$P = 1.40$ MPa (195.07)			
Sat.	.194 44	2583.6	2778.1	6.5865	.163 33	2588.8	2784.8	6.5233	.140 84	2592.8	2790.0	6.4693
200	.2060	2621.9	2827.9	6.6940	.169 30	2612.8	2815.9	6.5898	.143 02	2603.1	2803.3	6.4975
250	.2327	2709.9	2942.6	6.9247	.192 34	2704.2	2935.0	6.8294	.163 50	2698.3	2927.2	6.7467
300	.2579	2793.2	3051.2	7.1229	.2138	2789.2	3045.8	7.0317	.182 28	2785.2	3040.4	6.9534
350	.2825	2875.2	3157.7	7.3011	.2345	2872.2	3153.6	7.2121	.2003	2869.2	3149.5	7.1360
400	.3066	2957.3	3263.9	7.4651	.2548	2954.9	3260.7	7.3774	.2178	2952.5	3257.5	7.3026
500	.3541	3124.4	3478.5	7.7622	.2946	3122.8	3476.3	7.6759	.2521	3121.1	3474.1	7.6027
600	.4011	3296.8	3697.9	8.0290	.3339	3295.6	3696.3	7.9435	.2860	3294.4	3694.8	7.8710
700	.4478	3475.3	3923.1	8.2731	.3729	3474.4	3922.0	8.1881	.3195	3473.6	3920.8	8.1160
800	.4943	3660.4	4154.7	8.4996	.4118	3659.7	4153.8	8.4148	.3528	3659.0	4153.0	8.3431
900	.5407	3852.2	4392.9	8.7118	.4505	3851.6	4392.2	8.6272	.3861	3851.1	4391.5	8.5556
1000	.5871	4050.5	4637.6	8.9119	.4892	4050.0	4637.0	8.8274	.4192	4049.5	4636.4	8.7559
1100	.6335	4255.1	4888.6	9.1017	.5278	4254.6	4888.0	9.0172	.4524	4254.1	4887.5	8.9457
1200	.6798	4465.6	5145.4	9.2822	.5665	4465.1	5144.9	9.1977	.4855	4464.7	5144.4	9.1262
1300	.7261	4681.3	5407.4	9.4543	.6051	4680.9	5407.0	9.3698	.5186	4680.4	5406.5	9.2984
	$P = 1.60$ MPa (201.41)				$P = 1.80$ MPa (207.15)				$P = 2.00$ MPa (212.42)			
Sat.	.123 80	2596.0	2794.0	6.4218	.110 42	2598.4	2797.1	6.3794	.099 63	2600.3	2799.5	6.3409
225	.132 87	2644.7	2857.3	6.5518	.116 73	2636.6	2846.7	6.4808	.103 77	2628.3	2835.8	6.4147
250	.141 84	2692.3	2919.2	6.6732	.124 97	2686.0	2911.0	6.6066	.111 44	2679.6	2902.5	6.5453
300	.158 62	2781.1	3034.8	6.8844	.140 21	2776.9	3029.2	6.8226	.125 47	2772.6	3023.5	6.7664
350	.174 56	2866.1	3145.4	7.0694	.154 57	2863.0	3141.2	7.0100	.138 57	2859.8	3137.0	6.9563
400	.190 05	2950.1	3254.2	7.2374	.168 47	2947.7	3250.9	7.1794	.151 20	2945 2	3247.6	7.1271
500	.2203	3119.5	3472.0	7.5390	.195 50	3117.9	3469.8	7.4825	.175 68	3116.2	3467.6	7.4317
600	.2500	3293.3	3693.2	7.8080	.2220	3292.1	3691.7	7.7523	.199 60	3290.9	3690.1	7.7024
700	.2794	3472.7	3919.7	8.0535	.2482	3471.8	3918.5	7.9983	.2232	3470.9	3917.4	7.9487

Table A-3.3 (Continued)

T	v	u	h	s	v	u	h	s	v	u	h	s
	P = 1.60 MPa (201.41)				P = 1.80 MPa (207.15)				P = 2.00 MPa (212.42)			
800	.3086	3658.3	4152.1	8.2808	.2742	3657.6	4151.2	8.2258	.2467	3657.0	4150.3	8.1765
900	.3377	3850.5	4390.8	8.4935	.3001	3849.9	4390.1	8.4386	.2700	3849.3	4389.4	8.3895
1000	.3668	4049.0	4635.8	8.6938	.3260	4048.5	4635.2	8.6391	.2933	4048.0	4634.6	8.5901
1100	.3958	4253.7	4887.0	8.8837	.3518	4253.2	4886.4	8.8290	.3166	4252.7	4885.9	8.7800
1200	.4248	4464.2	5143.9	9.0643	.3776	4463.7	5143.4	9.0096	.3398	4463.3	5142.9	8.9607
1300	.4538	4679.9	5406.0	9.2364	.4034	4679.5	5405.6	9.1818	.3631	4679.0	5405.1	9.1329
	P = 2.50 MPa (223.99)				P = 3.00 MPa (233.90)				P = 3.50 MPa (242.60)			
Sat.	.079 98	2603.1	2803.1	6.2575	.066 68	2604.1	2804.2	6.1869	.057 07	2603.7	2803.4	6.1253
225	.080 27	2605.6	2806.3	6.2639								
250	.087 00	2662.6	2880.1	6.4085	.070 58	2644.0	2855.8	6.2872	.058 72	2623.7	2829.2	6.1749
300	.098 90	2761.6	3008.8	6.6438	.081 14	2750.1	2993.5	6.5390	.068 42	2738.0	2977.5	6.4461
350	.109 76	2851.9	3126.3	6.8403	.090 53	2843.7	3115.3	6.7428	.076 78	2835.3	3104.0	6.6579
400	.120 10	2939.1	3239.3	7.0148	.099 36	2932.8	3230.9	6.9212	.084 53	2926.4	3222.3	6.8405
450	.130 14	3025.5	3350.8	7.1746	.107 87	3020.4	3344.0	7.0834	.091 96	3015.3	3337.2	7.0052
500	.139 98	3112.1	3462.1	7.3234	.116 19	3108.0	3456.5	7.2338	.099 18	3103.0	3450.9	7.1572
600	.159 30	3288.0	3686.3	7.5960	.132 43	3285.0	3682.3	7.5085	.113 24	3282.1	3678.4	7.4339
700	.178 32	3468.7	3914.5	7.8435	.148 38	3466.5	3911.7	7.7571	.126 99	3464.3	3908.8	7.6837
800	.197 16	3655.3	4148.2	8.0720	.164 14	3653.5	4145.9	7.9862	.140 56	3651.8	4143.7	7.9134
900	.215 90	3847.9	4387.6	8.2853	.179 80	3846.5	4385.9	8.1999	.154 02	3845.0	4384.1	8.1276
1000	.2346	4046.7	4633.1	8.4861	.195 41	4045.4	4631.6	8.4009	.167 43	4044.1	4630.1	8.3288
1100	.2532	4251.5	4884.6	8.6762	.210 98	4250.3	4883.3	8.5912	.180 80	4249.2	4881.9	8.5192
1200	.2718	4462.1	5141.7	8.8569	.226 52	4460.9	5140.5	8.7720	.194 15	4459.8	5139.3	8.7000
1300	.2905	4677.8	5404.0	9.0291	.242 06	4676.6	5402.8	8.9442	.207 49	4675.5	5401.7	8.8723

	$P = 4.0$ MPa (250.40)				$P = 4.5$ MPa (257.49)				$P = 5.0$ MPa (263.99)			
Sat.	.049 78	2602.3	2801.4	6.0701	.044 06	2600.1	2798.3	6.0198	.039 44	2597.1	2794.3	5.9734
275	.054 57	2667.9	2886.2	6.2285	.047 30	2650.3	2863.2	6.1401	.041 41	2631.3	2838.3	6.0544
300	.058 84	2725.3	2960.7	6.3615	.051 35	2712.0	2943.1	6.2828	.045 32	2698.0	2924.5	6.2084
350	.066 45	2826.7	3092.5	6.5821	.058 40	2817.8	3080.6	6.5131	.051 94	2808.7	3068.4	6.4493
400	.073 41	2919.9	3213.6	6.7690	.064 75	2913.3	3204.7	6.7047	.057 81	2906.6	3195.7	6.6459
450	.080 02	3010.2	3330.3	6.9363	.070 74	3005.0	3323.3	6.8746	.063 30	2999.7	3316.2	6.8186
500	.086 43	3099.5	3445.3	7.0901	.076 51	3095.3	3439.6	7.0301	.068 57	3091.0	3433.8	6.9759
600	.098 85	3279.1	3674.4	7.3688	.087 65	3276.0	3670.5	7.3110	.078 69	3273.0	3666.5	7.2589
700	.110 95	3462.1	3905.9	7.6198	.098 47	3459.9	3903.0	7.5631	.088 49	3457.6	3900.1	7.5122
800	.122 87	3650.0	4141.5	7.8502	.109 11	3648.3	4139.3	7.7942	.098 11	3646.6	4137.1	7.7440
900	.134 69	3843.6	4382.3	8.0647	.119 65	3842.2	4380.6	8.0091	.107 62	3840.7	4378.8	7.9593
1000	.146 45	4042.9	4628.7	8.2662	.130 13	4041.6	4627.2	8.2108	.117 07	4040.4	4625.7	8.1612
1100	.158 17	4248.0	4880.6	8.4567	.140 56	4246.8	4879.3	8.4015	.126 48	4245.6	4878.0	8.3520
1200	.169 87	4458.6	5138.1	8.6376	.150 98	4457.5	5136.9	8.5825	.135 87	4456.3	5135.7	8.5331
1300	.181 56	4674.3	5400.5	8.8100	.161 39	4673.1	5399.4	8.7549	.145 26	4672.0	5398.2	8.7055

	$P = 6.0$ MPa (275.64)				$P = 7.0$ MPa (285.88)				$P = 8.0$ MPa (295.06)			
Sat.	.032 44	2589.7	2784.3	5.8892	.027 37	2580.5	2772.1	5.8133	.023 52	2569.8	2758.0	5.7432
300	.036 16	2667.2	2884.2	6.0674	.029 47	2632.2	2838.4	5.9305	.024 26	2590.9	2785.0	5.7906
350	.042 23	2789.6	3043.0	6.3335	.035 24	2769.4	3016.0	6.2283	.029 95	2747.7	2987.3	6.1301
400	.047 39	2892.9	3177.2	6.5408	.039 93	2878.6	3158.1	6.4478	.034 32	2863.8	3138.3	6.3634
450	.052 14	2988.9	3301.8	6.7193	.044 16	2978.0	3287.1	6.6327	.038 17	2966.7	3272.0	6.5551
500	.056 65	3082.2	3422.2	6.8803	.048 14	3073.4	3410.3	6.7975	.041 75	3064.3	3398.3	6.7240
550	.061 01	3174.6	3540.6	7.0288	.051 95	3167.2	3530.9	6.9486	.045 16	3159.8	3521.0	6.8778
600	.065 25	3266.9	3658.4	7.1677	.055 65	3260.7	3650.3	7.0894	.048 45	3254.4	3642.0	7.0206
700	.073 52	3453.1	3894.2	7.4234	.062 83	3448.5	3888.3	7.3476	.054 81	3443.9	3882.4	7.2812
800	.081 60	3643.1	4132.7	7.6566	.069 81	3639.5	4128.2	7.5822	.060 97	3636.0	4123.8	7.5173
900	.089 58	3837.8	4375.3	7.8727	.076 69	3835.0	4371.8	7.7991	.067 02	3832.1	4368.3	7.7351
1000	.097 49	4037.8	4622.7	8.0751	.083 50	4035.3	4619.8	8.0020	.073 01	4032.8	4616.9	7.9384
1100	.105 36	4243.3	4875.4	8.2661	.090 27	4240.9	4872.8	8.1933	.078 96	4238.6	4870.3	8.1300

Table A-3.3 (Continued)

T	v	u	h	s	v	u	h	s	v	u	h	s
	$P = 6.0$ MPa (275.64)				$P = 7.0$ MPa (285.88)				$P = 8.0$ MPa (295.06)			
1200	.113 21	4454.0	5133.3	8.4474	.097 03	4451.7	5130.9	8.3747	.084 89	4449.5	5128.5	8.3115
1300	.121 06	4669.6	5396.0	8.6199	.103 77	4667.3	5393.7	8.5473	.090 80	4665.0	5391.5	8.4842
	$P = 9.0$ MPa (303.40)				$P = 10.0$ MPa (311.06)				$P = 12.5$ MPa (327.89)			
Sat.	.020 48	2557.8	2742.1	5.6772	.018 026	2544.4	2724.7	5.6141	.013 495	2505.1	2673.8	5.4624
325	.023 27	2646.6	2856.0	5.8712	.019 861	2610.4	2809.1	5.7568				
350	.025 80	2724.4	2956.6	6.0361	.022 42	2699.2	2923.4	5.9443	.016 126	2624.6	2826.2	5.7118
400	.029 93	2848.4	3117.8	6.2854	.026 41	2832.4	3096.5	6.2120	.020 00	2789.3	3039.3	6.0417
450	.033 50	2955.2	3256.6	6.4844	.029 75	2943.4	3240.9	6.4190	.022 99	2912.5	3199.8	6.2719
500	.036 77	3055.2	3386.1	6.6576	.032 79	3045.8	3373.7	6.5966	.025 60	3021.7	3341.8	6.4618
550	.039 87	3152.2	3511.0	6.8142	.035 64	3144.6	3500.9	6.7561	.028 01	3125.0	3475.2	6.6290
600	.042 85	3248.1	3633.7	6.9589	.038 37	3241.7	3625.3	6.9029	.030 29	3225.4	3604.0	6.7810
650	.045 74	3343.6	3755.3	7.0943	.041 01	3338.2	3748.2	7.0398	.032 48	3324.4	3730.4	6.9218
700	.048 57	3439.3	3876.5	7.2221	.043 58	3434.7	3870.5	7.1687	.034 60	3422.9	3855.3	7.0536
800	.054 09	3632.5	4119.3	7.4596	.048 59	3628.9	4114.8	7.4077	.038 69	3620.0	4103.6	7.2965
900	.059 50	3829.2	4364.8	7.6783	.053 49	3826.3	4361.2	7.6272	.042 67	3819.1	4352.5	7.5182
1000	.064 85	4030.3	4614.0	7.8821	.058 32	4027.8	4611.0	7.8315	.046 58	4021.6	4603.8	7.7237
1100	.070 16	4236.3	4867.7	8.0740	.063 12	4234.0	4865.1	8.0237	.050 45	4228.2	4858.8	7.9165
1200	.075 44	4447.2	5126.2	8.2556	.067 89	4444.9	5123.8	8.2055	.054 30	4439.3	5118.0	8.0987
1300	.080 72	4662.7	5389.2	8.4284	.072 65	4460.5	5387.0	8.3783	.058 13	4654.8	5381.4	8.2717

Table A-3.3 (Continued)

	$P = 15.0$ MPa (342.24)				$P = 17.5$ MPa (354.75)				$P = 20.0$ MPa (365.81)			
Sat.	.010 337	2455.5	2610.5	5.3098	.007 920	2390.2	2528.8	5.1419	.005 834	2293.0	2409.7	4.9269
350	.011 470	2520.4	2692.4	5.4421								
400	.015 649	2740.7	2975.5	5.8811	.012 447	2685.0	2902.9	5.7213	.009 942	2619.3	2818.1	5.5540
450	.018 445	2879.5	3156.2	6.1404	.015 174	2844.2	3109.7	6.0184	.012 695	2806.2	3060.1	5.9017
500	.020 80	2996.6	3308.6	6.3443	.017 358	2970.3	3274.1	6.2383	.014 768	2942.9	3238.2	6.1401
550	.022 93	3104.7	3448.6	6.5199	.019 288	3083.9	3421.4	6.4230	.016 555	3062.4	3393.5	6.3348
600	.024 91	3208.6	3582.3	6.6776	.021 06	3191.5	3560.1	6.5866	.018 178	3174.0	3537.6	6.5048
650	.026 80	3310.3	3712.3	6.8224	.022 74	3296.0	3693.9	6.7357	.019 693	3281.4	3675.3	6.6582
700	.028 61	3410.9	3840.1	6.9572	.024 34	3398.7	3824.6	6.8736	.021 13	3386.4	3809.0	6.7993
800	.032 10	3610.9	4092.4	7.2040	.027 38	3601.8	4081.1	7.1244	.023 85	3592.7	4069.7	7.0544
900	.035 46	3811.9	4343.8	7.4279	.030 31	3804.7	4335.1	7.3507	.026 45	3797.5	4326.4	7.2830
1000	.038 75	4015.4	4596.6	7.6348	.033 16	4009.3	4589.5	7.5589	.028 97	4003.1	4582.5	7.4925
1100	.042 00	4222.6	4852.6	7.8283	.035 97	4216.9	4846.4	7.7531	.031 45	4211.3	4840.2	7.6874
1200	.045 23	4433.8	5112.3	8.0108	.038 76	4428.3	5106.6	7.9360	.033 91	4422.8	5101.0	7.8707
1300	.048 45	4649.1	5376.0	8.1840	.041 54	4643.5	5370.5	8.1093	.036 36	4638.0	5365.1	8.0442
	$P = 25.0$ MPa				$P = 30.0$ MPa				$P = 35.0$ MPa			
375	.001 973 1	1798.7	1848.0	4.0320	.001 789 2	1737.8	1791.5	3.9305	.001 700 3	1702.9	1762.4	3.8722
400	.006 004	2430.1	2580.2	5.1418	.002 790	2067.4	2151.1	4.4728	.002 100	1914.1	1987.6	4.2126
425	.007 881	2609.2	2806.3	5.4723	.005 303	2455.1	2614.2	5.1504	.003 428	2253.4	2373.4	4.7747
450	.009 162	2720.7	2949.7	5.6744	.006 735	2619.3	2821.4	5.4424	.004 961	2498.7	2672.4	5.1962
500	.011 123	2884.3	3162.4	5.9592	.008 678	2820.7	3081.1	5.7905	.006 927	2751.9	2994.4	5.6282
550	.012 724	3017.5	3335.6	6.1765	.010 168	2970.3	3275.4	6.0342	.008 345	2921.0	3213.0	5.9026
600	.014 137	3137.9	3491.4	6.3602	.011 446	3100.5	3443.9	6.2331	.009 527	3062.0	3395.5	6.1179
650	.015 433	3251.6	3637.4	6.5229	.012 596	3221.0	3598.9	6.4058	.010 575	3189.8	3559.9	6.3010
700	.016 646	3361.3	3777.5	6.6707	.013 661	3335.8	3745.6	6.5606	.011 533	3309.8	3713.5	6.4631
800	.018 912	3574.3	4047.1	6.9345	.015 623	3555.5	4024.2	6.8332	.013 278	3536.7	4001.5	6.7450
900	.021 045	3783.0	4309.1	7.1680	.017 448	3768.5	4291.9	7.0718	.014 883	3754.0	4274.9	6.9886
1000	.023 10	3990.9	4568.5	7.3802	.019 196	3978.8	4554.7	7.2867	.016 410	3966.7	4541.1	7.2064
1100	.025 12	4200.2	4828.2	7.5765	.020 903	4189.2	4816.3	7.4845	.017 895	4178.3	4804.6	7.4057

Table A-3.3 (Continued)

T	v	u	h	s	v	u	h	s	v	u	h	s
	$P = 25.0$ MPa				$P = 30.0$ MPa				$P = 35.0$ MPa			
1200	.027 11	4412.0	5089.9	7.7605	.022 589	4401.3	5079.0	7.6692	.019 360	4390.7	5068.3	7.5910
1300	.029 10	4626.9	5354.4	7.9342	.024 266	4616.0	5344.0	7.8432	.020 815	4605.1	5333.6	7.7653
	$P = 40.0$ MPa				$P = 50.0$ MPa				$P = 60.0$ MPa			
375	.001 640 7	1677.1	1742.8	3.8290	.001 559 4	1638.6	1716.6	3.7639	.001 502 8	1609.4	1699.5	3.7141
400	.001 907 7	1854.6	1930.9	4.1135	.001 730 9	1788.1	1874.6	4.0031	.001 633 5	1745.4	1843.4	3.9318
425	.002 532	2096.9	2198.1	4.5029	.002 007	1959.7	2060.0	4.2734	.001 816 5	1892.7	2001.7	4.1626
450	.003 693	2365.1	2512.8	4.9459	.002 486	2159.6	2284.0	4.5884	.002 085	2053.9	2179.0	4.4121
500	.005 622	2678.4	2903.3	5.4700	.003 892	2525.5	2720.1	5.1726	.002 956	2390.6	2567.9	4.9321
550	.006 984	2869.7	3149.1	5.7785	.005 118	2763.6	3019.5	5.5485	.003 956	2658.8	2896.2	5.3441
600	.008 094	3022.6	3346.4	6.0114	.006 112	2942.0	3247.6	5.8178	.004 834	2861.1	3151.2	5.6452
650	.009 063	3158.0	3520.6	6.2054	.006 966	3093.5	3441.8	6.0342	.005 595	3028.8	3364.5	5.8829
700	.009 941	3283.6	3681.2	6.3750	.007 727	3230.5	3616.8	6.2189	.006 272	3177.2	3553.5	6.0824
800	.011 523	3517.8	3978.7	6.6662	.009 076	3479.8	3933.6	6.5290	.007 459	3441.5	3889.1	6.4109
900	.012 962	3739.4	4257.9	6.9150	.010 283	3710.3	4224.4	6.7882	.008 508	3681.0	4191.5	6.6805
1000	.014 324	3954.6	4527.6	7.1356	.011 411	3930.5	4501.1	7.0146	.009 480	3906.4	4475.2	6.9127
1100	.015 642	4167.4	4793.1	7.3364	.012 496	4145.7	4770.5	7.2184	.010 409	4124.1	4748.6	7.1195
1200	.016 940	4380.1	5057.7	7.5224	.013 561	4359.1	5037.2	7.4058	.011 317	4338.2	5017.2	7.3083
1300	.018 229	4594.3	5323.5	7.6969	.014 616	4572.8	5303.6	7.5808	.012 215	4551.4	5284.3	7.4837

Table A-3.5 Saturated Solid-Vapor

Temp. °C T	Specific Volume: Press. kPa P	Sat. Solid $v_i \times 10^3$	Sat. Vapor v_g	Internal Energy: Sat. Solid u_i	Subl. u_{ig}	Sat. Vapor u_g	Enthalpy: Sat. Solid h_i	Subl. h_{ig}	Sat. Vapor h_g	Entropy: Sat. Solid s_i	Subl. s_{ig}	Sat. Vapor s_g
.01	.6113	1.0908	206.1	−333.40	2708.7	2375.3	−333.40	2834.8	2501.4	−1.221	10.378	9.156
0	.6108	1.0908	206.3	−333.43	2708.8	2375.3	−333.43	2834.8	2501.3	−1.221	10.378	9.157
−2	.5176	1.0904	241.7	−337.62	2710.2	2372.6	−337.62	2835.3	2497.7	−1.237	10.456	9.219
−4	.4375	1.0901	283.8	−341.78	2711.6	2369.8	−341.78	2835.7	2494.0	−1.253	10.536	9.283
−6	.3689	1.0898	334.2	−345.91	2712.9	2367.0	−345.91	2836.2	2490.3	−1.268	10.616	9.348
−8	.3102	1.0894	394.4	−350.02	2714.2	2364.2	−350.02	2836.6	2486.6	−1.284	10.698	9.414
−10	.2602	1.0891	466.7	−354.09	2715.5	2361.4	−354.09	2837.0	2482.9	−1.299	10.781	9.481
−12	.2176	1.0888	553.7	−358.14	2716.8	2358.7	−358.14	2837.3	2479.2	−1.315	10.865	9.550
−14	.1815	1.0884	658.8	−362.15	2718.0	2355.9	−362.15	2837.6	2475.5	−1.331	10.950	9.619
−16	.1510	1.0881	786.0	−366.14	2719.2	2353.1	−366.14	2837.9	2471.8	−1.346	11.036	9.690
−18	.1252	1.0878	940.5	−370.10	2720.4	2350.3	−370.10	2838.2	2468.1	−1.362	11.123	9.762
−20	.1035	1.0874	1128.6	−374.03	2721.6	2347.5	−374.03	2838.4	2464.3	−1.377	11.212	9.835
−22	.0853	1.0871	1358.4	−377.93	2722.7	2344.7	−377.93	2838.6	2460.6	−1.393	11.302	9.909
−24	.0701	1.0868	1640.1	−381.80	2723.7	2342.0	−381.80	2838.7	2456.9	−1.408	11.394	9.985
−26	.0574	1.0864	1986.4	−385.64	2724.8	2339.2	−385.64	2838.9	2453.2	−1.424	11.486	10.062
−28	.0469	1.0861	2413.7	−389.45	2725.8	2336.4	−389.45	2839.0	2449.5	−1.439	11.580	10.141
−30	.0381	1.0858	2943	−393.23	2726.8	2333.6	−393.23	2839.0	2445.8	−1.455	11.676	10.221
−32	.0309	1.0854	3600	−396.98	2727.8	2330.8	−396.98	2839.1	2442.1	−1.471	11.773	10.303
−34	.0250	1.0851	4419	−400.71	2728.7	2328.0	−400.71	2839.1	2438.4	−1.486	11.872	10.386
−36	.0201	1.0848	5444	−404.40	2729.6	2325.2	−404.40	2839.1	2434.7	−1.501	11.972	10.470
−38	.0161	1.0844	6731	−408.06	2730.5	2322.4	−408.06	2839.0	2430.9	−1.517	12.073	10.556
−40	.0129	1.0841	8354	−411.70	2731.3	2319.6	−411.70	2838.9	2427.2	−1.532	12.176	10.644

THERMODYNAMIC PROPERTIES OF AMMONIA[a]

Table A-4.1 Saturated Ammonia

		Specific Volume m³/kg			Enthalpy kJ/kg			Entropy kJ/kg K		
Temp. °C	Abs. Press. kPa P	Sat. Liquid v_f	Evap. v_{fg}	Sat. Vapor v_g	Sat. Liquid h_f	Evap. h_{fg}	Sat. Vapor h_g	Sat. Liquid s_f	Evap. s_{fg}	Sat. Vapor s_g
−50	40.88	0.001 424	2.6239	2.6254	−44.3	1416.7	1372.4	−0.1942	6.3502	6.1561
−48	45.96	0.001 429	2.3518	2.3533	−35.5	1411.3	1375.8	−0.1547	6.2696	6.1149
−46	51.55	0.001 434	2.1126	2.1140	−26.6	1405.8	1379.2	−0.1156	6.1902	6.0746
−44	57.69	0.001 439	1.9018	1.9032	−17.8	1400.3	1382.5	−0.0768	6.1120	6.0352
−42	64.42	0.001 444	1.7155	1.7170	−8.9	1394.7	1385.8	−0.0382	6.0349	5.9967
−40	71.77	0.001 449	1.5506	1.5521	0.0	1389.0	1389.0	0.0000	5.9589	5.9589
−38	79.80	0.001 454	1.4043	1.4058	8.9	1383.3	1392.2	0.0380	5.8840	5.9220
−36	88.54	0.001 460	1.2742	1.2757	17.8	1377.6	1395.4	0.0757	5.8101	5.8858
−34	98.05	0.001 465	1.1582	1.1597	26.8	1371.8	1398.5	0.1132	5.7372	5.8504
−32	108.37	0.001 470	1.0547	1.0562	35.7	1365.9	1401.6	0.1504	5.6652	5.8156
−30	119.55	0.001 476	0.9621	0.9635	44.7	1360.0	1404.6	0.1873	5.5942	5.7815
−28	131.64	0.001 481	0.8790	0.8805	53.6	1354.0	1407.6	0.2240	5.5241	5.7481
−26	144.70	0.001 487	0.8044	0.8059	62.6	1347.9	1410.5	0.2605	5.4548	5.7153
−24	158.78	0.001 492	0.7373	0.7388	71.6	1341.8	1413.4	0.2967	5.3864	5.6831
−22	173.93	0.001 498	0.6768	0.6783	80.7	1335.6	1416.2	0.3327	5.3188	5.6515
−20	190.22	0.001 504	0.6222	0.6237	89.7	1329.3	1419.0	0.3684	5.2520	5.6205
−18	207.71	0.001 510	0.5728	0.5743	98.8	1322.9	1421.7	0.4040	5.1860	5.5900
−16	226.45	0.001 515	0.5280	0.5296	107.8	1316.5	1424.4	0.4393	5.1207	5.5600
−14	246.51	0.001 521	0.4874	0.4889	116.9	1310.0	1427.0	0.4744	5.0561	5.5305
−12	267.95	0.001 528	0.4505	0.4520	126.0	1303.5	1429.5	0.5093	4.9922	5.5015
−10	290.85	0.001 534	0.4169	0.4185	135.2	1296.8	1432.0	0.5440	4.9290	5.4730
−8	315.25	0.001 540	0.3863	0.3878	144.3	1290.1	1434.4	0.5785	4.8664	5.4449
−6	341.25	0.001 546	0.3583	0.3599	153.5	1283.3	1436.8	0.6128	4.8045	5.4173
−4	368.90	0.001 553	0.3328	0.3343	162.7	1276.4	1439.1	0.6469	4.7432	5.3901
−2	398.27	0.001 559	0.3094	0.3109	171.9	1269.4	1441.3	0.6808	4.6825	5.3633
0	429.44	0.001 566	0.2879	0.2895	181.1	1262.4	1443.5	0.7145	4.6223	5.3369
2	462.49	0.001 573	0.2683	0.2698	190.4	1255.2	1445.6	0.7481	4.5627	5.3108
4	497.49	0.001 580	0.2502	0.2517	199.6	1248.0	1447.6	0.7815	4.5037	5.2852
6	534.51	0.001 587	0.2335	0.2351	208.9	1240.6	1449.6	0.8148	4.4451	5.2599
8	573.64	0.001 594	0.2182	0.2198	218.3	1233.2	1451.5	0.8479	4.3871	5.2350
10	614.95	0.001 601	0.2040	0.2056	227.6	1225.7	1453.3	0.8808	4.3295	5.2104
12	658.52	0.001 608	0.1910	0.1926	237.0	1218.1	1455.1	0.9136	4.2725	5.1861
14	704.44	0.001 616	0.1789	0.1805	246.4	1210.4	1456.8	0.9463	4.2159	5.1621
16	752.79	0.001 623	0.1677	0.1693	255.9	1202.6	1458.5	0.9788	4.1597	5.1385
18	803.66	0.001 631	0.1574	0.1590	265.4	1194.7	1460.0	1.0112	4.1039	5.1151
20	857.12	0.001 639	0.1477	0.1494	274.9	1186.7	1461.5	1.0434	4.0486	5.0920
22	913.27	0.001 647	0.1388	0.1405	284.4	1178.5	1462.9	1.0755	3.9937	5.0692
24	972.19	0.001 655	0.1305	0.1322	294.0	1170.3	1464.3	1.1075	3.9392	5.0467
26	1033.97	0.001 663	0.1228	0.1245	303.6	1162.0	1465.6	1.1394	3.8850	5.0244
28	1098.71	0.001 671	0.1156	0.1173	313.2	1153.6	1466.8	1.1711	3.8312	5.0023
30	1166.49	0.001 680	0.1089	0.1106	322.9	1145.0	1467.9	1.2028	3.7777	4.9805
32	1237.41	0.001 689	0.1027	0.1044	332.6	1136.4	1469.0	1.2343	3.7246	4.9589
34	1311.55	0.001 698	0.0969	0.0986	342.3	1127.6	1469.9	1.2656	3.6718	4.9374
36	1389.03	0.001 707	0.0914	0.0931	352.1	1118.7	1470.8	1.2969	3.6192	4.9161
38	1469.92	0.001 716	0.0863	0.0880	361.9	1109.7	1471.5	1.3281	3.5669	4.8950
40	1554.33	0.001 726	0.0815	0.0833	371.7	1100.5	1472.2	1.3591	3.5148	4.8740
42	1642.35	0.001 735	0.0771	0.0788	381.6	1091.2	1472.8	1.3901	3.4630	4.8530
44	1734.09	0.001 745	0.0728	0.0746	391.5	1081.7	1473.2	1.4209	3.4112	4.8322
46	1829.65	0.001 756	0.0689	0.0707	401.5	1072.0	1473.5	1.4518	3.3595	4.8113
48	1929.13	0.001 766	0.0652	0.0669	411.5	1062.2	1473.7	1.4826	3.3079	4.7905
50	2032.62	0.001 777	0.0617	0.0635	421.7	1052.0	1473.7	1.5135	3.2561	4.7696

[a] Adapted from National Bureau of Standards Circular No. 142, *Tables of Thermodynamic Properties of Ammonia*.

Table A-4.2 (Continued)

	m³/kg	kJ/kg	kJ/kg·Δ_1K	m³/kg	kJ/kg	kJ/kg·Δ_1K	m³/kg	kJ/kg	kJ/kg·Δ_1K
T	*v*	*h*	*s*	*v*	*h*	*s*	*v*	*h*	*s*
	P = 1400 kPa (36.28)			*P* = 1600 kPa (41.05)			*P* = 1800 kPa (45.39)		
Sat.	0.0924	1470.9	4.9131	0.0809	1472.5	4.8630	0.0719	1473.4	4.8177
40	0.0944	1483.4	4.9534						
50	0.0995	1515.1	5.0530	0.0851	1502.9	4.9584	0.0739	1490.0	4.8693
60	0.1042	1544.7	5.1434	0.0895	1534.4	5.0543	0.0781	1523.5	4.9715
70	0.1088	1573.0	5.2270	0.0937	1564.0	5.1419	0.0820	1554.6	5.0635
80	0.1132	1600.2	5.3053	0.0977	1592.3	5.2232	0.0856	1584.1	5.1482
100	0.1216	1652.8	5.4501	0.1053	1646.4	5.3722	0.0926	1639.8	5.3018
120	0.1297	1703.9	5.5836	0.1125	1698.5	5.5084	0.0992	1693.1	5.4409
140	0.1376	1754.3	5.7087	0.1195	1749.7	5.6355	0.1055	1745.1	5.5699
160	0.1452	1804.5	5.8273	0.1263	1800.5	5.7555	0.1116	1796.5	5.6914
180	0.1528	1854.7	5.9406	0.1330	1851.2	5.8699	0.1177	1847.7	5.8069
	P = 2000 kPa (49.38)								
Sat.	0.0646	1473.7	4.7761						
50	0.0648	1476.1	4.7834						
60	0.0688	1512.0	4.8930						
70	0.0725	1544.9	4.9902						
80	0.0760	1575.6	5.0786						
100	0.0824	1633.2	5.2371						
120	0.0885	1687.6	5.3793						
140	0.0943	1740.4	5.5104						
160	0.0999	1792.4	5.6333						
180	0.1054	1844.1	5.7499						

THERMODYNAMIC PROPERTIES OF FREON-12[a]

Table A-5.1 Saturated Freon-12

Temp. °C	Abs. Press. MPa P	Specific Volume m³/kg Sat. Liquid v_f	Evap. v_{fg}	Sat. Vapor v_g	Enthalpy kJ/kg Sat. Liquid h_f	Evap. h_{fg}	Sat. Vapor h_g	Entropy kJ/kg K Sat. Liquid s_f	Evap. s_{fg}	Sat. Vapor s_g
−90	0.0028	0.000 608	4.414 937	4.415 545	−43.243	189.618	146.375	−0.2084	1.0352	0.8268
−85	0.0042	0.000 612	3.036 704	3.037 316	−38.968	187.608	148.640	−0.1854	0.9970	0.8116
−80	0.0062	0.000 617	2.137 728	2.138 345	−34.688	185.612	150.924	−0.1630	0.9609	0.7979
−75	0.0088	0.000 622	1.537 030	1.537 651	−30.401	183.625	153.224	−0.1411	0.9266	0.7855
−70	0.0123	0.000 627	1.126 654	1.127 280	−26.103	181.640	155.536	−0.1197	0.8940	0.7744
−65	0.0168	0.000 632	0.840 534	0.841 166	−21.793	179.651	157.857	−0.0987	0.8630	0.7643
−60	0.0226	0.000 637	0.637 274	0.637 910	−17.469	177.653	160.184	−0.0782	0.8334	0.7552
−55	0.0300	0.000 642	0.490 358	0.491 000	−13.129	175.641	162.512	−0.0581	0.8051	0.7470
−50	0.0391	0.000 648	0.382 457	0.383 105	−8.772	173.611	164.840	−0.0384	0.7779	0.7396
−45	0.0504	0.000 654	0.302 029	0.302 682	−4.396	171.558	167.163	−0.0190	0.7519	0.7329
−40	0.0642	0.000 659	0.241 251	0.241 910	−0.000	169.479	169.479	−0.0000	0.7269	0.7269
−35	0.0807	0.000 666	0.194 732	0.195 398	4.416	167.368	171.784	0.0187	0.7027	0.7214
−30	0.1004	0.000 672	0.158 703	0.159 375	8.854	165.222	174.076	0.0371	0.6795	0.7165
−25	0.1237	0.000 679	0.130 487	0.131 166	13.315	163.037	176.352	0.0552	0.6570	0.7121
−20	0.1509	0.000 685	0.108 162	0.108 847	17.800	160.810	178.610	0.0730	0.6352	0.7082
−15	0.1826	0.000 693	0.090 326	0.091 018	22.312	158.534	180.846	0.0906	0.6141	0.7046
−10	0.2191	0.000 700	0.075 946	0.076 646	26.851	156.207	183.058	0.1079	0.5936	0.7014
−5	0.2610	0.000 708	0.064 255	0.064 963	31.420	153.823	185.243	0.1250	0.5736	0.6986
0	0.3086	0.000 716	0.054 673	0.055 389	36.022	151.376	187.397	0.1418	0.5542	0.6960
5	0.3626	0.000 724	0.046 761	0.047 485	40.659	148.859	189.518	0.1585	0.5351	0.6937
10	0.4233	0.000 733	0.040 180	0.040 914	45.337	146.265	191.602	0.1750	0.5165	0.6916
15	0.4914	0.000 743	0.034 671	0.035 413	50.058	143.586	193.644	0.1914	0.4983	0.6897
20	0.5673	0.000 752	0.030 028	0.030 780	54.828	140.812	195.641	0.2076	0.4803	0.6879
25	0.6516	0.000 763	0.026 091	0.026 854	59.653	137.933	197.586	0.2237	0.4626	0.6863
30	0.7449	0.000 774	0.022 734	0.023 508	64.539	134.936	199.475	0.2397	0.4451	0.6848
35	0.8477	0.000 786	0.019 855	0.020 641	69.494	131.805	201.299	0.2557	0.4277	0.6834
40	0.9607	0.000 798	0.017 373	0.018 171	74.527	128.525	203.051	0.2716	0.4104	0.6820
45	1.0843	0.000 811	0.015 220	0.016 032	79.647	125.074	204.722	0.2875	0.3931	0.6806
50	1.2193	0.000 826	0.013 344	0.014 170	84.868	121.430	206.298	0.3034	0.3758	0.6792
55	1.3663	0.000 841	0.011 701	0.012 542	90.201	117.565	207.766	0.3194	0.3582	0.6777
60	1.5259	0.000 858	0.010 253	0.011 111	95.665	113.443	209.109	0.3355	0.3405	0.6760
65	1.6988	0.000 877	0.008 971	0.009 847	101.279	109.024	210.303	0.3518	0.3224	0.6742
70	1.8858	0.000 897	0.007 828	0.008 725	107.067	104.255	211.321	0.3683	0.3038	0.6721
75	2.0874	0.000 920	0.006 802	0.007 723	113.058	99.068	212.126	0.3851	0.2845	0.6697
80	2.3046	0.000 946	0.005 875	0.006 821	119.291	93.373	212.665	0.4023	0.2644	0.6667
85	2.5380	0.000 976	0.005 029	0.006 005	125.818	87.047	212.865	0.4201	0.2430	0.6631
90	2.7885	0.001 012	0.004 246	0.005 258	132.708	79.907	212.614	0.4385	0.2200	0.6585
95	3.0569	0.001 056	0.003 508	0.004 563	140.068	71.658	211.726	0.4579	0.1946	0.6526
100	3.3440	0.001 113	0.002 790	0.003 903	148.076	61.768	209.843	0.4788	0.1655	0.6444
105	3.6509	0.001 197	0.002 045	0.003 242	157.085	49.014	206.099	0.5023	0.1296	0.6319
110	3.9784	0.001 364	0.001 098	0.002 462	168.059	28.425	196.484	0.5322	0.0742	0.6064
112	4.1155	0.001 792	0.000 005	0.001 797	174.920	0.151	175.071	0.5651	0.0004	0.5655

Table A-5.2 (Continued)

	1.00 MPa			1.20 MPa			1.40 MPa		
50.0	0.018 366	210.152	0.7021	0.014 483	206.661	0.6812			
60.0	0.019 410	217.810	0.7254	0.015 463	214.805	0.7060	0.012 579	211.457	0.6876
70.0	0.020 397	225.319	0.7476	0.016 368	222.687	0.7293	0.013 448	219.822	0.7123
80.0	0.021 341	232.739	0.7689	0.017 221	230.398	0.7514	0.014 247	227.891	0.7355
90.0	0.022 251	240.101	0.7895	0.018 032	237.995	0.7727	0.014 997	235.766	0.7575
100.0	0.023 133	247.430	0.8094	0.018 812	245.518	0.7931	0.015 710	243.512	0.7785
110.0	0.023 993	254.743	0.8287	0.019 567	252.993	0.8129	0.016 393	251.170	0.7988
120.0	0.024 835	262.053	0.8475	0.020 301	260.441	0.8320	0.017 053	258.770	0.8183
130.0	0.025 661	269.369	0.8659	0.021 018	267.875	0.8507	0.017 695	266.334	0.8373
140.0	0.026 474	276.699	0.8839	0.021 721	275.307	0.8689	0.018 321	273.877	0.8558
150.0	0.027 275	284.047	0.9015	0.022 412	282.745	0.8867	0.018 934	281.411	0.8738
160.0	0.028 068	291.419	0.9187	0.023 093	290.195	0.9041	0.019 535	288.946	0.8914

	1.60 MPa			1.80 MPa			2.00 MPa		
70.0	0.011 208	216.650	0.6959	0.009 406	213.049	0.6794			
80.0	0.011 984	225.177	0.7204	0.010 187	222.198	0.7057	0.008 704	218.859	0.6909
90.0	0.012 698	233.390	0.7433	0.010 884	230.835	0.7298	0.009 406	228.056	0.7166
100.0	0.013 366	241.397	0.7651	0.011 526	239.155	0.7524	0.010 035	236.760	0.7402
110.0	0.014 000	249.264	0.7859	0.012 126	247.264	0.7739	0.010 615	245.154	0.7624
120.0	0.014 608	257.035	0.8059	0.012 697	255.228	0.7944	0.011 159	253.341	0.7835
130.0	0.015 195	264.742	0.8253	0.013 244	263.094	0.8141	0.011 676	261.384	0.8037
140.0	0.015 765	272.406	0.8440	0.013 772	270.891	0.8332	0.012 172	269.327	0.8232
150.0	0.016 320	280.044	0.8623	0.014 284	278.642	0.8518	0.012 651	277.201	0.8420
160.0	0.016 864	287.669	0.8801	0.014 784	286.364	0.8698	0.013 116	285.027	0.8603
170.9	0.017 398	295.290	0.8975	0.015 272	294.069	0.8874	0.013 570	292.822	0.8781
180.0	0.017 923	302.914	0.9145	0.015 752	301.767	0.9046	0.014 013	300.598	0.8955

Table A-5.2 (Continued)

Temp. °C	v m³/kg	h kJ/kg	s kJ/kg K	v m³/kg	h kJ/kg	s kJ/kg K	v m³/kg	h kJ/kg	s kJ/kg K
	2.50 MPa			3.00 MPa			3.50 MPa		
90.0	0.006 595	219.562	0.6823						
100.0	0.007 264	229.852	0.7103	0.005 231	220.529	0.6770			
110.0	0.007 837	239.271	0.7352	0.005 886	232.068	0.7075	0.004 324	222.121	0.6750
120.0	0.008 351	248.192	0.7582	0.006 419	242.208	0.7336	0.004 959	234.875	0.7078
130.0	0.008 827	256.794	0.7798	0.006 887	251.632	0.7573	0.005 456	245.661	0.7349
140.0	0.009 273	265.180	0.8003	0.007 313	260.620	0.7793	0.005 884	255.524	0.7591
150.0	0.009 697	273.414	0.8200	0.007 709	269.319	0.8001	0.006 270	264.846	0.7814
160.0	0.010 104	281.540	0.8390	0.008 083	277.817	0.8200	0.006 626	273.817	0.8023
170.0	0.010 497	289.589	0.8574	0.008 439	286.171	0.8391	0.006 961	282.545	0.8222
180.0	0.010 879	297.583	0.8752	0.008 782	294.422	0.8575	0.007 279	291.100	0.8413
190.0	0.011 250	305.540	0.8926	0.009 114	302.597	0.8753	0.007 584	299.528	0.8597
200.0	0.011 614	313.472	0.9095	0.009 436	310.718	0.8927	0.007 878	307.864	0.8775
	4.00 MPa								
120.0	0.003 736	224.863	0.6771						
130.0	0.004 325	238.443	0.7111						
140.0	0.004 781	249.703	0.7386						
150.0	0.005 172	259.904	0.7630						
160.0	0.005 522	269.492	0.7854						
170.0	0.005 845	278.684	0.8063						
180.0	0.006 147	287.602	0.8262						
190.0	0.006 434	296.326	0.8453						
200.0	0.006 708	304.906	0.8636						
210.0	0.006 972	313.380	0.8813						
220.0	0.007 228	321.774	0.8985						
230.0	0.007 477	330.108	0.9152						

Table A-6 Thermodynamic Properties of Normal Saturated Butane

T°C	P	Specific Volume (v)		Specific Enthalpy (h)		Specific Entropy (s)	
		m^3/kg		kJ/kg		$kJ/kg \cdot \Delta_1 K$	
	MPa abs.	Liquid v_f	Vapor v_g	h_f	h_g	s_f	s_g
−1.1	0.0993	1.66×10^{-3}	0.368	277.26	662.68	1.088	2.391
+1.7	0.1103	1.67×10^{-3}	0.335	283.31	666.17	1.105	2.391
4.4	0.1220	1.68×10^{-3}	0.305	288.89	669.89	1.130	2.391
7.2	0.1351	1.69×10^{-3}	0.279	294.94	673.38	1.156	2.391
10.0	0.1489	1.70×10^{-3}	0.254	301.45	677.33	1.181	2.391
12.8	0.1641	1.70×10^{-3}	0.233	307.26	680.82	1.202	2.387
15.6	0.1813	1.71×10^{-3}	0.212	313.54	684.31	1.227	2.387
18.3	0.1993	1.72×10^{-3}	0.195	319.36	688.26	1.248	2.387
21.1	0.2179	1.73×10^{-3}	0.180	325.87	691.99	1.273	2.387
23.9	0.2379	1.74×10^{-3}	0.165	331.69	695.24	1.298	2.387
26.7	0.2593	1.75×10^{-3}	0.154	337.27	698.50	1.319	2.387
29.4	0.2820	1.76×10^{-3}	0.142	343.78	702.45	1.348	2.387
32.2	0.3068	1.77×10^{-3}	0.131	350.06	705.94	1.365	2.391
35.0	0.3323	1.78×10^{-3}	0.122	356.11	709.43	1.390	2.391
37.8	0.3599	1.79×10^{-3}	0.113	363.32	713.38	1.424	2.391
40.6	0.3889	1.80×10^{-3}	0.106	370.07	716.64	1.449	2.395
43.3	0.4192	1.81×10^{-3}	0.099	376.58	719.90	1.474	2.395
46.1	0.4523	1.83×10^{-3}	0.092	383.79	723.62	1.503	2.395
48.9	0.4882	1.84×10^{-3}	0.086	390.30	727.34	1.528	2.395
51.7	0.5240	1.85×10^{-3}	0.080	397.75	730.36	1.558	2.399
54.4	0.5613	1.86×10^{-3}	0.074	404.72	733.85	1.583	2.399
57.2	0.5999	1.87×10^{-3}	0.069	411.93	737.34	1.612	2.399
60.0	0.6385	1.89×10^{-3}	0.065	418.45	740.13	1.637	2.403
62.8	0.6895	1.90×10^{-3}	0.062	425.89	743.16	1.666	2.403
65.6	0.7447	1.91×10^{-3}	0.056	433.80	746.65	1.696	2.403
68.3	0.7929	1.93×10^{-3}	0.052	440.31	749.67	1.721	2.403
71.1	0.8412	1.94×10^{-3}	0.049	448.22	753.16	1.750	2.408
73.9	0.8964	1.96×10^{-3}	0.046	455.43	755.95	1.779	2.408
76.7	0.9653	1.98×10^{-3}	0.043	463.34	758.51	1.813	2.408
79.4	1.0343	1.99×10^{-3}	0.040	470.08	761.07	1.838	2.408
82.2	1.1032	2.01×10^{-3}	0.038	476.83	763.86	1.863	2.408

Source: Adapted from the *Handbook Butane-Propane Gases*, 3rd ed., Table 4, p. 27. By permission from the *Butane-Propane News*.

Table A-6 Thermodynamic Properties of Normal Saturated Butane (Continued)

T°C	P	Internal Energy (u)		
	MPa abs.	kJ/kg u_f	kJ/kg u_{fg}	kJ/kg u_g
−1.1	0.0993	277.26	385.38	662.64
+1.7	0.1103	283.31	382.82	666.13
4.4	0.1220	288.89	380.96	669.85
7.2	0.1351	294.94	378.40	673.34
10.0	0.1489	301.45	375.84	677.29
12.8	0.1641	307.26	373.52	680.78
15.6	0.1813	313.54	370.73	684.27
18.3	0.1993	319.36	368.86	688.22
21.1	0.2179	325.87	366.08	691.95
23.9	0.2379	331.69	363.51	695.20
26.7	0.2593	337.27	361.19	698.46
29.4	0.2820	343.78	358.63	702.41
32.2	0.3068	350.06	355.84	705.90
35.0	0.3323	356.11	353.28	709.39
37.8	0.3599	363.32	350.02	713.34
40.6	0.3889	370.07	346.53	716.60
43.3	0.4192	376.58	343.28	719.86
46.1	0.4523	383.79	339.79	723.58
48.9	0.4882	390.30	337.00	727.30
51.7	0.5240	397.75	332.57	730.32
54.4	0.5613	404.72	329.09	733.81
57.2	0.5999	411.93	325.37	737.30
60.0	0.6385	418.45	321.64	740.09
62.8	0.6895	425.89	317.23	743.12
65.6	0.7447	433.80	312.81	746.61
68.3	0.7929	440.31	309.32	749.63
71.1	0.8412	448.22	304.90	753.12
73.9	0.8964	455.43	300.48	755.91
76.7	0.9653	463.34	295.13	758.47
79.4	1.0343	470.08	290.95	761.03
82.2	1.1032	467.83	295.99	763.82

Table A-7 ARDC Standard Atmosphere (1959)

Altitude (z) [m]	Temperature (T) [K]	Pressure (p) [kPa]	Density (ρ) [kg/m^3]
−5,000	320.69	177.610	1.9296
−4,000	314.18	159.600	1,7698
−3,000	307.67	142.970	1.6189
−2,000	301.16	127.780	1.4782
−1,000	294.66	113.930	1.3470
0	288.16	101.325	1.2250
500	284.91	95.461	1.1673
1,000	281.66	89.876	1.1117
1,500	278.41	84.560	1.0581
2,000	275.16	79.501	1.0066
2,500	271.92	74.692	0.9570
3,000	268.67	70.121	0.9093
3,500	265.42	65.780	0.8634
4,000	262.18	61.660	0.8194
4,500	258.93	57.752	0.7770
5,000	255.69	54.048	0.7364
6,000	249.20	47.217	0.6601
7,000	242.71	41.105	0.5900
8,000	236.23	35.651	0.5258
9,000	229.74	30.800	0.4671
10,000	223.26	26.500	0.4135
11,000	216.78	22.700	0.3648
12,000	216.66	19.399	0.3119
13,000	216.66	16.579	0.2666
14,000	216.66	14.170	0.2279
15,000	216.66	12.112	0.1948
16,000	216.66	10.353	0.1665
17,000	216.66	8.850	0.1423
18,000	216.66	7.565	0.1217
19,000	216.66	6.467	0.1040
20,000	216.66	5.529	0.0889
22,000	216.66	4.042	0.0650
24,000	216.66	2.955	0.0475
26,000	219.34	2.163	0.0344
28,000	225.29	1.595	0.0247
30,000	231.24	1.186	0.0179
35,000	246.09	0.584	0.0083
40,000	260.91	0.300	0.0040
45,000	275.71	0.160	0.0020
50,000	282.66	0.088	0.0011

Table A-8 Critical Constants

Substance	Formula	Molecular Weight (mol^{-1})	Temp. K	Pressure MPa	Volume m^3/kmol
Ammonia	NH_3	17.03	405.5	11.28	0.0724
Argon	Ar	39.948	151	4.86	0.0749
Bromine	Br_2	159.808	584	10.34	0.1355
Carbon dioxide	CO_2	44.01	304.2	7.39	0.0943
Carbon monoxide	CO	28.011	133	3.50	0.0930
Chlorine	Cl_2	70.906	417	7.71	0.1242
Deuterium (normal)	D_2	4.00	38.4	1.66	—
Helium	He	4.003	5.3	0.23	0.0578
Helium^3	He	3.00	3.3	0.12	—
Hydrogen (normal)	H_2	2.016	33.3	1.30	0.0649
Krypton	Kr	83.80	209.4	5.50	0.0924
Neon	Ne	20.183	44.5	2.73	0.0417
Nitrogen	N_2	28.013	126.2	3.39	0.0899
Nitrous oxide	N_2O	44.013	309.7	7.27	0.0961
Oxygen	O_2	31.999	154.8	5.08	0.0780
Sulfur dioxide	SO_2	64.063	430.7	7.88	0.1217
Water	H_2O	18.015	647.3	22.09	0.0568
Xenon	Xe	131.30	289.8	5.88	0.1186
Benzene	C_6H_6	78.115	562	4.92	0.2603
n-Butane	C_4H_{10}	58.124	425.2	3.80	0.2547
Carbon tetrachloride	CCl_4	153.82	556.4	4.56	0.2759
Chloroform	$CHCl_3$	119.38	536.6	5.47	0.2403
Dichlorodifluoromethane	CCl_2F_2	120.91	384.7	4.01	0.2179
Dichlorofluoromethane	$CHCl_2F$	102.92	451.7	5.17	0.1973
Ethane	C_2H_6	30.070	305.5	4.88	0.1480
Ethyl alcohol	C_2H_5OH	46.07	516	6.38	0.1673
Ethylene	C_2H_4	28.054	282.4	5.12	0.1242
n-Hexane	C_6H_{14}	86.178	507.9	3.03	0.3677
Methane	CH_4	16.043	191.1	4.64	0.0993
Methyl alcohol	CH_3OH	32.042	513.2	7.95	0.1180
Methyl chloride	CH_3Cl	50.488	416.3	6.68	0.1430
Propane	C_3H_8	44.097	370	4.26	0.1998
Propene	C_3H_6	42.081	365	4.62	0.1810
Propyne	C_3H_4	40.065	401	5.35	—
Trichlorofluoromethane	CCl_3F	137.37	471.2	4.38	0.2478

Source: K. A. Kobe and R. E. Lynn, Jr., *Chem. Rev.*, **52**:117–236 (1953).

Table A-11 (Continued)

T [K]	h [kJ/kg]	P_r [—]	u [kJ/kg]	v_r [—]	s° [kJ/kg·Δ_1K]
600	607.02	16.278	434.80	24.58	3.2223
610	617.53	17.297	442.43	23.51	3.2397
620	628.07	18.360	450.13	22.52	3.2569
630	638.65	19.475	457.83	21.57	3.2738
640	649.21	20.64	465.55	20.674	3.2905
650	659.84	21.86	473.32	19.828	3.3069
660	670.47	23.13	481.06	19.026	3.3232
670	681.15	24.46	488.88	18.266	3.3392
680	691.82	25.85	496.65	17.543	3.3551
690	702.52	27.29	504.51	16.857	3.3707
700	713.27	28.80	512.37	16.205	3.3861
710	724.01	30.38	520.26	15.585	3.4014
720	734.20	31.92	527.72	15.027	3.4156
730	745.62	33.72	536.12	14.434	3.4314
740	756.44	35.50	544.05	13.900	3.4461
750	767.30	37.35	552.05	13.391	3.4607
760	778.21	39.27	560.08	12.905	3.4751
770	789.10	41.27	568.10	12.440	3.4894
780	800.03	43.35	576.15	11.998	3.5035
790	810.98	45.51	584.22	11.575	3.5174
800	821.94	47.75	592.34	11.172	3.5312
810	832.96	50.08	600.46	10.785	3.5449
820	843.97	52.49	608.62	10.416	3.5584
830	855.01	55.00	616.79	10.062	3.5718
840	866.09	57.60	624.97	9.724	3.5850
850	877.16	60.29	633.21	9.400	3.5981
860	888.28	63.09	641.44	9.090	3.6111
870	899.42	65.98	649.70	8.792	3.6240
880	910.56	68.98	658.00	8.507	3.6367
890	921.75	72.08	666.31	8.233	3.6493
900	932.94	75.29	674.63	7.971	3.6619
910	944.15	78.61	682.98	7.718	3.6743
920	955.38	82.05	691.33	7.476	3.6865
930	966.64	85.60	699.73	7.244	3.6987
940	977.92	89.28	708.13	7.020	3.7108
950	989.22	93.08	716.57	6.805	3.7227
960	1000.53	97.00	725.01	6.599	3.7346
970	1011.88	101.06	733.48	6.400	3.7463
980	1023.25	105.24	741.99	6.209	3.7580
990	1034.63	109.57	750.48	6.025	3.7695
1000	1046.03	114.03	759.02	5.847	3.7810
1020	1068.89	123.12	775.67	5.521	3.8030
1040	1091.85	133.34	793.35	5.201	3.8259
1060	1114.85	143.91	810.61	4.911	3.8478
1080	1137.93	155.15	827.94	4.641	3.8694
1100	1161.07	167.07	845.34	4.390	3.8906
1120	1184.28	179.71	862.85	4.156	3.9116
1140	1207.54	193.07	880.37	3.937	3.9322
1160	1230.90	207.24	897.98	3.732	3.9525
1180	1254.34	222.2	915.68	3.541	3.9725

Table A-11 (Continued)

T [K]	h [kJ/kg]	P_r [—]	u [kJ/kg]	v_r [—]	$s°$ [kJ/kg·Δ_1K]
1200	1277.79	238.0	933.40	3.362	3.9922
1220	1301.33	254.7	951.19	3.194	4.0117
1240	1324.89	272.3	969.01	3.037	4.0308
1260	1348.55	290.8	986.92	2.889	4.0497
1280	1372.25	310.4	1004.88	2.750	4.0684
1300	1395.97	330.9	1022.88	2.619	4.0868
1320	1419.77	352.5	1040.93	2.497	4.1049
1340	1443.61	375.3	1059.03	2.381	4.1229
1360	1467.50	399.1	1077.17	2.272	4.1406
1380	1491.43	424.2	1095.36	2.169	4.1580
1400	1515.41	450.5	1113.62	2.072	4.1753
1420	1539.44	478.0	1131.90	1.9808	4.1923
1440	1563.49	506.9	1150.23	1.8942	4.2092
1460	1587.61	537.1	1168.61	1.8124	4.2258
1480	1611.80	568.8	1187.03	1.7350	4.2422
1500	1635.99	601.9	1205.47	1.6617	4.2585

Source: Adapted from Table 1 in *Gas Tables*, by Joseph H. Keenan and Joseph Kaye. Copyright 1948, by Joseph H. Keenan and Joseph Daye. Published by John Wiley & Sons, Inc., New York.

Table A-12 Psychrometric Chart (SI Version) for $P_b = 0.1013$ MPa

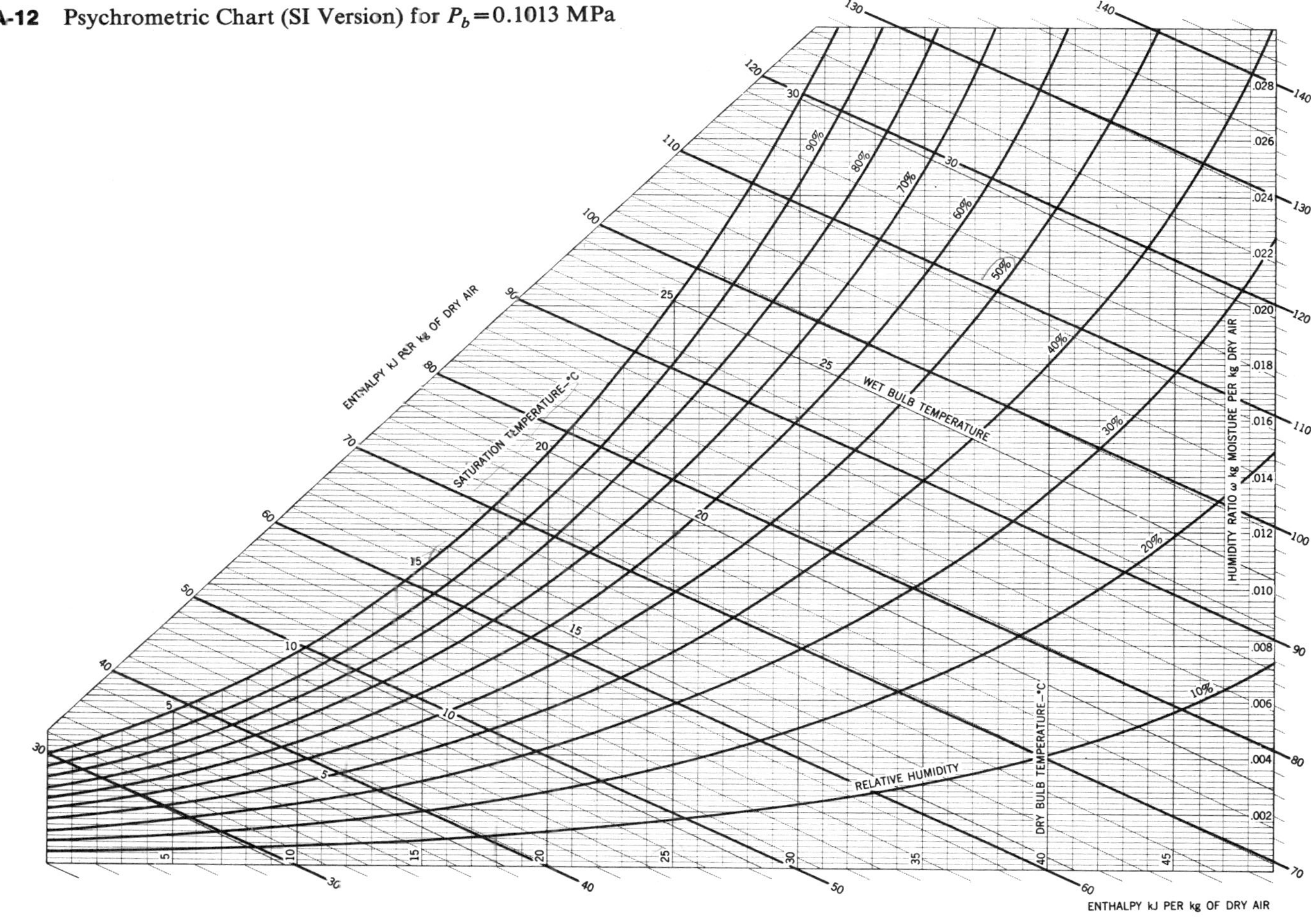

APPENDIX B
HEAT TRANSFER DATA

Table B-1 Properties of Air at Atmospheric Pressure

The values of μ, k, c_p, and P_r are not strongly pressure-dependent and may be used over a fairly wide range of pressures.

T(°K)	ρ (kg/m^3)	c_p (kJ/kg·Δ_1°C)	μ (Pa·s×10^5)	v (m^2/S ×10^6)	k (W/m·°C)	α_d (m^2/S ×10^4)	Pr (−)	α* [(1/m^3·°C)×10^{-6}]
100	3.6010	1.0266	0.6924	1.923	0.009246	0.02501	0.770	20396.
150	2.3675	1.0099	1.0283	4.343	0.013735	0.05745	0.753	2622
200	1.7684	1.0061	1.3289	7.490	0.01809	0.10165	0.739	641.8
250	1.4128	1.0053	1.488	9.49	0.02227	0.13161	0.722	282.9
300	1.1774	1.0057	1.983	15.68	0.02624	0.22160	0.708	87.61
350	0.9980	1.0090	2.075	20.76	0.03003	0.2983	0.697	45.24
400	0.8826	1.0140	2.286	25.90	0.03365	0.3760	0.689	25.13
450	0.7833	1.0207	2.484	28.86	0.03707	0.4222	0.683	16.31
500	0.7048	1.0295	2.671	37.90	0.04038	0.5564	0.680	9.286
550	0.6423	1.0392	2.848	44.34	0.04360	0.6532	0.680	6.182
600	0.5879	1.0551	3.018	51.34	0.04659	0.7512	0.680	4.234
650	0.5430	1.0635	3.177	58.51	0.04953	0.8578	0.682	3.009
700	0.5030	1.0752	3.332	66.25	0.05230	0.9672	0.684	2.188
750	0.4709	1.0856	3.481	73.91	0.05509	1.0774	0.686	1.638
800	0.4405	1.0978	3.625	82.29	0.05779	1.1951	0.689	1.246
850	0.4149	1.1095	3.765	90.75	0.06028	1.3097	0.692	0.9734
900	0.3925	1.1212	3.899	99.3	0.06279	1.4271	0.696	0.7674
950	0.3716	1.1321	4.023	108.2	0.06525	1.5510	0.699	0.6132
1000	0.3524	1.1417	4.152	117.8	0.06752	1.6779	0.702	0.4961
1100	0.3204	1.160	4.44	138.6	0.0732	1.969	0.704	0.3270
1200	0.2947	1.179	4.69	159.1	0.0782	2.251	0.707	0.2272
1300	0.2707	1.197	4.93	182.1	0.0837	2.583	0.705	0.1605
1400	0.2515	1.214	5.17	205.5	0.0891	2.920	0.705	0.1160
1500	0.2355	1.230	5.40	229.1	0.0946	3.262	0.705	0.0878
1600	0.2211	1.248	5.63	254.5	0.100	3.609	0.705	0.0672
1700	0.2082	1.267	5.85	280.5	0.105	3.977	0.705	0.0518
1800	0.1970	1.287	6.07	308.1	0.111	4.379	0.704	0.0407
1900	0.1858	1.309	6.29	338.5	0.117	4.811	0.704	0.0319
2000	0.1762	1.338	6.50	369.0	0.124	5.260	0.702	0.0253
2100	0.1682	1.372	6.72	399.6	0.131	5.715	0.700	0.0206
2200	0.1602	1.419	6.93	432.6	0.139	6.120	0.707	0.0167
2300	0.1538	1.482	7.14	464.0	0.149	6.540	0.710	0.0139
2400	0.1458	1.574	7.35	504.0	0.161	7.020	0.718	0.0116
2500	0.1394	1.688	7.57	543.5	0.175	7.441	0.730	0.0097

Source: From *Natl. Bur Stand. (U.S.) Circ. 564*, 1955.

$$^*\alpha = \left[\frac{g\beta\rho^2 C_p}{\mu k}\right]\left[\frac{1}{\text{m}^3\,°\text{C}}\right]$$

Table B-2 Properties of Gases at Atmospheric Pressure

Values of μ, k, C_p, and P_r are not strongly pressure-dependent for He, H_2, O_2, and N_2, and may be used over a fairly wide range of pressures.

T, °K	ρ (kg/m^3)	C_p (kJ/ kg·Δ_1°C)	μ (Pa·s)	ν (m^2/s)	k, (W/m·Δ_1°C)	α_d (m^2/s)	Pr (−)
				Helium			
144	0.3379	5.200	125.5×10^{-7}	37.11×10^{-6}	0.0928	0.5275×10^{-4}	0.70
200	0.2435	5.200	156.6	64.38	0.1177	0.9288	0.694
255	0.1906	5.200	181.7	95.50	0.1357	1.3675	0.70
366	0.13280	5.200	230.5	173.6	0.1691	2.449	0.71
477	0.10204	5.200	275.0	269.3	0.197	3.716	0.72
589	0.08282	5.200	311.3	375.8	0.225	5.215	0.72
700	0.07032	5.200	347.5	494.2	0.251	6.661	0.72
800	0.06023	5.200	381.7	634.1	0.275	8.774	0.72
				Hydrogen			
150	0.16371	12.602	5.595×10^{-6}	34.18×10^{-5}	0.0981	0.475×10^{-4}	0.718
200	0.12270	13.540	6.813	55.53	0.1282	0.772	0.719
250	0.09819	14.059	7.919	80.64	0.1561	1.130	0.713
300	0.08185	14.314	8.963	109.5	0.182	1.554	0.706
350	0.07016	14.436	9.954	141.9	0.206	2.031	0.697
400	0.06135	14.491	10.864	177.1	0.228	2.568	0.690
450	0.05462	14.499	11.779	215.6	0.251	3.164	0.682
500	0.04918	14.507	12.636	257.0	0.272	3.817	0.675
550	0.04469	14.532	13.475	301.6	0.292	4.516	0.668
600	0.04085	14.537	14.285	349.7	0.315	5.306	0.664
700	0.03492	14.574	15.89	455.1	0.351	6.903	0.659
800	0.03060	14.675	17.40	569.	0.384	8.563	0.664
900	0.02723	14.821	18.78	690.	0.412	10.217	0.676
				Oxygen			
150	2.6190	0.9178	11.490×10^{-6}	4.387×10^{-6}	0.01367	0.05688×10^{-4}	0.773
200	1.9559	0.9131	14.850	7.593	0.01824	0.10214	0.745
250	1.5618	0.9157	17.87	11.45	0.02259	0.15794	0.725
300	1.3007	0.9203	20.63	15.86	0.02676	0.22353	0.709
350	1.1133	0.9291	23.16	20.80	0.03070	0.2968	0.702
400	0.9755	0.9420	25.54	26.18	0.03461	0.3768	0.695
450	0.8682	0.9567	27.77	31.99	0.03828	0.4609	0.694
500	0.7801	0.9722	29.91	38.34	0.04173	0.5502	0.697
550	0.7096	0.9881	31.97	45.05	0.04517	0.6441	0.700
				Nitrogen			
200	1.7108	1.0429	12.947×10^{-6}	7.568×10^{-6}	0.01824	0.10224×10^{-4}	0.747
300	1.1421	1.0408	17.84	15.63	0.02620	0.22044	0.713
400	0.8538	1.0459	21.98	25.74	0.03335	0.3734	0.691
500	0.6824	1.0555	25.70	37.66	0.03984	0.5530	0.684
600	0.5687	1.0756	29.11	51.19	0.04580	0.7486	0.686
700	0.4934	1.0969	32.13	65.13	0.05123	0.9466	0.691
800	0.4277	1.1225	34.84	81.46	0.05609	1.1685	0.700
900	0.3796	1.1464	37.49	91.06	0.06070	1.3946	0.711
1000	0.3412	1.1677	40.00	117.2	0.06475	1.6250	0.724
1100	0.3108	1.1857	42.28	136.0	0.06850	1.8591	0.736
1200	0.2851	1.2037	44.50	156.1	0.07184	2.0932	0.748

Table B-2 (Continued)

T, °K	ρ (kg/m^3)	C_p (kJ/kg·Δ_1°C)	μ (Pa·s)	ν (m^2/s)	k, (W/m·Δ_1°C)	α_d (m^2/s)	Pr (−)
			Carbon Dioxide				
220	2.4733	0.783	11.105×10^{-6}	4.490×10^{-6}	0.010805	0.05920×10^{-4}	0.818
250	2.1657	0.804	12.590	5.813	0.012884	0.07401	0.793
300	1.7973	0.871	14.958	8.321	0.016572	0.10588	0.770
350	1.5362	0.900	17.205	11.19	0.02047	0.14808	0.755
400	1.3424	0.942	19.32	14.39	0.02461	0.19463	0.738
450	1.1918	0.980	21.34	17.90	0.02897	0.24813	0.721
500	1.0732	1.013	23.26	21.67	0.03352	0.3084	0.702
550	0.9739	1.047	25.08	25.74	0.03821	0.3750	0.685
600	0.8938	1.076	26.83	30.02	0.04311	0.4483	0.668
			Ammonia, NH_3				
273	0.7929	2.177	9.353×10^{-6}	1.18×10^{-5}	0.0220	0.1308×10^{-4}	0.90
323	0.6487	2.177	11.035	1.70	0.0270	0.1920	0.88
373	0.5590	2.236	12.886	2.30	0.0327	0.2619	0.87
423	0.4934	2.315	14.672	2.97	0.0391	0.3432	0.87
473	0.4405	2.395	16.49	3.74	0.0467	0.4421	0.84
			Water Vapor				
380	0.5863	2.060	12.71×10^{-6}	2.16×10^{-5}	0.0246	0.2036×10^{-4}	1.060
400	0.5542	2.014	13.44	2.42	0.0261	0.2338	1.040
450	0.4902	1.980	15.25	3.11	0.0299	0.307	1.010
500	0.4405	1.985	17.04	3.86	0.0339	0.387	0.996
550	0.4005	1.997	18.84	4.70	0.0379	0.475	0.991
600	0.3652	2.026	20.67	5.66	0.0422	0.573	0.986
650	0.3380	2.056	22.47	6.64	0.0464	0.666	0.995
700	0.3140	2.085	24.26	7.72	0.0505	0.772	1.000
750	0.2931	2.119	26.04	8.88	0.0549	0.883	1.005
800	0.2739	2.152	27.86	10.20	0.0592	1.001	1.010
850	0.2579	2.186	29.69	11.52	0.0637	1.130	1.019

Source: Adapted to SI units from E. R. G. Eckert and R. M. Drake, *Heat and Mass Transfer*, 2nd ed., McGraw-Hill Book Company, New York, 1959.

Table B-3 Properties of Water (Saturated Liquid)

°F	°C	c_p (kJ/kg·Δ_1°C)	ρ (kg/m³)	μ (kg/m·s)	k (W/m·Δ_1°C)	Pr (−)	$\left[\frac{g\beta\rho^2 C_p}{\mu k}\right] = \alpha$ (1/m³·Δ_1°C)
32	0.	4.225	999.8	1.79×10^{-3}	0.566	13.25	
40	4.44	4.208	999.8	1.55	0.575	11.35	1.91×10^{9}
50	10.	4.195	999.2	1.31	0.585	9.40	6.34×10^{9}
60	15.56	4.186	998.6	1.12	0.595	7.88	1.08×10^{10}
70	21.11	4.179	997.4	9.8×10^{-4}	0.604	6.78	1.46×10^{10}
80	26.67	4.179	995.8	8.6	0.614	5.85	1.91×10^{10}
90	32.22	4.174	994.9	7.65	0.623	5.12	2.48×10^{10}
100	37.78	4.174	993.0	6.82	0.630	4.53	3.3×10^{10}
110	43.33	4.174	990.6	6.16	0.637	4.04	4.19×10^{10}
120	48.89	4.174	988.8	5.62	0.644	3.64	4.89×10^{10}
130	54.44	4.179	985.7	5.13	0.649	3.30	5.66×10^{10}
140	60.	4.179	983.3	4.71	0.654	3.01	6.48×10^{10}
150	65.55	4.183	980.3	4.3	0.659	2.73	7.62×10^{10}
160	71.11	4.186	977.3	4.01	0.665	2.53	8.84×10^{10}
170	76.67	4.191	973.7	3.72	0.668	2.33	9.85×10^{10}
180	82.22	4.195	970.2	3.47	0.673	2.16	1.09×10^{11}
190	87.78	4.199	966.7	3.27	0.675	2.03	
200	93.33	4.204	963.2	3.06	0.678	1.90	
220	104.4	4.216	955.1	2.67	0.684	1.66	
240	115.6	4.229	946.7	2.44	0.685	1.51	
260	126.7	4.250	937.2	2.19	0.685	1.36	
280	137.8	4.271	928.1	1.98	0.685	1.24	
300	148.9	4.296	918.0	1.86	0.684	1.17	
350	176.7	4.371	890.4	1.57	0.677	1.02	
400	204.4	4.467	859.4	1.36	0.665	1.00	
450	232.2	4.585	825.7	1.20	0.646	0.85	
500	260.	4.731	785.2	1.07	0.616	0.83	
550	287.7	5.024	735.5	9.51×10^{-5}			
600	315.6	5.703	678.7	8.68			

Source: Adapted from A. I. Brown and S. M. Marco, *Introduction to Heat Transfer*, 3rd ed., McGraw-Hill Book Company, New York, 1958.

Table B-4 Properties of Saturated Liquids

t °C	ρ (kg/m³)	c_p (kJ/kg·Δ_1°C)	ν (m²/s)	k (W/m·Δ_1°C)	α_d (m²/s)	Pr (−)	β (°K⁻¹)
			Ammonia, NH_3				
−50	703.69	4.463	0.435×10^{-6}	0.547	1.742×10^{-7}	2.60	
−40	691.68	4.467	0.406	0.547	1.775	2.28	
−30	679.34	4.476	0.387	0.549	1.801	2.15	
−20	666.69	4.509	0.381	0.547	1.819	2.09	
−10	653.55	4.564	0.378	0.543	1.825	2.07	
0	640.10	4.635	0.373	0.540	1.819	2.05	
10	626.16	4.714	0.368	0.531	1.801	2.04	
20	611.75	4.798	0.359	0.521	1.775	2.02	2.45
30	596.37	4.890	0.349	0.507	1.742	2.01	$\times10^{-3}$
40	580.99	4.999	0.340	0.493	1.701	2.00	
50	564.33	5.116	0.330	0.476	1.654	1.99	

Table B-4 Properties of Saturated Liquids (Continued)

t °C	ρ (kg/m^3)	C_p (kJ/kg·Δ_1°C)	ν (m^2/s)	k, (W/m·Δ_1°C)	α_d (m^2/s)	Pr (–)	β (°K^{-1})
			Carbon Dioxide, CO_2				
−50	1,156.34	1.84	0.119×10^{-6}	0.0855	0.4021×10^{-7}	2.96	
−40	1,117.77	1.88	0.118	0.1011	0.4810	2.46	
−30	1,076.76	1.97	0.117	0.1116	0.5272	2.22	
−20	1,032.39	2.05	0.115	0.1151	0.5445	2.12	
−10	983.38	2.18	0.113	0.1099	0.5133	2.20	
0	926.99	2.47	0.108	0.1045	0.4578	2.38	
10	860.03	3.14	0.101	0.0971	0.3608	2.80	
20	772.57	5.0	0.091	0.0872	0.2219	4.10	14.00
30	597.81	36.4	0.080	0.0703	0.0279	28.7	$\times10^{-3}$
			Sulfur Dioxide, SO_2				
−50	1,560.84	1.3595	0.484×10^{-6}	0.242	1.141×10^{-7}	4.24	
−40	1,536.81	1.3607	0.424	0.235	1.130	3.74	
−30	1,520.64	1.3616	0.371	0.230	1.117	3.31	
−20	1,488.60	1.3624	0.324	0.225	1.107	2.93	
−10	1,463.61	1.3628	0.288	0.218	1.097	2.62	
0	1,438.46	1.3636	0.257	0.211	1.081	2.38	
10	1,412.51	1.3645	0.232	0.204	1.066	2.18	
20	1,386.40	1.3653	0.210	0.199	1.050	2.00	1.94
30	1,359.33	1.3662	0.190	0.192	1.035	1.83	$\times10^{-3}$
40	1,329.22	1.3674	0.173	0.185	1.019	1.70	
50	1,299.10	1.3683	0.162	0.177	0.999	1.61	
			Dichlorodifluoromethane (Freon), CCl_2F_2				
−50	1,546.75	0.8750	0.310×10^{-6}	0.067	0.501×10^{-7}	6.2	2.63
−40	1,518.71	0.8847	0.279	0.069	0.514	5.4	$\times10^{-3}$
−30	1,489.56	0.8956	0.253	0.069	0.526	4.8	
−20	1,460.57	0.9073	0.235	0.071	0.539	4.4	
−10	1,429.49	0.9203	0.221	0.073	0.550	4.0	
0	1,397.45	0.9345	0.214	0.073	0.557	3.8	
10	1,364.30	0.9496	0.203	0.073	0.560	3.6	
20	1,330.18	0.9659	0.198	0.073	0.560	3.5	
30	1,295.10	0.9835	0.194	0.071	0.560	3.5	
40	1,257.13	1.0019	0.191	0.169	0.555	3.5	
50	1,215.96	1.0216	0.190	0.067	0.545	3.5	
			Glycerin, $C_3H_5(OH)_3$				
0	1,276.03	2.261	0.00831	0.282	0.983×10^{-7}	84.7×10^3	
10	1,270.11	2.319	0.00300	0.284	0.965	31.0	
20	1,264.02	2.386	0.00118	0.286	0.947	12.5	0.50
30	1,258.09	2.445	0.00050	0.286	0.929	5.38	$\times10^{-3}$
40	1,252.01	2.512	0.00022	0.286	0.914	2.45	
50	1,244.96	2.583	0.00015	0.287	0.893	1.63	
			Ethylene Glycol, $C_2H_4(OH)_2$				
0	1,130.75	2.294	57.53×10^{-6}	0.242	0.934×10^{-7}	615	
20	1,116.65	2.382	19.18	0.249	0.939	204	0.65
40	1,101.43	2.474	8.69	0.256	0.939	93	$\times10^{-3}$
60	1,087.66	2.562	4.75	0.260	0.932	51	
80	1,077.56	2.650	2.98	0.261	0.921	32.4	
100	1,058.50	2.742	2.03	0.263	0.908	22.4	

Table B-4 (Continued)

t °C	ρ (kg/m^3)	c_p (kJ/kg·Δ_1°C)	ν (m^2/s)	k (W/m·Δ_1°C)	α_d (m^2/s)	Pr (−)	β (°K^{-1})
			Engine Oil (Unused)				
0	899.12	1.796	0.00428	0.147	0.911×10^{-7}	47,100	
20	888.23	1.880	0.00090	0.145	0.872	10,400	0.70
40	876.05	1.964	0.00024	0.144	0.834	2,870	$\times10^{-3}$
60	864.04	2.047	0.839×10^{-4}	0.140	0.800	1,050	
80	852.02	2.131	0.375	0.138	0.769	490	
100	840.01	2.219	0.203	0.137	0.738	276	
120	828.96	2.307	0.124	0.135	0.710	175	
140	816.94	2.395	0.080	0.133	0.686	116	
160	805.89	2.483	0.056	0.132	0.663	84	
			Mercury, Hg				
0	13,628.22	0.1403	0.124×10^{-6}	8.20	42.99×10^{-7}	0.0288	
20	13,579.04	0.1394	0.114	8.69	46.06	0.0249	1.82
50	13,505.84	0.1386	0.104	9.40	50.22	0.0207	$\times10^{-4}$
100	13,384.58	0.1373	0.0928	10.51	57.16	0.0162	
150	13,264.28	0.1365	0.0853	11.49	63.54	0.0134	
200	13,144.94	0.1570	0.0802	12.34	69.08	0.0116	
250	13,025.60	0.1357	0.0765	13.07	74.06	0.0103	
315.5	12,847.	0.134	0.0673	14.02	81.5	0.0083	

Source: Adapted to SI Units from E. R. G. Eckert and R. M. Drake, *Heat and Mass Transfer*, 2nd ed., McGraw-Hill Book Company, New York, 1959.

Table B-5. Property Values for Metals

Nickel steel														
Ni ≈ 0%	7,897	0.452	73	2.026										
20%	7,933	0.46	19	0.526										
40%	8,169	0.46	10	0.279										
80%	8,618	0.46	35	0.872										
Invar 36% Ni	8,137	0.46	10.7	0.286										
Chrome steel														
Cr = 0%	7,897	0.452	73	2.026	87	73	67	62	55	48	40	36	35	36
1%	7,865	0.46	61	1.665		62	55	52	47	42	36	33	33	
5%	7,833	0.46	40	1.110		40	38	36	36	33	29	29	29	
20%	7,689	0.46	22	0.635		22	22	22	22	24	24	26	29	
Cr-Ni (chrome-nickel): 15% Cr, 10% Ni	7,865	0.46	19	0.526										
18% Cr, 8% Ni (V2A)	7,817	0.46	16.3	0.444		16.3	17	17	19	19	22	26	31	
20% Cr, 15% Ni	7,833	0.46	15.1	0.415										
25% Cr, 20% Ni	7,865	0.46	12.8	0.361										
Tungsten steel														
W = 0%	7,897	0.452	73	2.026										
1%	7,913	0.448	66	1.858										
5%	8,073	0.435	54	1.525										
10%	8,314	0.419	48	1.391										
Copper:														
Pure	8,954	0.3831	386	11.234	407	386	379	374	369	363	353			
Aluminum bronze 95% Cu, 5%Al	8,666	0.410	83	2.330										
Bronze 75% Cu, 25% Sn	8,666	0.343	26	0.859										
Red brass 85% Cu, 9% Sn, 6% Zn	8,714	0.385	61	1.804		59	71							

Table B-5 (Continued)

Metal	Properties at 20°C				Thermal conductivity k, W/m·°C									
	ρ, kg/m^3	c_p, kJ/kg·°C	k, W/m·°C	α, m^2/s × 10^5	−100°C −148°F	0°C 32°F	100°C 212°F	200°C 392°F	300°C 572°F	400°C 752°F	600°C 1112°F	800°C 1472°F	1000°C 1832°F	1200°C 2192°F
Aluminum:														
Pure	2,707	0.896	204	8.418	215	202	206	215	228	249				
Al-Cu (Duralumin) 94–96% Al, 3–5% Cu, trace Mg	2,787	0.883	164	6.676	126	159	182	194						
Al-Si (Silumin, copper-bearing) 86.5% Al, 1% Cu	2,659	0.867	137	5.933	119	137	144	152	161					
Al-Si (Alusil) 78–80% Al, 20–22% Si	2,627	0.854	161	7.172	144	157	168	175	178					
Al-Mg-Si 97% Al, 1% Mg, 1% Si, 1% Mn	2,707	0.892	177	7.311		175	189	204						
Lead	11,373	0.130	35	2.343	36.9	35.1	33.4	31.5	29.8					
Iron:														
Pure	7,897	0.452	73	2.034	87	73	67	62	55	48	40	36	35	36
Wrought iron 0.5% C	7,849	0.46	59	1.626		59	57	52	48	45	36	33	33	33
Steel (C max ≈ 1.5%):														
Carbon steel														
C ≈ 0.5%	7,833	0.465	54	1.474		55	52	48	45	42	35	31	29	31
1.0%	7,801	0.473	43	1.172		43	43	42	40	36	33	29	28	29
1.5%	7,753	0.486	36	0.970		36	36	36	35	33	31	28	28	29

Table B-5 (Continued)

Metal	Properties at 20°C				Thermal Conductivity k (W/m·Δ, °C)									
	ρ (kg/m^3)	c_p (kJ/kg·°C)	k (W/m·°C)	α (m^2/s ×10^5)	−100°C −148°F	0°C 32°F	100°C 212°F	200°C 392°F	300°C 572°F	400°C 752°F	600°C 1112°F	800°C 1472°F	1000°C 1832°F	1200°C 2192°F
Brass, 70% Cu, 30% Zn	8,522	0.385	111	3.412	88		128	144	147	147				
German silver 62% Cu, 15% Ni, 22% Zn	8,618	0.394	24.9	0.733	19.2		31	40	45	48				
Constantan 60% Cu, 40% Ni	8,922	0.410	22.7	0.612	21		22.2	26						
Magnesium:														
Pure	1,746	1.013	171	9.708	178	171	168	163	157					
Mg-Al (Electrolytic) 6–8% Al, 1–2% Zn	1,810	1.00	66	3.605		52	62	74	83					
Molybdenum	10,220	0.251	123	4.790	138	125	118	114	111	109	106	102	99	92
Nickel:														
Pure (99.9%)	8,906	0.4459	90	2.266	104	93	83	73	64	59				
Ni-Cr 90% Ni, 10% Cr	8,666	0.444	17	0.444		17.1	18.9	20.9	22.8	24.6				
80% Ni, 20% Cr	8,314	0.444	12.6	0.343		12.3	13.8	15.6	17.1	18.0	22.5			
Silver:														
Purest	10,524	0.2340	419	17.004	419	417	415	412						
Pure (99.9%)	10,524	0.2340	407	16.563	419	410	415	374	362	360				
Tin, pure	7,304	0.2265	64	3.884	74	65.9	59	57						
Tungsten	19,350	0.1344	163	6.271		166	151	142	133	126	112	76		
Zinc, pure	7,144	0.3843	112.2	4.106	114	112	109	106	100	93				
Gold, pure	19,300	0.1289	315	12.662	331	318	312	310	305	299	286			

Table B-6 Physical Properties of Some Common Low-Melting-Point Metals

Metal	Melting point, °C	Normal boiling point, °C	Temperature, °C	Density, kg/m³ × 10⁻³	Viscosity, kg/m·s × 10³	Heat capacity, kJ/kg·°C	Thermal conductivity, W/m·°C	Prandtl number
Bismuth	271	1477	316	10.01	1.62	0.144	16.4	0.014
			760	9.47	0.79	0.165	15.6	0.0084
Lead	327	1737	371	10.5	2.40	0.159	16.1	0.024
			704	10.1	1.37	0.155	14.9	0.016
Lithium	179	1317	204	0.51	0.60	4.19	38.1	0.065
			982	0.44	0.42	4.19		
Mercury	−39	357	10	13.6	1.59	0.138	8.1	0.027
			316	12.8	0.86	0.134	14.0	0.0084
Potassium	63.8	760	149	0.81	0.37	0.796	45.0	0.0066
			704	0.67	0.14	0.754	33.1	0.0031
Sodium	97.8	883	204	0.90	0.43	1.34	80.3	0.0072
			704	0.78	0.18	1.26	59.7	0.0038
Sodium potassium:								
22% Na	19	826	93.3	0.848	0.49	0.946	24.4	0.019
			760	0.69	0.146	0.883		
56% Na	−11	784	93.3	0.89	0.58	1.13	25.6	0.026
			760	0.74	0.16	1.04	28.9	0.058
Lead bismuth, 44.5% Pb	125	1670	288	10.3	1.76	0.147	10.7	0.024
			649	9.84	1.15			

Table B-7 Properties of Nonmetals

Substance	Temperature, °C	k, W/m·°C	ρ, kg/m³	C, kJ/kg·°C	α, m²/s × 10^7
Structural and heat-resistant materials					
Asphalt	20–55	0.74–0.76			
Brick:					
Building brick, common	20	0.69	1600	0.84	5.2
Face		1.32	2000		
Carborundum brick	600	18.5			
	1400	11.1			
Chrome brick	200	2.32	3000	0.84	9.2
	550	2.47			9.8
	900	1.99			7.9
Diatomaceous earth, molded and fired	200	0.24			
	870	0.31			
Fireclay brick, burnt 2426°F	500	1.04	2000	0.96	5.4
	800	1.07			
	1100	1.09			
Burnt 2642°F	500	1.28	2300	0.96	5.8
	800	1.37			
	1100	1.40			
Missouri	200	1.00	2600	0.96	4.0
	600	1.47			
	1400	1.77			
Magnesite	200	3.81		1.13	
	650	2.77			
	1200	1.90			
Cement, portland		0.29	1500		
Mortar	23	1.16			
Concrete, cinder	23	0.76			
Stone 1-2-4 mix	20	1.37	1900–2300	0.88	8.2–6.8
Glass, window	20	0.78 (avg)	2700	0.84	3.4
Corosilicate	30–75	1.09	2200		
Plaster, gypsum	20	0.48	1440	0.84	4.0
Metal lath	20	0.47			
Wood lath	20	0.28			
Stone:					
Granite		1.73–3.98	2640	0.82	8–18
Limestone	100–300	1.26–1.33	2500	0.90	5.6–5.9
Marble		2.07–2.94	2500–2700	0.80	10–13.6
Sandstone	40	1.83	2160–2300	0.71	11.2–11.9
Wood (across the grain):					
Balsa, 8.8 lb/ft³	30	0.055	140		
Cypress	30	0.097	460		
Fir	23	0.11	420	2.72	0.96
Maple or oak	30	0.166	540	2.4	1.28
Yellow pine	23	0.147	640	2.8	0.82
White pine	30	0.112	430		

Table B-7 (continued).

Substance	Temperature, °C	k, W/m·°C	ρ, kg/m³	c_p, kJ/kg·°C	α, m²/s $\times 10^7$
Insulating material					
Asbestos:					
Loosely packed	−45	0.149			
	0	0.154	470–570	0.816	3.3–4
	100	0.161			
Asbestos-cement boards	20	0.74			
Sheets	51	0.166			
Felt, 40 laminations/in	38	0.057			
	150	0.069			
	260	0.083			
20 laminations/in	38	0.078			
	150	0.095			
	260	0.112			
Corrugated, 4 plies/in	38	0.087			
	93	0.100			
	150	0.119			
Asbestos cement	...	2.08			
Balsam wool, 2.2 lb/ft³	32	0.04	35		
Cardboard, corrugated	...	0.064			
Celotex	32	0.048			
Corkboard, 10 lb/ft³	30	0.043	160		
Cork, regranulated	32	0.045	45–120	1.88	2–5.3
Ground	32	0.043	150		
Diatomaceous earth (Sil-o-cel)	0	0.061	320		
Felt, hair	30	0.036	130–200		
Wool	30	0.052	330		
Fiber, insulating board	20	0.048	240		
Glass wool, 1.5 lb/ft³	23	0.038	24	0.7	22.6
Insulex, dry	32	0.064			
		0.144			
Kapok	30	0.035			
Magnesia, 85%	38	0.067	270		
	93	0.071			
	150	0.074			
	204	0.080			
Rock wool, 10 lb/ft³	32	0.040	160		
Loosely packed	150	0.067	64		
	260	0.087			
Sawdust	23	0.059			
Silica aerogel	32	0.024	140		
Wood shavings	23	0.059			

Table B-8 Normal Total Emissivity of Various Surfaces[†]

Surface	T, °C	Emissivity ϵ, [−]
Metals and their oxides		
Aluminum		
Highly polished plate, 98.3% pure	227–577	0.039–0.057
Commercial sheet	100	0.09
Heavily oxidized	148–504	0.20–0.31
Al-surfaced roofing	43	0.216
Brass:		
Highly polished:		
73.2% Cu, 26.7% Zn	890–1250	0.028–0.031
62.4% Cu, 36.8% Zn, 0.4% Pb, 0.3% Al	920–377	0.033–0.037
82.9% Cu, 17.0% Zn	277	0.030
Hard-rolled, polished, but direction of polishing visible	21	0.038
Dull plate	50–350	0.22
Chromium (see nickel alloys for Ni-Cr steels), polished	38–1093	0.08–0.36
Copper:		
Polished	117	0.023
	100	0.052
Plate, heated long time, covered with thick oxide layer	25	0.78
Gold, pure, highly polished	227–632	0.018–0.035
Iron and steel (not including stainless):		
Steel, polished	100	0.066
Iron, polished	427–1027	0.14–0.38
Cast iron, newly turned	22	0.44
turned and heated	882–987	0.60–0.70
Mild steel; A	232–1067	0.20–0.32
Oxidized surfaces:		
Iron plate, pickled, then rusted red	20	0.61
Iron, dark-gray surface	100	0.31
Rough ingot iron	927–1117	0.87–0.95
Sheet steel with strong, rough oxide layer	24	0.80
Lead:		
Unoxidized, 99.96% pure	127–227	0.057–0.075
Gray oxidized	24	0.28
Oxidized at 300°F	197	0.63
Magnesium, magnesium oxide	277–827	0.55–0.20
Molybdenum:		
Filament	727–2600	0.096–0.202
Massive, polished	100	0.071
Monel metal, oxidized at 1110°F	197–597	0.41–0.46
Nickel:		
Polished	100	0.072
Nickel oxide	647–1257	0.59–0.86
Nickel alloys:		
Copper nickel, polished	100	0.059
Nichrome wire, bright	50–997	0.65–0.79
Nichrome wire, oxidized	50–497	0.95–0.98
Platinum; polished plate, pure	227–627	0.054–0.104
Silver:		
Polished, pure	227–627	0.020–0.032
Polished	38–371	0.022–0.031
Stainless steels:		
Poished	100	0.074
Type 301; B	232–941	0.54–0.63
Tin, bright tinned iron	24	0.043–0.064

Table B-8 Normal Total Emissivity of Various Surfaces† (Continued)

Surface	T, °C	Emissivity ϵ, [–]
Tungsten, filament	3317	0.39
Zinc, galvanized sheet iron, fairly bright	28	0.23
Refractories, building materials, paints, and miscellaneous		
Alumina (85–99.5% Al_2O_3, 0–12% SiO_2, 0–1% Ge_2O_3); effect of mean grain size, microns (μm):		
10μm		0.30–0.18
50μm		0.39–0.28
100 μm		0.50–0.40
Asbestos, board	23	0.96
Brick:		
Red, rough, but no gross irregularities	21	0.93
Fireclay	1000	0.75
Carbon:		
T-carbon (Gebrüder Siemens) 0.9% ash, started with emissivity of 0.72 at 260°F but on heating changed to values given	127–627	0.81–0.79
Filament	1037–1407	0.526
Rough plate	100–320	0.77
Lampblack, rough deposit	100–500	0.84–0.78
Concrete tiles	1000	0.63
Enamel, white fused, on iron	19	0.90
Glass:		
Smooth	22	0.94
Pyrex, lead, and soda	267–537	0.95–0.85
Paints, lacquers, varnishes:		
Snow-white enamel varnish on rough iron plate	23	0.906
Black shiny lacquer, sprayed on iron	24	0.875
Black shiny shellac on tinned iron sheet	21	0.821
Black matte shellac	77–147	0.91
Black or white lacquer	37–93	0.80–0.95
Flat black lacquer	37–93	0.96–0.98
Paints, lacquers, varnishes:		
Aluminum paints and lacquers:		
10% Al, 22% lacquer body, on rough or smooth surface	100	0.52
Other Al paints, varying age and Al content	100	0.27–0.67
Porcelain, glazed	22	0.92
Quartz, rough, fused	21	0.93
Roofing paper	21	0.91
Rubber, hard, glossy plate	23	0.94
Water	0–100	0.95–0.963

†Courtesy of H. C. Hottel, from W. H. McAdams, "Heat Transmission," 3d ed., McGraw-Hill Book Company, New York, 1954.

APPENDIX C
CONVERSION FACTORS (EQUATIONS OF EQUALITY)

Table C-1. Equations of Equality.

C-1.1 Dimensions

Length (L)

1 in. = 0.0254 m
1 in. = 2.540 cm
1 ft = 0.3048 m
1 mi = 1.609 Km

Area (L^2)

1 in^2 = 0.64516×10^{-3} m^2
1 in^2 = 6.4516 cm^2
1 ft^2 = 0.09290 m^2
1 m^2 = 10,000 cm^2

Volume (L^3)

1 in^3 = 16.387 cm^3
1 ft^3 = 0.0283 m^3
1 gal = 3.7854 liters
1 gal = 0.0037854 m^3
1 m^3 = 1,000,000 cm^3
1 m^3 = 1000 liters
1 liter = 1000 cm^3

C-1.2 Fluid Flow, Mass, Density, and Pressure

Flow Velocity (V)

1 ft/sec = 0.3048 m/s
1 ft/sec = 1097 m/hr
1 mi/hr = 1.609 km/hr

Flow Rates—Volume

1 ft^3/sec = 0.0283 m^3/s
1 ft^3/sec = 101.88 m^3/hr
1 gal/min = 0.2271 m^3/hr

Flow Rates—Mass

1 lb_m/sec = 0.4536 kg/s
1 lb_m/sec = 1632.96 kg/hr
1 lb_m/hr = 0.000126 kg/s

Mass

1 lb_m = 0.4536 kg
1 metric ton = 1000 kg

Force

1 lb_f = 4.4482 N

Density

1 lb_m/ft^3 = 16.0185 kg/m^3
1 lb_m/ft^3 = 0.016019 gm/cm^3

Pressure

1 lb_f/in^2 = 6894.76 Pa
1 lb_f/ft^2 = 47.88 Pa
1 atm = 101352 Pa = 0.10135 MPa

Viscosity—Absolute (μ)

1 lb_m/ft sec = 1.488 Pa·s
1 lb_f sec/ft^2 = 47.88 Pa·s
1 centripoise = 1×10^{-3} Pa·s

Viscosity—Kinematic (ν)

1 ft^2/sec = 0.0929 m^2/s
1 centistoke – 1×10^{-6} m^2/s

C-1.3 Latent and Specific Heats, and Temperature

Latent Heat

1 BTU/lb = 2.3254 J/gm = 2.324 kJ/kg
1 gm cal/gm = 4.1865 J/gm

Specific Heat

1 BTU/lb Δ_1°F = 4186 J/kg Δ_1°C
1 gm cal/gm Δ_1°C = 4186 J/kg Δ_1°C

APPENDIX D
INTERPOLATION AND PROBLEM SOLVING

D-1 LINEAR INTERPOLATION

When values of parameters are sought that are between values listed in thermodynamic tables, interpolation is used. In most cases a linear (or "straight line") interpolation is used, as illustrated in the following examples.

PROBLEM 1

What is the saturation temperature T_s of steam at a pressure of 0.9384 kPa?

An inspection of the steam tables (Appendix A, Table A-1.1) shows that the temperature will be between 5 and 10°C. To determine the temperature, the following steps must be taken:

Step 1
Write the equation you plan to use, such as:

$$T_s = 5 + \Delta T \qquad [°\mathrm{C}] \tag{D-1}$$

$$\Delta T = \left(\frac{\Delta p}{\Delta p^1}\right)(\Delta T') \qquad [\Delta °\mathrm{C}] \tag{D-2}$$

Step 2
Prepare a table of T and p.

T [°C]	p [kPa]	
5	0.8721	(steam table)
T_s	0.9384	(given)
10	1.2276	(steam table)

Step 3
Define and evaluate parameters.

$$\Delta p = p_s - p_5 = 0.9384 - 0.8721 = 0.0663 \quad [\text{kPa}]$$

$$\Delta p' = p_{10} - p_5 = 1.2276 - 0.8721 = 0.3555 \quad [\text{kPa}]$$

$$\Delta T' = 10 - 5 = 5 \quad [\Delta°\text{C}]$$

Step 4
Solve for ΔT (Eq. D-2).

$$\Delta T = \frac{(0.0663)}{(0.3555)}(5) \quad \frac{[\text{kPa}]}{[\text{kPa}]}[\Delta°\text{C}] = [\Delta°\text{C}]$$

$$= 0.94 \quad [\Delta°\text{C}]$$

Step 5
Solve for T_s (Eq. D-1).

$$T_s = 5 + 0.94 \quad [°\text{C}] + [\Delta°\text{C}] = [°\text{C}]$$

Answer

$$\mathbf{T_s = 5.94} \quad [°\text{C}]$$

PROBLEM 2

If steam expands to a pressure of 0.050 MPa at a specific entropy s value of 8.000, what is the specific enthalpy h?

Following the steps outlined in Problem 1, Problem 2 can be solved as follows: An inspection shows that the value of h will be between that for 150 and 200°C in Steam Table A-1.3. The problem can be solved by determining the temperature of the steam for $s = 8.000$ kJ/kg in the manner of Problem 1, and then in a similar manner determining the value of h for the determined temperature.

Instead of taking these two steps, Problem 2 can be interpolated directly between h and s as follows.

Step 1
Write the equation you plan to use such as:

$$h = 2780.1 + \Delta h \quad [\text{kJ/kg}] \tag{D-3}$$

$$\Delta h = \left(\frac{\Delta s}{\Delta s'}\right)(\Delta h') \quad [\text{kJ/kg}] \tag{D-4}$$

Step 2
Make a table of h and s.

T [°C]	h [kJ/kg]	s [kJ/kg·Δ_1K]	
150	2780.1	7.9401	(steam table)
	h	8.000	(given)
200	2877.7	8.1580	(steam table)

Step 3
Define and evaluate parameters.

$$\Delta s = 8.000 - 7.9401 = 0.0599 \qquad [\text{kJ/kg}\cdot\Delta_1\text{K}]$$

$$\Delta s' = 8.1580 - 7.9401 = 0.2179 \qquad [\text{kJ/kg}\cdot\Delta_1\text{K}]$$

$$\Delta h' = 2877.7 - 2780.1 = 97.6 \qquad [\text{kJ/kg}]$$

Step 4
Solve for Δh (Eq. D-4).

$$\Delta h = \frac{(0.0599)}{(0.2179)}(97.6) \qquad \frac{[\text{kJ/kg}\cdot\Delta_1\text{K}]}{[\text{kJ/kg}\cdot\Delta_1\text{K}]}[\text{kJ/kg}]$$

$$= 26.8 \qquad [\text{kJ/kg}]$$

Step 5
Solve for h (Eq. D-3).

$$h = 2780.1 + 26.8 \qquad [\text{kJ/kg}] + [\text{kJ/kg}]$$

Answer

$$\mathbf{h = 2806.9} \qquad [\textbf{kJ/kg}]$$

Cases may be encountered where a double interpolation must be made. An example of such a case is given in Problem 3 where the two parameters involved are between the listed values in the thermodynamic tables.

PROBLEM 3

What is the temperature of superheat steam with a specific entrophy s of 7.200 $\text{kJ/kg}\cdot\Delta_1\text{K}$ at a pressure of 0.36 MPa?

There are two approaches to this interpolation: One would be to determine the temperature at 0.30 and 0.40 MPa of steam with $s = 7.200$ $\text{kJ/kg}\cdot\Delta_1\text{K}$, and then interpolate the temperature for the 0.36 MPa pressure. The other would be to determine the value of s at 200 and 150°C at 0.36 MPa pressure, and then interpolate for the temperature for $s = 7.200$ $\text{kJ/kg}\cdot\Delta_1\text{K}$.

Either way will give the same answer so that the choice is a matter of personal preference.

If the second method is chosen, the interpolation can be accomplished as follows.

1. Determine the value of s at 200°C, 0.36 MPa.

(a) $s_{200} = 7.3115 - \Delta s \qquad [\text{kJ/kg}\cdot\Delta_1\text{K}]$

$$\Delta s = \frac{(\Delta p)}{(\Delta p')}(\Delta s')$$

(b)

p [MPa]	s [kJ/kg·Δ_1K]
0.30	7.3115
0.36	s_{200}
0.40	7.1706

(c) $\Delta p = 0.36 - 0.30 = 0.06$ MPa

$\Delta p' = 0.40 - 0.30 = 0.10$ MPa

$\Delta s' = 7.3115 - 7.1706 = 0.1409$ kJ/kg·Δ_1K

(d) $\Delta s = \dfrac{(0.06)}{(0.10)}(0.1409) \quad \dfrac{[\text{MPa}]}{[\text{MPa}]}[\text{kJ/kg}\cdot\Delta_1\text{K}]$

$= 0.0845 \quad [\text{kJ/kg}\cdot\text{K}]$

(e) $s_{200} = 7.3115 - 0.0845 \quad [\text{kJ/kg}\cdot\Delta_1\text{K}] - [\text{kJ/kg}\cdot\Delta_1\text{K}]$

$= 7.2270$ kJ/kg·Δ_1K

2. Determine the value of s at 150°C, 0.36 MPa.

(a) $s_{150} = 7.0778 - \Delta s \quad [\text{kJ/kg}\cdot\Delta_1\text{K}]$

$\Delta s = \dfrac{(\Delta p)}{(\Delta p')}(\Delta s')$

(b)

p [MPa]	s [kJ/kg·Δ_1K]
0.30	7.0778
0.36	s_{150}
0.40	6.9299

(c) $\Delta p = = 0.36 - 0.30 = 0.06 \quad$ [MPa]

$\Delta p' = 0.40 - 0.30 = 0.10 \quad$ [MPa]

$\Delta s' = 7.0778 - 6.9299 = 0.1479 \quad$ [kJ/kg·Δ_1K]

(d) $\Delta s = \dfrac{(0.06)}{(0.10)}(0.1479) \quad \dfrac{[\text{MPa}]}{[\text{MPa}]}[\text{kJ/kg}\cdot\Delta_1\text{K}]$

$= 0.0887 \quad [\text{kJ/kg}\cdot\text{K}]$

(e) $s_{150} = 7.0778 - 0.0887 \quad [\text{kJ/kg}\cdot\Delta_1\text{K}] - [\text{kJ/kg}\cdot\Delta_1\text{K}]$

$= 6.9891 \quad [\text{kJ/kg}\cdot\Delta_1\text{K}]$

3. Determine the Temperature at 0.36 MPa and $s=7.2000$ [kJ/kg·Δ_1K].

(a) $T=200-\Delta T$ [°C]

$$\Delta T=\frac{(\Delta s)}{(\Delta s')}(\Delta T') \qquad [\Delta°C]$$

(b)

T [°C]	s [kJ/kg·Δ_1K]
150	6.9891
T	7.2000
200	7.2270

(c) $\Delta s = 7.2270-7.2000=0.0270$ [kJ/kg·Δ_1K]

$\Delta s' = 7.2270-6.9891=0.2379$ [kJ/kg·Δ_1K]

$\Delta T' = 200-150=50\ \Delta°C$

(d) $\Delta T=\frac{(0.0270)}{(0.2379)}(50) \qquad \frac{[kJ/kg\cdot\Delta_1 K]}{[kJ/kg\cdot\Delta_1 K]}[\Delta°C]$

$=5.67\ \Delta°C$

(e) $T=200-5.67$ [°C]−[Δ°C]=[°C]

Answer

$T=194.33$ [°C]

D-2 AN APPROACH TO PROBLEM SOLVING

Problem solving in engineering technology or applied engineering courses is significantly different from the usual problem-solving methods in subject engineering courses. In most engineering subject courses, the student *knows* the formula involved for the particular aspect of the subject being studied. The usual scope of the problems constitutes manipulations of various permutations of the known equation parameters.

In an applied engineering course, such as Applied Thermodynamics and Heat Transfer (for which this book was written), there is no way of immediately knowing the sequence or equation(s) that will solve the problem. Therefore, beginning students in an applied course are perplexed and may feel lost because they do not immediately know the exact equation to use in the manner they have become accustomed to from previous experience with subject courses. Students are confronted with the need for visualizing and analyzing the problem and developing a workable plan for solving the problem, before the numerical solution of the problem is initiated.

A sequence of steps for solving applied thermodynamics and applied heat transfer problems is presented on the following page. This suggested sequence has evolved through a series of classes in these subjects at the California State Polytechnic University, Pomona, in the Engineering Technology Department, and has proved to be quite effective. The illustrative sample problems in this text have followed the outlined sequence and format.

Table D-2 An Approach to Problem Solving.

STEP 1—READ THE PROBLEM CAREFULLY

- Determine what the problem is about.
- Establish what is wanted.

STEP 2—IDENTIFY THE THERMODYNAMIC SYSTEM OR PROCESS(ES) INVOLVED

- Establish whether the system is OPEN or CLOSED.
- Make a diagram of the system indicating the STATE POINTS at the beginning and end of the processes involved and any work or heat involved with each process.

STEP 3—ORGANIZE A DETAILED SYSTEM DESCRIPTION (to the extent necessary)

- Make a TABLE OF STATE.
- List and identify each process involved.

STEP 4—DETERMINE OPTIONAL ROUTES FOR SOLUTION OF THE PROBLEM

- Review Steps 1, 2, and 3.
- Select candidate basic solution equations for obtaining the problem answer.
- Determine if and how the parameters required by the candidate equations can be evaluated.

STEP 5—ESTABLISH YOUR PLAN FOR SOLUTION OF THE PROBLEM

- Select the basic solution equation that is the most convenient one to use, and establish (list) your selected units for your answer.
- Mathematically solve the equation for the problem parameter(s) before using the equation.
- List the sequence of calculations and the equation for each required to evaluate the parameters of your selected basic solution equation, and indicate the units for each (in the order listed in the equation).

At the end of Step 5 you have "solved" the problem. You (or a computer) can start the numerical solution of the problem.

STEP 6—EVALUATION OF THE PARAMETERS OF YOUR BASIC SOLUTION EQUATION

- Evaluate the parameters required by your basic solution equation in the sequence outlined in Step 5.

- For each step in the outlined sequence:
  - Identify the calculation for later reference.
  - List the equation to be solved with units.
  - List equation parameter values, units, and sources in the same sequence as listed in the equation.
  - Substitute parameter numerical and unit values separately in the same sequential order as listed in the equation.
  - Solve separately for the parameter numerical value and units.

STEP 7—SOLVING YOUR BASIC SOLUTION EQUATION

- Substitute the parametric values obtained in Step 6 in your basic solution equation in the same sequential order as listed in your basic solution equation.
- Make a similar separate substitution of parameter units.
- Solve both the numerical and unit substitutions to obtain your answer.

STEP 8—CHECKING YOUR ANSWER

- Review your answer. Is it what the problem statement asked for?
- Does the numerical value seem to be what would reasonably be expected for the units indicated?

  Common sources of error include:
  - Erroneous (inverse) conversion to k (watts or whatever) resulting in the answer being off by a factor of 1,000,000.
  - A misplaced decimal point.
  - Errors in transposing numbers.
  - Misreading tables.
- Does the solution of the unit substitution give the answer units stated in Step 5?

  Common sources of errors include:
  - Errors in transposing units.
  - Not solving the unit substitution in the equation, and assuming the units are right.
  - Conversion factor errors (units inverse of what was intended).

STEP 9—CLEARLY LIST AND IDENTIFY YOUR ANSWER

- Numerical value and units.

D-3 SOLUTIONS BY ITERATION

When the values of the primary properties describing the state of the substance must be determined by iteration, two iterative steps combined with a graphical plot will suffice to determine the sought value. The coordinates of the graphical plot would be the two primary properties for the description of state. upon these graphical coordinates the two points would be plotted for the assumed primary property value for each of the two iterations.

If the value of one of the two primary properties were known, the intersection of the line defined by the two points with the coordinate line of the known value of the one primary property establishes the value of the second primary property.

If a value of neither primary property is known, there will be two points for each iterative selection of a primary value (one point for each of the two secondary properties). The intersection of the two lines defined by these two iteration points of each of the corresponding secondary properties evaluates the two primary properties.

EXAMPLE 1 ONE PRIMARY AND ONE SECONDARY PROPERTY KNOWN

Describe the state of steam with a specific entropy s of 7.500 [kJ/kgΔ_1K] and a quality x of 0.98 [—].

Solution

1. Select pressure as the second primary property to describe the state.

Table of State

Substance: H_2O	2ϕ	
Mass	M	[kg]1 (a)
Pressure	p	[kPa]43 (step 5)
Quality	x	[—]0.98 (g)

2. Make an iteration graph.
3. First iteration—select a pressure.
   (a) Select a pressure; if there is no obvious guiding basis, just guess. Try 25 kPa.
   (b) Calculate the quality. Equation:

$$x = \frac{s - s_f}{s_g - s_f} \quad [—] \qquad \text{(Eq. 2-26x)}$$

Selected pressure 25 *kPa*.

$$s = 7.500 \text{ kJ/kJ}\cdot\Delta_1\text{K} \qquad \text{(given)}$$

$$s_f = 0.8931 \text{ kJ/kg}\cdot\Delta_1\text{K} \qquad \text{(st)} \qquad \text{saturated liquid at 25 kPa}$$

$$(s_g - s_f) = 6.9383 \text{ kJ/kg}\cdot\Delta_1\text{K} \qquad \text{(st)}$$

$$x = \frac{7.500 - 0.8931}{6.9383} \qquad \frac{[\text{kJ/kg}\cdot\Delta_1\text{K} - \text{kJ/kg}\cdot\Delta_1\text{K}]}{[\text{kJ/kg}\cdot\Delta_1\text{K}]}$$

$$= \frac{6.6069}{6.9383} = 0.952 \qquad [—]$$

   (c) Plot the point on a graph.
4. Second iteration
   As the quality of the first iteration is low, it indicates too low a pressure. Select a higher pressure. *Try* 40 *kPa*.

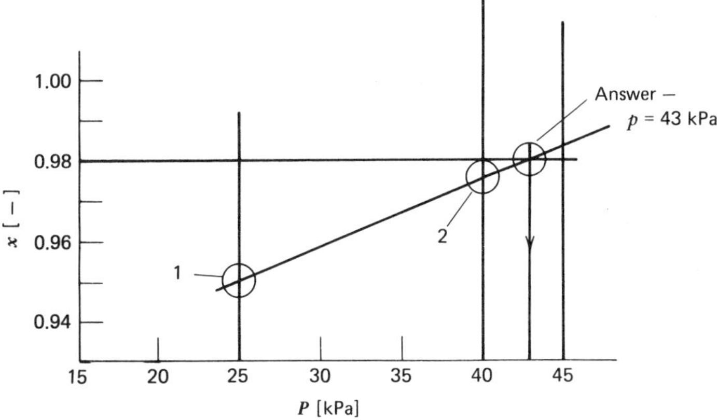

(a) Calculate the quality. Equation parameters are

$$s_f = 1.0259\ \text{kJ/kg}\cdot\Delta_1\text{K} \qquad \text{(st)} \qquad \text{saturated liquid at 40 kPa}$$

$$s_{f_g} = 6.6441\ \text{kJ/kg}\cdot\Delta_1\text{K} \qquad \text{(st)}$$

$$x = \frac{7.500 - 1.0259}{6.6441} = \frac{6.4741}{6.6441} = 0.974$$

(b) Plot the point on a graph.

5. Determine the pressure value. (If a higher accuracy is required, make additional refined calculations as appropriate.)
   (a) Draw a line through the two iteration points on the graph, extending across the quality line of 0.98.
   (b) Read the pressure at the intersection point of the two lines.

EXAMPLE 2 TWO SECONDARY PROPERTIES KNOWN

Describe the state of steam with a specific internal energy of 3200 kJ/kg and a specific volume of 0.155 m^3/kg.

Solution

1. Select the two primary properties to describe the state.

   To do this a decision must be made as to whether the substance is in the single-phase superheat region or in the two-phase saturation region.

   An inspection of the saturated steam tables indicates that the u of saturated vapor peaks at 2604 kJ/kg and the v ranges from 206 to 0.003 m^3/kg. The given 0.155 m^3/kg specific volume by itself is not indicative of the state of the steam. However, because the given specific internal energy of 3200 kJ/kg is higher than the peak internal energy of saturated vapor, the steam must be single-phase superheat. Therefore, temperature T and pressure p will describe the state.

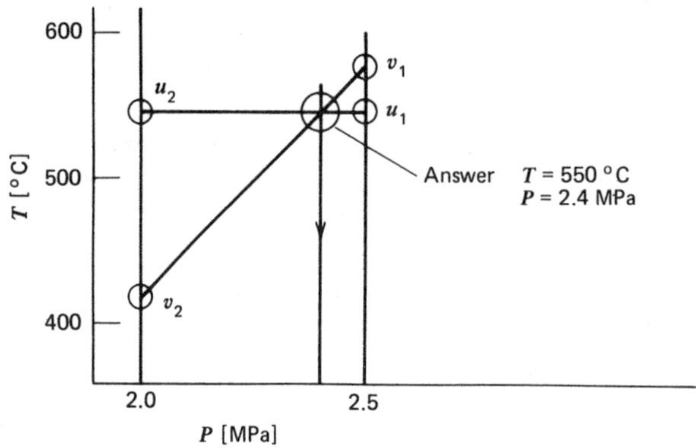

Table of State

Substance: H_2O	superheated
Mass M [kg]	1 (a)
Pressure p [Pa]	2.4 (step 5)
Temperature T [°C]	550 (step 5)

2. Make an iteration graph.
3. First iteration. For a selected pressure determine the temperatures corresponding to the given values of u and v.

   An inspection of the superheated steam tables indicates values of u and v close to those given at 2.50 MPa and 600°C. Therefore, try $p = 2.50$ MPa.

   (a) Determine the temperature for steam with a u of 3200 kJ/kg at 2.50 MPa.

$$T^1 = 500 + \Delta T \qquad °C$$

T [°C]	u [kJ/kg]
500	3112.1
T'	3200.0
600	3288.0

$$\Delta T = (600 - 500)\frac{(3200.0 - 3112.1)}{(3288.0 - 3112.1)}$$

$$= 100\frac{87.9}{175.9}$$

$$= 50.0\ \Delta°C$$

$$T' = 500 + 50.0 = 550°C$$

(b) Determine the temperature for steam with a v of 0.155 m^3/kg at 2.50 MPa.

$$T'' = 500 + \Delta T \qquad {}^\circ C$$

T [°C]	v [m³/kg]
500	0.1400
T''	0.155
600	0.1593

$$\Delta T = (600 - 500)\frac{(0.155 - 0.1400)}{(0.1593 - 0.1400)}$$

$$= 100\frac{0.015}{0.0193}$$

$$= 77.7\ \Delta{}^\circ C$$

$$T'' = 500 + 77.7 = 577.7{}^\circ C$$

(c) Plot the two points on the graph.

4. Second iteration. The two choices are to select $p = 3.0$ or 2.0 MPa. As the given values of u and v are closer to the same temperature at 2.0 MPa, *try 2.00 MPa*.
(a) Determine the temperature for steam with a u of 3200 kJ/kg at 2.00 MPa.

$$T' = 500 + \Delta T \qquad {}^\circ C$$

T [°C]	u [kJ/kg]
500	3116.2
T'	3200.0
600	3290.9

$$\Delta T = (600 - 500)\frac{(3200.0 - 3116.2)}{(3290.9 - 3116.2)}$$

$$= 100\frac{83.8}{174.7} = 48.0\ \Delta{}^\circ C$$

$$T' = 500 + 48.0 = 548.0{}^\circ C$$

(b) Determine the temperature for steam with a v of 0.155 m^3/kg at 2.00 MPa.

$$T'' = 400 + \Delta T \qquad {}^\circ C$$

T [°C]	v [m³/kg]
400	0.1512
T''	0.155
500	0.1757

$$\Delta T = (500 - 400)\frac{(0.1550 - 0.1512)}{(0.1757 - 0.1512)}$$

$$= (100)\frac{0.0038}{0.0245} = 15.5\ \Delta{}^\circ C$$

$$T'' = 400 + 15.5 = 415.5{}^\circ C$$

(c) Plot the two points on a graph.

5. Determine the pressure and temperature values. If a higher accuracy is required, make additional refining calculations as appropriate.

- Draw a line through the two iteration points for u.
- Draw a line through the two iteration points for v.
- Extend the lines if necessary to effect an intersection.
- Read the values of temperature T and pressure p at the intersection point of the two lines.

APPENDIX E
MATHEMATICAL TABLES

The natural logarithm of a number is the index of the power to which the base e (=2.7182818) must be raised in order to equal the number.

EXAMPLE: $\log_e 4.12 = \ln 4.12 = 1.4159$.

The table gives the natural logarithms of numbers from 1.00 to 9.99 directly, and permits the finding of the logarithms of numbers outside of that range by the addition or subtraction of the natural logarithms of powers of 10.

EXAMPLE: $\log_e 679. = \log_e 6.79 + \log_e 10^2 = 1.9155 + 4.6052 = 6.5207$.
$\log_e 0.679 = \log_e 6.79 - \log_e 10^2 = 1.9155 - 4.6052 = -2.6897$.

Natural Logarithms of Powers of 10

$\log_e 10 = 2.302\ 585$	$\log_e 10^4 = 9.210\ 340$	$\log_e 10^7 = 16.118\ 096$
$\log_e 10^2 = 4.605\ 170$	$\log_e 10^5 = 11.512\ 925$	$\log_e 10^8 = 18.420\ 681$
$\log_e 10^3 = 6.907\ 755$	$\log_e 10^6 = 13.815\ 511$	$\log_e 10^9 = 20.723\ 266$

To obtain the common logarithm, the natural logarithm is multiplied by $\log_{10} e$, which is 0.434 294, or $\log_{10} N = 0.434\ 294 \log_e N$.

A negative number or number less than zero has no real logarithm.

Table E-1 Natural Logarithms

N	0	1	2	3	4	5	6	7	8	9
1.0	**0.0000**	**0.0100**	**0.0198**	**0.0296**	**0.0392**	**0.0488**	**0.0583**	**0.0677**	**0.0770**	**0.0862**
1.1	0.0953	0.1044	0.1133	0.1222	0.1310	0.1398	0.1484	0.1570	0.1655	0.1740
1.2	0.1823	0.1906	0.1989	0.2070	0.2151	0.2231	0.2311	0.2390	0.2469	0.2546
1.3	0.2624	0.2700	0.2776	0.2852	0.2927	0.3001	0.3075	0.3148	0.3221	0.3293
1.4	0.3365	0.3436	0.3507	0.3577	0.3646	0.3716	0.3784	0.3853	0.3920	0.3988
1.5	0.4055	0.4121	0.4187	0.4253	0.4318	0.4383	0.4447	0.4511	0.4574	0.4637
1.6	0.4700	0.4762	0.4824	0.4886	0.4947	0.5008	0.5068	0.5128	0.5188	0.5247
1.7	0.5306	0.5365	0.5423	0.5481	0.5539	0.5596	0.5653	0.5710	0.5766	0.5822
1.8	0.5878	0.5933	0.5988	0.6043	0.6098	0.6152	0.6206	0.6259	0.6313	0.6366
1.9	0.6419	0.6471	0.6523	0.6575	0.6627	0.6678	0.6729	0.6780	0.6831	0.6881
2.0	**0.6931**	**0.6981**	**0.7031**	**0.7080**	**0.7129**	**0.7178**	**0.7227**	**0.7275**	**0.7324**	**0.7372**
2.1	0.7419	0.7467	0.7514	0.7561	0.7608	0.7655	0.7701	0.7747	0.7793	0.7839
2.2	0.7885	0.7930	0.7975	0.8020	0.8065	0.8109	0.8154	0.8198	0.8242	0.8286
2.3	0.8329	0.8372	0.8416	0.8459	0.8502	0.8544	0.8587	0.8629	0.8671	0.8713
2.4	0.8755	0.8796	0.8838	0.8879	0.8920	0.8961	0.9002	0.9042	0.9083	0.9123
2.5	0.9163	0.9203	0.9243	0.9282	0.9322	0.9361	0.9400	0.9439	0.9478	0.9517
2.6	0.9555	0.9594	0.9632	0.9670	0.9708	0.9746	0.9783	0.9821	0.9858	0.9895
2.7	0.9933	0.9969	1.0006	1.0043	1.0080	1.0116	1.0152	1.0188	1.0225	1.0260
2.8	1.0296	1.0332	1.0367	1.0403	1.0438	1.0473	1.0508	1.0543	1.0578	1.0613
2.9	1.0647	1.0682	1.0716	1.0750	1.0784	1.0818	1.0852	1.0886	1.0919	1.0953
3.0	**1.0986**	**1.1019**	**1.1053**	**1.1086**	**1.1119**	**1.1151**	**1.1184**	**1.1217**	**1.1249**	**1.1282**
3.1	1.1314	1.1346	1.1378	1.1410	1.1442	1.1474	1.1506	1.1537	1.1569	1.1600
3.2	1.1632	1.1663	1.1694	1.1725	1.1756	1.1787	1.1817	1.1848	1.1878	1.1909
3.3	1.1939	1.1969	1.2000	1.2030	1.2060	1.2090	1.2119	1.2149	1.2179	1.2208
3.4	1.2238	1.2267	1.2296	1.2326	1.2355	1.2384	1.2413	1.2442	1.2470	1.2499
3.5	1.2528	1.2556	1.2585	1.2613	1.2641	1.2669	1.2698	1.2726	1.2754	1.2782
3.6	1.2809	1.2837	1.2865	1.2892	1.2920	1.2947	1.2975	1.3002	1.3029	1.3056
3.7	1.3083	1.3110	1.3137	1.3164	1.3191	1.3218	1.3244	1.3271	1.3297	1.3324
3.8	1.3350	1.3376	1.3403	1.3429	1.3455	1.3481	1.3507	1.3533	1.3558	1.3584
3.9	1.3610	1.3635	1.3661	1.3686	1.3712	1.3737	1.3762	1.3788	1.3813	1.3838
4.0	**1.3863**	**1.3888**	**1.3913**	**1.3938**	**1.3962**	**1.3987**	**1.4012**	**1.4036**	**1.4061**	**1.4085**
4.1	1.4110	1.4134	1.4159	1.4183	1.4207	1.4231	1.4255	1.4279	1.4303	1.4327
4.2	1.4351	1.4375	1.4398	1.4422	1.4446	1.4469	1.4493	1.4516	1.4540	1.4563
4.3	1.4586	1.4609	1.4633	1.4656	1.4679	1.4702	1.4725	1.4748	1.4770	1.4793
4.4	1.4816	1.4839	1.4861	1.4884	1.4907	1.4929	1.4951	1.4974	1.4996	1.5019
4.5	1.5041	1.5063	1.5085	1.5107	1.5129	1.5151	1.5173	1.5195	1.5217	1.5239
4.6	1.5261	1.5282	1.5304	1.5326	1.5347	1.5369	1.5390	1.5412	1.5433	1.5454
4.7	1.5476	1.5497	1.5518	1.5539	1.5560	1.5581	1.5602	1.5623	1.5644	1.5665
4.8	1.5686	1.5707	1.5728	1.5748	1.5769	1.5790	1.5810	1.5831	1.5851	1.5872
4.9	1.5892	1.5913	1.5933	1.5953	1.5974	1.5994	1.6014	1.6034	1.6054	1.6074

Table E-1 Natural Logarithms (Continued)

N	0	1	2	3	4	5	6	7	8	9
5.0	**1.6094**	**1.6114**	**1.6134**	**1.6154**	**1.6174**	**1.6194**	**1.6214**	**1.6233**	**1.6253**	**1.6273**
5.1	1.6292	1.6312	1.6332	1.6351	1.6371	1.6390	1.6409	1.6429	1.6448	1.6467
5.2	1.6487	1.6506	1.6525	1.6544	1.6563	1.6582	1.6601	1.6620	1.6639	1.6658
5.3	1.6677	1.6696	1.6715	1.6734	1.6752	1.6771	1.6790	1.6808	1.6827	1.6845
5.4	1.6864	1.6882	1.6901	1.6919	1.6938	1.6956	1.6974	1.6993	1.7011	1.7029
5.5	1.7047	1.7066	1.7084	1.7102	1.7120	1.7138	1.7156	1.7174	1.7192	1.7210
5.6	1.7228	1.7246	1.7263	1.7281	1.7299	1.7317	1.7334	1.7352	1.7370	1.7387
5.7	1.7405	1.7422	1.7440	1.7457	1.7475	1.7492	1.7509	1.7527	1.7544	1.7561
5.8	1.7579	1.7596	1.7613	1.7630	1.7647	1.7664	1.7681	1.7699	1.7716	1.7733
5.9	1.7750	1.7766	1.7783	1.7800	1.7817	1.7834	1.7851	1.7867	1.7884	1.7901
6.0	**1.7918**	**1.7934**	**1.7951**	**1.7967**	**1.7984**	**1.8001**	**1.8017**	**1.8034**	**1.8050**	**1.8066**
6.1	1.8083	1.8099	1.8116	1.8132	1.8148	1.8165	1.8181	1.8197	1.8213	1.8229
6.2	1.8245	1.8262	1.8278	1.8294	1.8310	1.8326	1.8342	1.8358	1.8374	1.8390
6.3	1.8405	1.8421	1.8437	1.8453	1.8469	1.8485	1.8500	1.8516	1.8532	1.8547
6.4	1.8563	1.8579	1.8594	1.8610	1.8625	1.8641	1.8656	1.8672	1.8687	1.8703
6.5	1.8718	1.8733	1.8749	1.8764	1.8779	1.8795	1.8810	1.8825	1.8840	1.8856
6.6	1.8871	1.8886	1.8901	1.8916	1.8931	1.8946	1.8961	1.8976	1.8991	1.9006
6.7	1.9021	1.9036	1.9051	1.9066	1.9081	1.9095	1.9110	1.9125	1.9140	1.9155
6.8	1.9169	1.9184	1.9199	1.9213	1.9228	1.9242	1.9257	1.9272	1.9286	1.9301
6.9	1.9315	1.9330	1.9344	1.9359	1.9373	1.9387	1.9402	1.9416	1.9430	1.9445
7.0	**1.9459**	**1.9473**	**1.9488**	**1.9502**	**1.9516**	**1.9530**	**1.9544**	**1.9559**	**1.9573**	**1.9587**
7.1	1.9601	1.9615	1.9629	1.9643	1.9657	1.9671	1.9685	1.9699	1.9713	1.9727
7.2	1.9741	1.9755	1.9769	1.9782	1.9796	1.9810	1.9824	1.9838	1.9851	1.9865
7.3	1.9879	1.9892	1.9906	1.9920	1.9933	1.9947	1.9961	1.9974	1.9988	2.0001
7.4	2.0015	2.0028	2.0042	2.0055	2.0069	2.0082	2.0096	2.0109	2.0122	2.0136
7.5	2.0149	2.0162	2.0176	2.0189	2.0202	2.0215	2.0229	2.0242	2.0255	2.0268
7.6	2.0281	2.0295	2.0308	2.0321	2.0334	2.0347	2.0360	2.0373	2.0386	2.0399
7.7	2.0412	2.0425	2.0438	2.0451	2.0464	2.0477	2.0490	2.0503	2.0516	2.0528
7.8	2.0541	2.0554	2.0567	2.0580	2.0592	2.0605	2.0618	2.0631	2.0643	2.0656
7.9	2.0669	2.0681	2.0694	2.0707	2.0719	2.0732	2.0744	2.0757	2.0769	2.0782
8.0	**2.0794**	**2.0807**	**2.0819**	**2.0832**	**2.0844**	**2.0857**	**2.0869**	**2.0882**	**2.0894**	**2.0906**
8.1	2.0919	2.0931	2.0943	2.0956	2.0968	2.0980	2.0992	2.1005	2.1017	2.1029
8.2	2.1041	2.1054	2.1066	2.1078	2.1090	2.1102	2.1114	2.1126	2.1138	2.1150
8.3	2.1163	2.1175	2.1187	2.1199	2.1211	2.1223	2.1235	2.1247	2.1258	2.1270
8.4	2.1282	2.1294	2.1306	2.1318	2.1330	2.1342	2.1353	2.1365	2.1377	2.1389
8.5	2.1401	2.1412	2.1424	2.1436	2.1448	2.1459	2.1471	2.1483	2.1494	2.1506
8.6	2.1518	2.1529	2.1541	2.1552	2.1564	2.1576	2.1587	2.1599	2.1610	2.1622
8.7	2.1633	2.1645	2.1656	2.1668	2.1679	2.1691	2.1702	2.1713	2.1725	2.1736
8.8	2.1748	2.1759	2.1770	2.1782	2.1793	2.1804	2.1815	2.1827	2.1038	2.1849
8.9	2.1861	2.1872	2.1883	2.1894	2.1905	2.1917	2.1928	2.1939	2.1950	2.1961
9.0	**2.1972**	**2.1983**	**2.1994**	**2.2006**	**2.2017**	**2.2028**	**2.2039**	**2.2050**	**2.2061**	**2.2072**
9.1	2.2083	2.2094	2.2105	2.2116	2.2127	2.2138	2.2148	2.2159	2.2170	2.2181
9.2	2.2192	2.2203	2.2214	2.2225	2.2235	2.2246	2.2257	2.2268	2.2279	2.2289
9.3	2.2300	2.2311	2.2322	2.2332	2.2343	2.2354	2.2364	2.2375	2.2386	2.2396
9.4	2.2407	2.2418	2.2428	2.2439	2.2450	2.2460	2.2471	2.2481	2.2492	2.2502
9.5	2.2513	2.2523	2.2534	2.2544	2.2555	2.2565	2.2576	2.2586	2.2597	2.2607
9.6	2.2618	2.2628	2.2638	2.2649	2.2659	2.2670	2.2680	2.2690	2.2701	2.2711
9.7	2.2721	2.2732	2.2742	2.2752	2.2762	2.2773	2.2783	2.2793	2.2803	2.2814
9.8	2.2824	2.2834	2.2844	2.2854	2.2865	2.2875	2.2885	2.2895	2.2905	2.2915
9.9	2.2925	2.2935	2.2946	2.2956	2.2966	2.2976	2.2986	2.2996	2.3006	2.3016

Table E-2 Hyperbolic Functions

x	Natural Values					x	Natural Values				
	e^x	e^{-x}	Sinh x	Cosh x	Tanh x		e^x	e^{-x}	Sinh x	Cosh x	Tanh x
0.00	1.0000	1.0000	0.0000	1.0000	.00000	0.60	1.8221	.54881	0.6367	1.1855	.53705
0.01	1.0101	.99005	0.0100	1.0001	.01000	0.61	1.8404	.54335	0.6485	1.1919	.54413
0.02	1.0202	.98020	0.0200	1.0002	.02000	0.62	1.8589	.53794	0.6605	1.1984	.55113
0.03	1.0305	.97045	0.0300	1.0005	.02999	0.63	1.8776	.53259	0.6725	1.2051	.55805
0.04	1.0408	.96079	0.0400	1.0008	.03998	0.64	1.8965	.52729	0.6846	1.2119	.56490
0.05	1.0513	.95123	0.0500	1.0013	.04996	0.65	1.9155	.52205	0.6967	1.2188	.57167
0.06	1.0618	.94176	0.0600	1.0018	.05993	0.66	1.9348	.51685	0.7090	1.2258	.57836
0.07	1.0725	.93239	0.0701	1.0025	.06989	0.67	1.9542	.51171	0.7213	1.2330	.58498
0.08	1.0833	.92312	0.0801	1.0032	.07983	0.68	1.9739	.50662	0.7336	1.2402	.59152
0.09	1.0942	.91393	0.0901	1.0041	.08976	0.69	1.9937	.50158	0.7461	1.2476	.59798
0.10	1.1052	.90484	0.1002	1.0050	.09967	0.70	2.0138	.49659	0.7586	1.2552	.60437
0.11	1.1163	.89583	0.1102	1.0061	.10956	0.71	2.0340	.49164	0.7712	1.2628	.61068
0.12	1.1275	.88692	0.1203	1.0072	.11943	0.72	2.0544	.48675	0.7838	1.2706	.61691
0.13	1.1388	.87810	0.1304	1.0085	.12927	0.73	2.0751	.48191	0.7966	1.2785	.62307
0.14	1.1503	.86936	0.1405	1.0098	.13909	0.74	2.0959	.47711	0.8094	1.2865	.62915
0.15	1.1618	.86071	0.1506	1.0113	.14889	0.75	2.1170	.47237	0.8223	1.2947	.63515
0.16	1.1735	.85214	0.1607	1.0128	.15865	0.76	2.1383	.46767	0.8353	1.3030	.64108
0.17	1.1853	.84366	0.1708	1.0145	.16838	0.77	2.1598	.46301	0.8484	1.3114	.64693
0.18	1.1972	.83527	0.1810	1.0162	.17808	0.78	2.1815	.45841	0.8615	1.3199	.65271
0.19	1.2092	.82696	0.1911	1.0181	.18775	0.79	2.2034	.45384	0.8748	1.3286	.65841
0.20	1.2214	.81873	0.2013	1.0201	.19738	0.80	2.2255	.44933	0.8881	1.3374	.66404
0.21	1.2337	.81058	0.2115	1.0221	.20697	0.81	2.2479	.44486	0.9015	1.3464	.66959
0.22	1.2461	.80252	0.2218	1.0243	.21652	0.82	2.2705	.44043	0.9150	1.3555	.67507
0.23	1.2586	.79453	0.2320	1.0266	.22603	0.83	2.2933	.43605	0.9286	1.3647	.68048
0.24	1.2712	.78663	0.2423	1.0289	.23550	0.84	2.3164	.43171	0.9423	1.3740	.68581
0.25	1.2840	.77880	0.2526	1.0314	.24492	0.85	2.3396	.42741	0.9561	1.3835	.69107
0.26	1.2969	.77105	0.2629	1.0340	.25430	0.86	2.3632	.42316	0.9700	1.3932	.69626
0.27	1.3100	.76338	0.2733	1.0367	.26362	0.87	2.3869	.41895	0.9840	1.4029	.70137
0.28	1.3231	.75578	0.2837	1.0395	.27291	0.88	2.4109	.41478	0.9981	1.4128	.70642
0.29	1.3364	.74826	0.2941	1.0423	.28213	0.89	2.4351	.41066	1.0122	1.4229	.71139
0.30	1.3499	.74082	0.3045	1.0453	.29131	0.90	2.4596	.40657	1.0265	1.4331	.71630
0.31	1.3634	.73345	0.3150	1.0484	.30044	0.91	2.4843	.40252	1.0409	1.4434	.72113
0.32	1.3771	.72615	0.3255	1.0516	.30951	0.92	2.5093	.39852	1.0554	1.4539	.72590
0.33	1.3910	.71892	0.3360	1.0549	.31852	0.93	2.5345	.39455	1.0700	1.4645	.73059
0.34	1.4049	.71177	0.3466	1.0584	.32748	0.94	2.5600	.39063	1.0847	1.4753	.73522
0.35	1.4191	.70469	0.3572	1.0619	.33638	0.95	2.5857	.38674	1.0995	1.4862	.73978
0.36	1.4333	.69768	0.3678	1.0655	.34521	0.96	2.6117	.38289	1.1144	1.4973	.74428
0.37	1.4477	.69073	0.3785	1.0692	.35399	0.97	2.6379	.37908	1.1294	1.5085	.74870
0.38	1.4623	.68386	0.3892	1.0731	.36271	0.98	2.6645	.37531	1.1446	1.5199	.75307
0.39	1.4770	.67706	0.4000	1.0770	.37136	0.99	2.6912	.37158	1.1598	1.5314	.75736
0.40	1.4918	.67032	0.4108	1.0811	.37995	1.00	2.7183	.36788	1.1752	1.5431	.76159
0.41	1.5068	.66365	0.4216	1.0852	.38847	1.01	2.7456	.36422	1.1907	1.5549	.76576
0.42	1.5220	.65705	0.4325	1.0895	.39693	1.02	2.7732	.36059	1.2063	1.5669	.76987
0.43	1.5373	.65051	0.4434	1.0939	.40532	1.03	2.8011	.35701	1.2220	1.5790	.77391
0.44	1.5527	.64404	0.4543	1.0984	.41364	1.04	2.8292	.35345	1.2379	1.5913	.77789
0.45	1.5683	.63763	0.4653	1.1030	.42190	1.05	2.8577	.34994	1.2539	1.6038	.78181
0.46	1.5841	.63128	0.4764	1.1077	.43008	1.06	2.8864	.34646	1.2700	1.6164	.78566
0.47	1.6000	.62500	0.4875	1.1125	.43820	1.07	2.9154	.34301	1.2862	1.6292	.78946
0.48	1.6161	.61878	0.4986	1.1174	.44624	1.08	2.9447	.33960	1.3025	1.6421	.79320
0.49	1.6323	.61263	0.5098	1.1225	.45422	1.09	2.9743	.33622	1.3190	1.6552	.79688
0.50	1.6487	.60653	0.5211	1.1276	.46212	1.10	3.0042	.33287	1.3356	1.6685	.80050
0.51	1.6653	.60050	0.5324	1.1329	.46995	1.11	3.0344	.32956	1.3524	1.6820	.80406
0.52	1.6820	.59452	0.5438	1.1383	.47770	1.12	3.0649	.32628	1.3693	1.6956	.80757
0.53	1.6989	.58860	0.5552	1.1438	.48538	1.13	3.0957	.32303	1.3863	1.7093	.81102
0.54	1.7160	.58275	0.5666	1.1494	.49299	1.14	3.1268	.31982	1.4035	1.7233	.81441
0.55	1.7333	.57695	0.5782	1.1551	.50052	1.15	3.1582	.31664	1.4208	1.7374	.81775
0.56	1.7507	.57121	0.5897	1.1609	.50798	1.16	3.1899	.31349	1.4382	1.7517	.82104
0.57	1.7683	.56553	0.6014	1.1669	.51536	1.17	3.2220	.31037	1.4558	1.7662	.82427
0.58	1.7860	.55990	0.6131	1.1730	.52267	1.18	3.2544	.30728	1.4735	1.7808	.82745
0.59	1.8040	.55433	0.6248	1.1792	.52990	1.19	3.2871	.30422	1.4914	1.7957	.83058
0.60	1.8221	.54881	0.6367	1.1855	.53705	1.20	3.3201	.30119	1.5095	1.8107	.83365

Table E-2 Hyperbolic Functions (Continued)

x	Natural Values					x	Natural Values				
	e^x	e^{-x}	Sinh x	Cosh x	Tanh x		e^x	e^{-x}	Sinh x	Cosh x	Tanh x
1.20	**3.3201**	**.30119**	**1.5095**	**1.8107**	**.83365**	**1.80**	**6.0496**	**.16530**	**2.9422**	**3.1075**	**.94681**
1.21	3.3535	.29820	1.5276	1.8258	.83668	1.81	6.1104	.16365	2.9734	3.1371	.94783
1.22	3.3872	.29523	1.5460	1.8412	.83965	1.82	6.1719	.16203	3.0049	3.1669	.94884
1.23	3.4212	.29229	1.5645	1.8568	.84258	1.83	6.2339	.16041	3.0367	3.1972	.94983
1.24	3.4556	.28938	1.5831	1.8725	.84546	1.84	6.2965	.15882	3.0689	3.2277	.95080
1.25	3.4903	.28650	1.6019	1.8884	.84828	1.85	6.3598	.15724	3.1013	3.2585	.95175
1.26	3.5254	.28365	1.6209	1.9045	.85106	1.86	6.4237	.15567	3.1340	3.2897	.95268
1.27	3.5609	.28083	1.6400	1.9208	.85380	1.87	6.4883	.15412	3.1671	3.3212	.95359
1.28	3.5966	.27804	1.6593	1.9373	.85648	1.88	6.5535	.15259	3.2005	3.3530	.95449
1.29	3.6328	.27527	1.6788	1.9540	.85913	1.89	6.6194	.15107	3.2341	3.3852	.95537
1.30	**3.6693**	**.27253**	**1.6984**	**1.9709**	**.86172**	**1.90**	**6.6859**	**.14957**	**3.2682**	**3.4177**	**.95624**
1.31	3.7062	.26982	1.7182	1.9880	.86428	1.91	6.7531	.14808	3.3025	3.4506	.95709
1.32	3.7434	.26714	1.7381	2.0053	.86678	1.92	6.8210	.14661	3.3372	3.4838	.95792
1.33	3.7810	.26448	1.7583	2.0228	.86925	1.93	6.8895	.14515	3.3722	3.5173	.95873
1.34	3.8190	.26185	1.7786	2.0404	.87167	1.94	6.9588	.14370	3.4075	3.5512	.95953
1.35	3.8574	.25924	1.7991	2.0583	.87405	1.95	7.0287	.14227	3.4432	3.5855	.96032
1.36	3.8962	.25666	1.8198	2.0764	.87639	1.96	7.0993	.14086	3.4792	3.6201	.96109
1.37	3.9354	.25411	1.8406	2.0947	.87869	1.97	7.1707	.13946	3.5156	3.6551	.96185
1.38	3.9749	.25158	1.8617	2.1132	.88095	1.98	7.2427	.13807	3.5523	3.6904	.96259
1.39	4.0149	.24908	1.8829	2.1320	.88317	1.99	7.3155	.13670	3.5894	3.7261	.96331
1.40	**4.0552**	**.24660**	**1.9043**	**2.1509**	**.88535**	**2.00**	**7.3891**	**.13534**	**3.6269**	**3.7622**	**.96403**
1.41	4.0960	.24414	1.9259	2.1700	.88749	2.01	7.4633	.13399	3.6647	3.7987	.96473
1.42	4.1371	.24171	1.9477	2.1894	.88960	2.02	7.5383	.13266	3.7028	3.8355	.96541
1.43	4.1787	.23931	1.9697	2.2090	.89167	2.03	7.6141	.13134	3.7414	3.8727	.96609
1.44	4.2207	.23693	1.9919	2.2288	.89370	2.04	7.6906	.13003	3.7803	3.9103	.96675
1.45	4.2631	.23457	2.0143	2.2488	.89569	2.05	7.7679	.12873	3.8196	3.9483	.96740
1.46	4.3060	.23224	2.0369	2.2691	.89765	2.06	7.8460	.12745	3.8593	3.9867	.96803
1.47	4.3492	.22993	2.0597	2.2896	.89958	2.07	7.9248	.12619	3.8993	4.0255	.96865
1.48	4.3929	.22764	2.0827	2.3103	.90147	2.08	8.0045	.12493	3.9398	4.0647	.96926
1.49	4.4371	.22537	2.1059	2.3312	.90332	2.09	8.0849	.12369	3.9806	4.1043	.96986
1.50	**4.4817**	**.22313**	**2.1293**	**2.3524**	**.90515**	**2.10**	**8.1662**	**.12246**	**4.0219**	**4.1443**	**.97045**
1.51	4.5267	.22091	2.1529	2.3738	.90694	2.11	8.2482	.12124	4.0635	4.1847	.97103
1.52	4.5722	.21871	2.1768	2.3955	.90870	2.12	8.3311	.12003	4.1056	4.2256	.97159
1.53	4.6182	.21654	2.2008	2.4174	.91042	2.13	8.4149	.11884	4.1480	4.2669	.97215
1.54	4.6646	.21438	2.2251	2.4395	.91212	2.14	8.4994	.11765	4.1909	4.3085	.97269
1.55	4.7115	.21225	2.2496	2.4619	.91379	2.15	8.5849	.11648	4.2342	4.3507	.97323
1.56	4.7588	.21014	2.2743	2.4845	.91542	2.16	8.6711	.11533	4.2779	4.3932	.97375
1.57	4.8066	.20805	2.2993	2.5073	.91703	2.17	8.7583	.11418	4.3221	4.4362	.97426
1.58	4.8550	.20598	2.3245	2.5305	.91860	2.18	8.8463	.11304	4.3666	4.4797	.97477
1.59	4.9037	.20393	2.3499	2.5538	.92015	2.19	8.9352	.11192	4.4116	4.5236	.97526
1.60	**4.9530**	**.20190**	**2.3756**	**2.5775**	**.92167**	**2.20**	**9.0250**	**.11080**	**4.4571**	**4.5679**	**.97574**
1.61	5.0028	.19989	2.4015	2.6013	.92316	2.21	9.1157	.10970	4.5030	4.6127	.97622
1.62	5.0531	.19790	2.4276	2.6255	.92462	2.22	9.2073	.10861	4.5494	4.6580	.97668
1.63	5.1039	.19593	2.4540	2.6499	.92606	2.23	9.2999	.10753	4.5962	4.7037	.97714
1.64	5.1552	.19398	2.4806	2.6746	.92747	2.24	9.3933	.10646	4.6434	4.7499	.97759
1.65	5.2070	.19205	2.5075	2.6995	.92886	2.25	9.4877	.10540	4.6912	4.7966	.97803
1.66	5.2593	.19014	2.5346	2.7247	.93022	2.26	9.5831	.10435	4.7394	4.8437	.97846
1.67	5.3122	.18825	2.5620	2.7502	.93155	2.27	9.6794	.10331	4.7880	4.8914	.97888
1.68	5.3656	.18637	2.5896	2.7760	.93286	2.28	9.7767	.10228	4.8372	4.9395	.97929
1.69	5.4195	.18452	2.6175	2.8020	.93415	2.29	9.8749	.10127	4.8868	4.9881	.97970
1.70	**5.4739**	**.18268**	**2.6456**	**2.8283**	**.93541**	**2.30**	**9.9742**	**.10026**	**4.9370**	**5.0372**	**.98010**
1.71	5.5290	.18087	2.6740	2.8549	.93665	2.31	10.074	.09926	4.9876	5.0868	.98049
1.72	5.5845	.17907	2.7027	2.8818	.93786	2.32	10.176	.09827	5.0387	5.1370	.98087
1.73	5.6407	.17728	2.7317	2.9090	.93906	2.33	10.278	.09730	5.0903	5.1876	.98124
1.74	5.6973	.17552	2.7609	2.9364	.94023	2.34	10.381	.09633	5.1425	5.2388	.98161
1.75	5.7546	.17377	2.7904	2.9642	.94138	2.35	10.486	.09537	5.1951	5.2905	.98197
1.76	5.8124	.17204	2.8202	2.9922	.94250	2.36	10.591	.09442	5.2483	5.3427	.98233
1.77	5.8709	.17033	2.8503	3.0206	.94361	2.37	10.697	.09348	5.3020	5.3954	.98267
1.78	5.9299	.16864	2.8806	3.0492	.94470	2.38	10.805	.09255	5.3562	5.4487	.98301
1.79	5.9895	.16696	2.9112	3.0782	.94576	2.39	10.913	.09163	5.4109	5.5026	.98335
1.80	**6.0496**	**.16530**	**2.9422**	**3.1075**	**.94681**	**2.40**	**11.023**	**.09072**	**5.4662**	**5.5569**	**.98367**

Table E-2 Hyperbolic Functions (Continued)

x	Natural Values					x	Natural Values				
	e^x	e^{-x}	Sinh x	Cosh x	Tanh x		e^x	e^{-x}	Sinh x	Cosh x	Tanh x
2.40	**11.023**	**.09072**	**5.4662**	**5.5569**	**.98367**	**3.00**	**20.086**	**.04979**	**10.018**	**10.068**	**.99505**
2.41	11.134	.08982	5.5221	5.6119	.98400	3.01	20.287	.04929	10.119	10.168	.99515
2.42	11.246	.08892	5.5785	5.6674	.98431	3.02	20.491	.04880	10.221	10.270	.99525
2.43	11.359	.08804	5.6354	5.7235	.98462	3.03	20.697	.04832	10.325	10.373	.99534
2.44	11.473	.08716	5.6929	5.7801	.98492	3.04	20.905	.04783	10.429	10.477	.99543
2.45	11.588	.08629	5.7510	5.8373	.98522	3.05	21.115	.04736	10.534	10.581	.99552
2.46	11.705	.08543	5.8097	5.8951	.98551	3.06	21.328	.04689	10.640	10.687	.99561
2.47	11.822	.08458	5.8689	5.9535	.98579	3.07	21.542	.04642	10.748	10.794	.99570
2.48	11.941	.08374	5.9288	6.0125	.98607	3.08	21.758	.04596	10.856	10.902	.99578
2.49	12.061	.08291	5.9892	6.0721	.98635	3.09	21.977	.04550	10.966	11.011	.99587
2.50	**12.182**	**.08208**	**6.0502**	**6.1323**	**.98661**	**3.10**	**22.198**	**.04505**	**11.077**	**11.122**	**.99595**
2.51	12.305	.08127	6.1118	6.1931	.98688	3.11	22.421	.04460	11.188	11.233	.99603
2.52	12.429	.08046	6.1741	6.2545	.98714	3.12	22.646	.04416	11.301	11.345	.99611
2.53	12.554	.07966	6.2369	6.3166	.98739	3.13	22.874	.04372	11.415	11.459	.99618
2.54	12.680	.07887	6.3004	6.3793	.98764	3.14	23.104	.04328	11.530	11.574	.99626
2.55	12.807	.07808	6.3645	6.4426	.98788	3.15	23.336	.04285	11.647	11.689	.99633
2.56	12.936	.07730	6.4293	6.5066	.98812	3.16	23.571	.04243	11.764	11.807	.99641
2.57	13.066	.07654	6.4946	6.5712	.98835	3.17	23.807	.04200	11.883	11.925	.99648
2.58	13.197	.07577	6.5607	6.6365	.98858	3.18	24.047	.04159	12.003	12.044	.99655
2.59	13.330	.07502	6.6274	6.7024	.98881	3.19	24.288	.04117	12.124	12.165	.99662
2.60	**13.464**	**.07427**	**6.6947**	**6.7690**	**.98903**	**3.20**	**24.533**	**.04076**	**12.246**	**12.287**	**.99668**
2.61	13.599	.07353	6.7628	6.8363	.98924	3.21	24.779	.04036	12.369	12.410	.99675
2.62	13.736	.07280	6.8315	6.9043	.98946	3.22	25.028	.03996	12.494	12.534	.99681
2.63	13.874	.07208	6.9008	6.9729	.98966	3.23	25.280	.03956	12.620	12.660	.99688
2.64	14.013	.07136	6.9709	7.0423	.98987	3.24	25.534	.03916	12.747	12.786	.99694
2.65	14.154	.07065	7.0417	7.1123	.99007	3.25	25.790	.03877	12.876	12.915	.99700
2.66	14.296	.06995	7.1132	7.1831	.99026	3.26	26.050	.03839	13.006	13.044	.99706
2.67	14.440	.06925	7.1854	7.2546	.99045	3.27	26.311	.03801	13.137	13.175	.99712
2.68	14.585	.06856	7.2583	7.3268	.99064	3.28	26.576	.03763	13.269	13.307	.99717
2.69	14.732	.06788	7.3319	7.3998	.99083	3.29	26.843	.03725	13.403	13.440	.99723
2.70	**14.880**	**.06721**	**7.4063**	**7.4735**	**.99101**	**3.30**	**27.113**	**.03688**	**13.538**	**13.575**	**.99728**
2.71	15.029	.06654	7.4814	7.5479	.99118	3.31	27.385	.03652	13.674	13.711	.99734
2.72	15.180	.06587	7.5572	7.6231	.99136	3.32	27.660	.03615	13.812	13.848	.99739
2.73	15.333	.06522	7.6338	7.6991	.99153	3.33	27.938	.03579	13.951	13.987	.99744
2.74	15.487	.06457	7.7112	7.7758	.99170	3.34	28.219	.03544	14.092	14.127	.99749
2.75	15.643	.06393	7.7894	7.8533	.99186	3.35	28.503	.03508	14.234	14.269	.99754
2.76	15.800	.06329	7.8683	7.9316	.99202	3.36	28.789	.03474	14.377	14.412	.99759
2.77	15.959	.06266	7.9480	8.0106	.99218	3.37	29.079	.03439	14.522	14.556	.99764
2.78	16.119	.06204	8.0285	8.0905	.99233	3.38	29.371	.03405	14.668	14.702	.99768
2.79	16.281	.06142	8.1098	8.1712	.99248	3.39	29.666	.03371	14.816	14.850	.99773
2.80	**16.445**	**.06081**	**8.1919**	**8.2527**	**.99263**	**3.40**	**29.964**	**.03337**	**14.965**	**14.999**	**.99777**
2.81	16.610	.06020	8.2749	8.3351	.99278	3.41	30.265	.03304	15.116	15.149	.99782
2.82	16.777	.05961	8.3586	8.4182	.99292	3.42	30.569	.03271	15.268	15.301	.99786
2.83	16.945	.05901	8.4432	8.5022	.99306	3.43	30.877	.03239	15.422	15.455	.99790
2.84	17.116	.05843	8.5287	8.5871	.99320	3.44	31.187	.03206	15.577	15.610	.99795
2.85	17.288	.05784	8.6150	8.6728	.99333	3.45	31.500	.03175	15.734	15.766	.99799
2.86	17.462	.05727	8.7021	8.7594	.99346	3.46	31.817	.03143	15.893	15.924	.99803
2.87	17.637	.05670	8.7902	8.8469	.99359	3.47	32.137	.03112	16.053	16.084	.99807
2.88	17.814	.05613	8.8791	8.9352	.99372	3.48	32.460	.03081	16.215	16.245	.99810
2.89	17.993	.05558	8.9689	9.0244	.99384	3.49	32.786	.03050	16.378	16.408	.99814
2.90	**18.174**	**.05502**	**9.0596**	**9.1146**	**.99396**	**3.50**	**33.115**	**.03020**	**16.543**	**16.573**	**.99818**
2.91	18.357	.05448	9.1512	9.2056	.99408	3.51	33.448	.02990	16.709	16.739	.99821
2.92	18.541	.05393	9.2437	9.2976	.99420	3.52	33.784	.02960	16.877	16.907	.99825
2.93	18.728	.05340	9.3371	9.3905	.99431	3.53	34.124	.02930	17.047	17.077	.99828
2.94	18.916	.05287	9.4315	9.4844	.99443	3.54	34.467	.02901	17.219	17.248	.99832
2.95	19.106	.05234	9.5268	9.5791	.99454	3.55	34.813	.02872	17.392	17.421	.99835
2.96	19.298	.05182	9.6231	9.6749	.99464	3.56	35.163	.02844	17.567	17.596	.99838
2.97	19.492	.05130	9.7203	9.7716	.99475	3.57	35.517	.02816	17.744	17.772	.99842
2.98	19.688	.05079	9.8185	9.8693	.99485	3.58	35.874	.02788	17.923	17.951	.99845
2.99	19.886	.05029	9.9177	9.9680	.99496	3.59	36.234	.02760	18.103	18.131	.99848
3.00	**20.086**	**.04979**	**10.018**	**10.068**	**.99505**	**3.60**	**36.598**	**.02732**	**18.285**	**18.313**	**.99851**

Table E-2 Hyperbolic Functions (Continued)

x	e^x	e^{-x}	Sinh x	Cosh x	Tanh x
	Natural Values				
3.60	**36.598**	**.02732**	**18.285**	**18.313**	**.99851**
3.61	36.966	.02705	18.470	18.497	.99854
3.62	37.338	.02678	18.655	18.682	.99857
3.63	37.713	.02652	18.843	18.870	.99859
3.64	38.092	.02625	19.033	19.059	.99862
3.65	38.475	.02599	19.224	19.250	.99865
3.66	38.861	.02573	19.418	19.444	.99868
3.67	39.252	.02548	19.613	19.639	.99870
3.68	39.646	.02522	19.811	19.836	.99873
3.69	40.045	.02497	20.010	20.035	.99875
3.70	**40.447**	**.02472**	**20.211**	**20.236**	**.99878**
3.71	40.854	.02448	20.415	20.439	.99880
3.72	41.264	.02423	20.620	20.644	.99883
3.73	41.679	.02399	20.828	20.852	.99885
3.74	42.098	.02375	21.037	21.061	.99887
3.75	42.521	.02352	21.249	21.272	.99889
3.76	42.948	.02328	21.463	21.486	.99892
3.77	43.380	.02305	21.679	21.702	.99894
3.78	43.816	.02282	21.897	21.919	.99896
3.79	44.256	.02260	22.117	22.140	.99898
3.80	**44.701**	**.02237**	**22.339**	**22.362**	**.99900**
3.81	45.150	.02215	22.564	22.586	.99902
3.82	45.604	.02193	22.791	22.813	.99904
3.83	46.063	.02171	23.020	23.042	.99906
3.84	46.525	.02149	23.252	23.274	.99908
3.85	46.993	.02128	23.486	23.507	.99909
3.86	47.465	.02107	23.722	23.743	.99911
3.87	47.942	.02086	23.961	23.982	.99913
3.88	48.424	.02065	24.202	24.222	.99915
3.89	48.911	.02045	24.445	24.466	.99916
3.90	**49.402**	**.02024**	**24.691**	**24.711**	**.99918**
3.91	49.899	.02004	24.939	24.960	.99920
3.92	50.400	.01984	25.190	25.210	.99921
3.93	50.907	.01964	25.444	25.463	.99923
3.94	51.419	.01945	25.700	25.719	.99924
3.95	51.935	.01925	25.958	25.977	.99926
3.96	52.457	.01906	26.219	26.238	.99927
3.97	52.985	.01887	26.483	26.502	.99929
3.98	53.517	.01869	26.749	26.768	.99930
3.99	54.055	.01850	27.018	27.037	.99932
4.00	**54.598**	**.01832**	**27.290**	**27.308**	**.99933**
4.01	55.147	.01813	27.564	27.583	.99934
4.02	55.701	.01795	27.842	27.860	.99936
4.03	56.261	.01777	28.122	28.139	.99937
4.04	56.826	.01760	28.404	28.422	.99938
4.05	57.397	.01742	28.690	28.707	.99939
4.06	57.974	.01725	28.979	28.996	.99941
4.07	58.557	.01708	29.270	29.287	.99942
4.08	59.145	.01691	29.564	29.581	.99943
4.09	59.740	.01674	29.862	29.878	.99944
4.10	**60.340**	**.01657**	**30.162**	**30.178**	**.99945**
4.11	60.947	.01641	30.465	30.482	.99946
4.12	61.559	.01624	30.772	30.788	.99947
4.13	62.178	.01608	31.081	31.097	.99948
4.14	62.803	.01592	31.393	31.409	.99949
4.15	63.434	.01576	31.709	31.725	.99950
4.16	64.072	.01561	32.028	32.044	.99951
4.17	64.715	.01545	32.350	32.365	.99952
4.18	65.366	.01530	32.675	32.691	.99953
4.19	66.023	.01515	33.004	33.019	.99954
4.20	**66.686**	**.01500**	**33.336**	**33.351**	**.99955**

x	e^x	e^{-x}	Sinh x	Cosh x	Tanh x
	Natural Values				
4.20	**66.686**	**.01500**	**33.336**	**33.351**	**.99955**
4.21	67.357	.01485	33.671	33.686	.99956
4.22	68.033	.01470	34.009	34.024	.99957
4.23	68.717	.01455	34.351	34.366	.99958
4.24	69.408	.01441	34.697	34.711	.99958
4.25	70.105	.01426	35.046	35.060	.99959
4.26	70.810	.01412	35.398	35.412	.99960
4.27	71.522	.01398	35.754	35.768	.99961
4.28	72.240	.01384	36.113	36.127	.99962
4.29	72.966	.01370	36.476	36.490	.99962
4.30	**73.700**	**.01357**	**36.843**	**36.857**	**.99963**
4.31	74.440	.01343	37.214	37.227	.99964
4.32	75.189	.01330	37.588	37.601	.99965
4.33	75.944	.01317	37.966	37.979	.99965
4.34	76.708	.01304	38.347	38.360	.99966
4.35	77.478	.01291	38.733	38.746	.99967
4.36	78.257	.01278	39.122	39.135	.99967
4.37	79.044	.01265	39.515	39.528	.99968
4.38	79.838	.01253	39.913	39.925	.99969
4.39	80.640	.01240	40.314	40.326	.99969
4.40	**81.451**	**.01228**	**40.719**	**40.732**	**.99970**
4.41	82.269	.01216	41.129	41.141	.99970
4.42	83.096	.01203	41.542	41.554	.99971
4.43	83.931	.01191	41.960	41.972	.99972
4.44	84.775	.01180	42.382	42.393	.99972
4.45	85.627	.01168	42.808	42.819	.99973
4.46	86.488	.01156	43.238	43.250	.99973
4.47	87.357	.01145	43.673	43.684	.99974
4.48	88.235	.01133	44.112	44.123	.99974
4.49	89.121	.01122	44.555	44.566	.99975
4.50	**90.017**	**.01111**	**45.003**	**45.014**	**.99975**
4.51	90.922	.01100	45.455	45.466	.99976
4.52	91.836	.01089	45.912	45.923	.99976
4.53	92.759	.01078	46.374	46.385	.99977
4.54	93.691	.01067	46.840	46.851	.99977
4.55	94.632	.01057	47.311	47.321	.99978
4.56	95.583	.01046	47.787	47.797	.99978
4.57	96.544	.01036	48.267	48.277	.99979
4.58	97.514	.01025	48.752	48.762	.99979
4.59	98.494	.01015	49.242	49.252	.99979
4.60	**99.484**	**.01005**	**49.737**	**49.747**	**.99980**
4.61	100.48	.00995	50.237	50.247	.99980
4.62	101.49	.00985	50.742	50.752	.99981
4.63	102.51	.00975	51.252	51.262	.99981
4.64	103.54	.00966	51.767	51.777	.99981
4.65	104.58	.00956	52.288	52.297	.99982
4.66	105.64	.00947	52.813	52.823	.99982
4.67	106.70	.00937	53.344	53.354	.99982
4.68	107.77	.00928	53.880	53.890	.99983
4.69	108.85	.00919	54.422	54.431	.99983
4.70	**109.95**	**.00910**	**54.969**	**54.978**	**.99983**
4.71	111.05	.00900	55.522	55.531	.99984
4.72	112.17	.00892	56.080	56.089	.99984
4.73	113.30	.00883	56.643	56.652	.99984
4.74	114.43	.00874	57.213	57.222	.99985
4.75	115.58	.00865	57.788	57.796	.99985
4.76	116.75	.00857	58.369	58.377	.99985
4.77	117.92	.00848	58.955	58.964	.99986
4.78	119.10	.00840	59.548	59.556	.99986
4.79	120.30	.00831	60.147	60.155	.99986
4.80	**121.51**	**.00823**	**60.751**	**60.759**	**.99986**

Table E-2 Hyperbolic Functions (Continued)

x	Natural Values					x	Natural Values				
	e^x	e^{-x}	Sinh x	Cosh x	Tanh x		e^x	e^{-x}	Sinh x	Cosh x	Tanh x
4.80	**121.51**	**.00823**	**60.751**	**60.760**	**.99986**	**5.40**	**221.41**	**.00452**	**110.70**	**110.71**	**.99996**
4.81	122.73	.00815	61.362	61.370	.99987	5.41	223.63	.00447	111.81	111.82	.99996
4.82	123.97	.00807	61.979	61.987	.99987	5.42	225.88	.00443	112.94	112.94	.99996
4.83	125.21	.00799	62.601	62.609	.99987	5.43	228.15	.00438	114.07	114.08	.99996
4.84	126.47	.00791	63.231	63.239	.99987	5.44	230.44	.00434	115.22	115.22	.99996
4.85	127.74	.00783	63.866	63.874	.99988	5.45	232.76	.00430	116.38	116.38	.99996
4.86	129.02	.00775	64.508	64.516	.99988	5.46	235.10	.00425	117.55	117.55	.99996
4.87	130.32	.00767	65.157	65.164	.99988	5.47	237.46	.00421	118.73	118.73	.99996
4.88	131.63	.00760	65.812	65.819	.99988	5.48	239.85	.00417	119.92	119.93	.99997
4.89	132.95	.00752	66.473	66.481	.99989	5.49	242.26	.00413	121.13	121.13	.99997
4.90	**134.29**	**.00745**	**67.141**	**67.149**	**.99989**	**5.50**	**244.69**	**.00409**	**122.34**	**122.35**	**.99997**
4.91	135.64	.00737	67.816	67.823	.99989	5.51	247.15	.00405	123.57	123.58	.99997
4.92	137.00	.00730	68.498	68.505	.99989	5.52	249.64	.00401	124.82	124.82	.99997
4.93	138.38	.00723	69.186	69.193	.99990	5.53	252.14	.00397	126.07	126.07	.99997
4.94	139.77	.00715	69.882	69.889	.99990	5.54	254.68	.00393	127.34	127.34	.99997
4.95	141.17	.00708	70.584	70.591	.99990	5.55	257.24	.00389	128.62	128.62	.99997
4.96	142.59	.00701	71.293	71.300	.99990	5.56	259.82	.00385	129.91	129.91	.99997
4.97	144.03	.00694	72.010	72.017	.99990	5.57	262.43	.00381	131.22	131.22	.99997
4.98	145.47	.00687	72.734	72.741	.99991	5.58	265.07	.00377	132.53	132.54	.99997
4.99	146.94	.00681	73.465	73.472	.99991	5.59	267.74	.00374	133.87	133.87	.99997
5.00	**148.41**	**.00674**	**74.203**	**74.210**	**.99991**	**5.60**	**270.43**	**.00370**	**135.21**	**135.22**	**.99997**
5.01	149.90	.00667	74.949	74.956	.99991	5.61	273.14	.00366	136.57	136.57	.99997
5.02	151.41	.00660	75.702	75.710	.99991	5.62	275.89	.00362	137.94	137.95	.99997
5.03	152.93	.00654	76.463	76.470	.99991	5.63	278.66	.00359	139.33	139.33	.99997
5.04	154.47	.00647	77.232	77.238	.99992	5.64	281.46	.00355	140.73	140.73	.99997
5.05	156.02	.00641	78.008	78.014	.99992	5.65	284.29	.00352	142.14	142.15	.99998
5.06	157.59	.00635	78.792	78.798	.99992	5.66	287.15	.00348	143.57	143.58	.99998
5.07	159.17	.00628	79.584	79.590	.99992	5.67	290.03	.00345	145.02	145.02	.99998
5.08	160.77	.00622	80.384	80.390	.99992	5.68	292.95	.00341	146.47	146.48	.99998
5.09	162.39	.00616	81.192	81.198	.99992	5.69	295.89	.00338	147.95	147.95	.99998
5.10	**164.02**	**.00610**	**82.008**	**82.014**	**.99993**	**5.70**	**298.87**	**.00335**	**149.43**	**149.44**	**.99998**
5.11	165.67	.00604	82.832	82.838	.99993	5.71	301.87	.00331	150.93	150.94	.99998
5.12	167.34	.00598	83.665	83.671	.99993	5.72	304.90	.00328	152.45	152.45	.99998
5.13	169.02	.00592	84.506	84.512	.99993	5.73	307.97	.00325	153.98	153.99	.99998
5.14	170.72	.00586	85.355	85.361	.99993	5.74	311.06	.00321	155.53	155.53	.99998
5.15	172.43	.00580	86.213	86.219	.99993	5.75	314.19	.00318	157.09	157.10	.99998
5.16	174.16	.00574	87.079	87.085	.99993	5.76	317.35	.00315	158.67	158.68	.99998
5.17	175.91	.00568	87.955	87.960	.99994	5.77	320.54	.00312	160.27	160.27	.99998
5.18	177.68	.00563	88.839	88.844	.99994	5.78	323.76	.00309	161.88	161.88	.99998
5.19	179.47	.00557	89.732	89.737	.99994	5.79	327.01	.00306	163.51	163.51	.99998
5.20	**181.27**	**.00552**	**90.633**	**90.639**	**.99994**	**5.80**	**330.30**	**.00303**	**165.15**	**165.15**	**.99998**
5.21	183.09	.00546	91.544	91.550	.99994	5.81	333.62	.00300	166.81	166.81	.99998
5.22	184.93	.00541	92.464	92.470	.99994	5.82	336.97	.00297	168.48	168.49	.99998
5.23	186.79	.00535	93.394	93.399	.99994	5.83	340.36	.00294	170.18	170.18	.99998
5.24	188.67	.00530	94.332	94.338	.99994	5.84	343.78	.00291	171.89	171.89	.99998
5.25	190.57	.00525	95.281	95.286	.99994	5.85	347.23	.00288	173.62	173.62	.99998
5.26	192.48	.00520	96.238	96.243	.99995	5.86	350.72	.00285	175.36	175.36	.99998
5.27	194.42	.00514	97.205	97.211	.99995	5.87	354.25	.00282	177.12	177.13	.99998
5.28	196.37	.00509	98.182	98.188	.99995	5.88	357.81	.00279	178.90	178.91	.99998
5.29	198.34	.00504	99.169	99.174	.99995	5.89	361.41	.00277	180.70	180.70	.99998
5.30	**200.34**	**.00499**	**100.17**	**100.17**	**.99995**	**5.90**	**365.04**	**.00274**	**182.52**	**182.52**	**.99998**
5.31	202.35	.00494	101.17	101.18	.99995	5.91	368.71	.00271	184.35	184.35	.99999
5.32	204.38	.00489	102.19	102.19	.99995	5.92	372.41	.00269	186.20	186.21	.99999
5.33	206.44	.00484	103.22	103.22	.99995	5.93	376.15	.00266	188.08	188.08	.99999
5.34	208.51	.00480	104.25	104.26	.99995	5.94	379.93	.00263	189.97	189.97	.99999
5.35	210.61	.00475	105.30	105.31	.99995	5.95	383.75	.00261	191.88	191.88	.99999
5.36	212.72	.00470	106.36	106.36	.99996	5.96	387.61	.00258	193.80	193.81	.99999
5.37	214.86	.00465	107.43	107.43	.99996	5.97	391.51	.00255	195.75	195.75	.99999
5.38	217.02	.00461	108.51	108.51	.99996	5.98	395.44	.00253	197.72	197.72	.99999
5.39	219.20	.00456	109.60	109.60	.99996	5.99	399.41	.00250	199.71	199.71	.99999
5.40	**221.41**	**.00452**	**110.70**	**110.71**	**.99996**	**6.00**	**403.43**	**.00248**	**201.71**	**201.72**	**.99999**

Table E-3 Mensuration

Solids Having Curved Surfaces

31. Right Circular Cylinder (and Truncated Right Circular Cylinder) 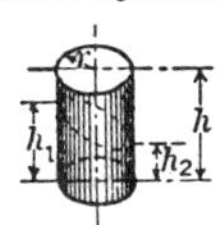	*For Right Circular Cylinder:* $A_l = 2\pi rh$; $A_t = 2\pi r(r + h)$; $V = \pi r^2 h$. *For Truncated Right Circular Cylinder:* $A_l = \pi r(h_1 + h_2)$; $A_t = \pi r\left[h_1 + h_2 + r + \sqrt{r^2 + \left(\frac{h_1 - h_2}{2}\right)^2}\right]$; $V = \frac{\pi r^2}{2}(h_1 + h_2)$.
32. Ungula (Wedge) of Right Circular Cylinder 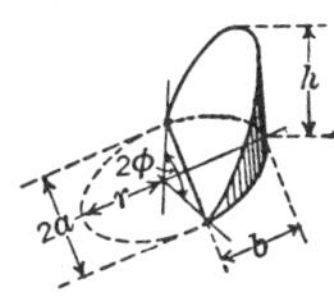	$A_l = \frac{2rh}{b}[a + (b - r)\phi]$; $V = \frac{h}{3b}[a(3r^2 - a^2) + 3r^2(b - r)\phi]$ $= \frac{hr^3}{b}\left[\sin\phi - \frac{\sin^3\phi}{3} - \phi\cos\phi\right]$. *For Semicircular Base* (letting $a = b = r$): $A_l = 2rh$; $V = \frac{2r^2h}{3}$.
33. General Cylinder 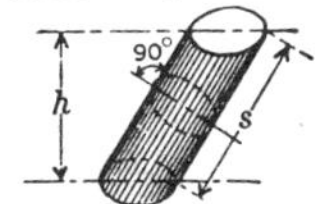	$A_l = p_b h = p_r s$; $V = A_b h = A_r s$.
34. Right Circular Cone (and Frustum of Right Circular Cone) 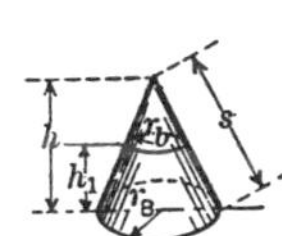	*For Right Circular Cone:* $A_l = \pi r_B s = \pi r_B\sqrt{r_B^2 + h^2}$; $A_t = \pi r_B(r_B + s)$; $V = \frac{\pi r_B^2 h}{3}$. *For Frustum of Right Circular Cone:* $s = \sqrt{h_1^2 + (r_B - r_b)^2}$; $A_l = \pi s(r_B + r_b)$; $V = \frac{\pi h_1}{3}(r_B^2 + r_b^2 + r_B r_b)$.
35. General Cone (and Frustum of General Cone) 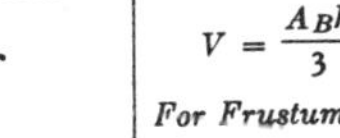	*For General Cone:* $V = \frac{A_B h}{3}$. *For Frustum of General Cone:* $V = \frac{h_1}{3}(A_B + A_b + \sqrt{A_B A_b})$.
36. Sphere 	Let diameter $= d$. $A_t = 4\pi r^2 = \pi d^2$; $V = \frac{4\pi r^3}{3} = \frac{\pi d^3}{6}$.
37. Spherical Sector (and Hemisphere) 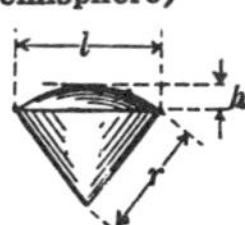	*For Spherical Sector:* $A_t = \frac{\pi r}{2}(4h + l)$; $V = \frac{2\pi r^2 h}{3}$. *For Hemisphere* $\left(\text{letting } h = \frac{l}{2} = r\right)$: $A_t = 3\pi r^2$; $V = \frac{2\pi r^3}{3}$.

Notation. Lines $a, b, c, \cdots$; altitude (perpendicular height), $h, h_1, \cdots$; slant height, s; radius, r; perimeter of base, p_b; perimeter of a right section, p_r; angle in radians, ϕ; arc, s; chord of segment, l; rise, h; area of base, A_b or A_B; area of a right section, A_r; total area of convex surface, A_l; total area of all surfaces, A_t; volume, V.

REFERENCES

1. G. P. Sutton. *Rocket Propulsion Elements*, 2nd ed., John Wiley & Sons, Inc., New York, 1956.
2. K. C. Rolle. *Introduction to Thermodynamics*. Charles E. Merrill Publishing Company, Columbus, Ohio, 1973.
3. R. D. Bent, and J. L. McKinley. *Aircraft Powerplants*, 4th ed. McGraw-Hill Book Company, New York, 1978.
4. I. Granet. *Thermodynamics and Heat Power*. Reston, Va. 1974.
5. "The International System of Units (SI)," *NBS Special Publication 330*, 1974 ed. U.S. Department of Commerce, National Bureau of Standards, Washington, D.C.
6. J. H. Keenan, F. G. Keyes, P. G. Hill, and J. G. Moore. *Steam Tables*. John Wiley & Sons, Inc., New York, 1969.
7. G. J. Van Wylen, and R. E. Sonntag. *Fundamentals of Classical Thermodynamics*, 2nd ed. (SI Version). John Wiley & Sons, Inc., New York, 1976.
8. J. H. Keenan, and J. Kaye. *Gas Tables*. John Wiley & Sons, Inc., New York, 1948.
9. P. J. Potter. *Power Plant Theory and Design*, 2nd ed. The Ronald Press Company, New York, 1959.
10. C. F. Taylor and E. S. Taylor. *The Internal-Combustion Engine*. International Textbook Company, Scranton, Pa., 1962.
11. G. P. Sutton. *Rocket Propulsion Elements*, 2nd ed. John Wiley & Sons, Inc., New York, 1956.
12. M. J. Zucrow. *Aircraft and Missile Propulsion*, Vol. 2. John Wiley & Sons, Inc., New York, 1958.
13. C. W. Smith. *Aircraft Gas Turbines*. John Wiley, & Sons, Inc., New York, 1956.
14. W. H. Severns, and J. R. Fellows. *Air Conditioning and Refrigeration*. John Wiley & Sons, Inc., New York, 1958.
15. N. C. Harris, and D. F. Conde. *Modern Air Conditioning Practice*, 2nd ed. McGraw-Hill Book Company, New York, 1974.

16. Frank, Kreith. *Principles of Heat Transfer*. International Textbook Company, Scranton, Pa., 1958.
17. J. P. Holman. *Heat Transfer*, 4th ed. McGraw-Hill Book Company, 1976.
18. W. H. McAdams. *Heat Transmission*, 3rd ed. McGraw-Hill Book Company, 1954.
19. R. A. Bowman, A. C. Mueller, and W. M. Nagle. "Mean Temperature Difference in Design." *Trans. ASME*, **62**:283–294 (1940).
20. W. M. Kays, and A. L. London. *Compact Heat Exchangers*, 2nd ed. McGraw-Hill Book Company, New York, 1964.

7.	6267 kW	
15.	130 kN	
19.	660 kJ/kg	

CHAPTER 12

4(a)	0.00589	—
(b)	6°C	
5(a)	0.0070	—
(b)	0.388	—
6.	0.807 kg/hr	
9.	0.527	—
11.	0.11	—
12.	0.00077 kg/kg	
18(a)	16.6°C	
(b)	No change	
19(a)	−16.5 kJ/kg	
(b)	8.5 kJ/kg	
22.	Below 0.29	—
25.	Approx. −8°C	

CHAPTER 13

6.	0.50 $W/\Delta_1°C$	
7.	10 W	
10.	0.070 W	
12.	0.055 W	
20.	11 000 $W/m^2\Delta_1°C$	
25.	1324 W/m^2	
27.	6236 W	
30.	0.80	—
31.	0.78	—

CHAPTER 14

5.	27.3 $\Delta_1°C$
7.	Water = 57.4°C
	Oil = 51.1°C
9.	146.3 $W/\Delta_1°C$
11.	51°C

INDEX

HEAT TRANSFER PROBLEM:

$$\frac{q}{A} = \frac{\Delta T}{\Sigma R}$$

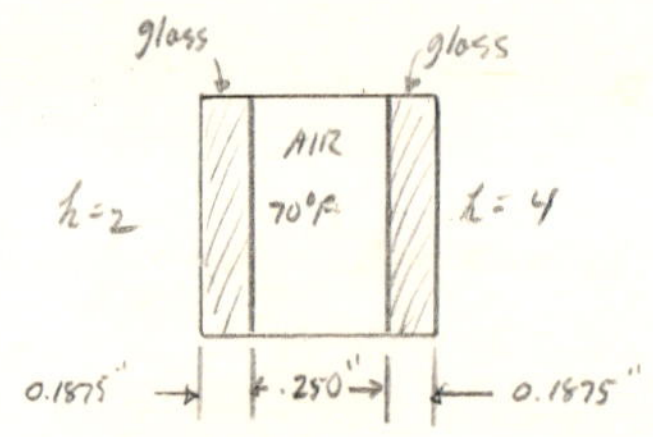

$$k_{glass} = 0.78 \frac{W}{m\,^{\circ}C} \left(.451 \frac{BTU}{hr \cdot ft \cdot {}^{\circ}F}\right)$$

$$k_{AIR} = 0.015 \frac{BTU}{hr \cdot ft\,^{\circ}F}$$

$$\frac{q}{A} = \frac{\Delta T}{\frac{1}{2} + 2\left(\frac{.0156\ ft}{.451 \frac{BTU}{hr \cdot ft \cdot {}^{\circ}F}}\right) + \left(\frac{0.0208\ ft}{0.015 \frac{BTU}{hr \cdot ft \cdot {}^{\circ}F}}\right) + \frac{1}{4}}$$

$$\frac{q}{A} = \frac{\Delta T}{2.2} \left[\frac{BTU}{hr \cdot ft^2}\right]$$

CRANKCASE PROBLEM:

$$h_c = \frac{Nu\,k}{L}\ ; \quad Nu = .036 (P_r)^{0.3} (R_N)^{0.8}$$

$$P_r =$$

$$R_N = \frac{V L \rho}{\mu}$$

$$q = h_c A \Delta T$$